POWER SHIFT

ENERGY AND SOCIETY
Brian Black, Series Editor

Saharan Winds: Energy Systems and Aeolian Imaginaries in Western Sahara
Joanna Allan

American Energy Cinema
Edited by Robert Lifset, Raechel Lutz, and Sarah Stanford-McIntyre

Engaging the Atom: The History of Nuclear Energy and Society in Europe from the 1950s to the Present
Edited by Arne Kaijser, Markku Lehtonen, Jan-Henrik Meyer, and Mar Rubio-Varas

Transportation and the Culture of Climate Change: Accelerating Ride to Global Crisis
Edited by Tatiana Prorokova-Konrad

Energy Culture: Art and Theory on Oil and Beyond
Edited by Imre Szeman and Jeff Diamanti

On Petrocultures: Globalization, Culture, and Energy
Imre Szeman

Oil and Urbanization on the Pacific Coast: Ralph Bramel Lloyd and the Shaping of the Urban West
Michael R. Adamson

Oil and Nation: A History of Bolivia's Petroleum Sector
Stephen C. Cote

Power Shift

Keywords for a New Politics of Energy

Edited by Imre Szeman and Jennifer Wenzel

Foreword by David E. Nye

WEST VIRGINIA UNIVERSITY PRESS
MORGANTOWN

Copyright © 2025 West Virginia University Press
All rights reserved
First edition published 2025 by West Virginia University Press
Printed in the United States of America

ISBN 978-1-959000-49-5 (paperback) / ISBN 978-1-959000-50-1 (ebook) /
978-1-959000-51-8 (cloth)

Library of Congress Control Number: 2024046634

Cover design by Daniel Benneworth-Gray
Book design by Than Saffel

NO AI TRAINING: Without in any way limiting the author's exclusive rights under copyright, any use of this publication to train generative artificial intelligence (AI) technologies to generate text is expressly prohibited. The author reserves all rights to license uses of this work for generative AI training and development of machine learning language models.

For EU safety/GPSR concerns, please direct inquiries to WVUPress@mail.wvu.edu or our physical mailing address at West Virginia University Press / PO Box 6295 / West Virginia University / Morgantown, WV, 26508, USA.

CONTENTS

FOREWORD

We urgently need *Power Shift* if we are to find our way forward toward a decarbonized future. For the world is making a full-scale energy transition, a relatively rare historical event that will change everything in society. *Full-scale* means that the transition involves central aspects of both production and consumption, including transformations in work, domestic life, infrastructure, social organization, and values. These changes are not inevitable, nor are they identical in every society. Energy systems are heterogeneous, based not only on technologies but also on local resources, geography, and the culture of those living there. Some quick examples. The Standing Rock Sioux oppose an oil pipeline and prefer different energy technologies than the state. France makes most of its electricity in atomic reactors. Denmark absolutely rejects nuclear plants but uses windmills and other renewable energy sources to produce two-thirds of its electricity. Norway relies almost entirely on hydro. Germany has invested heavily in solar and wind energies and is seeking to escape its dependence on fossil fuels. There are also striking contrasts in energy use. In the Netherlands, one-third of commuters ride bicycles; in Belgium, only one in eight does. Australians use twice as much energy per capita as the British. Floridians use twice as much electricity as Californians. What explains these variations? Geography and culture, not technology.

The world has been undergoing a full-scale energy transition away from fossil fuels that began slowly in the 1970s and started to seem urgent in 1990s but has really gotten well underway only in the 2020s. That transition will vary from one place to another, but it must be made. The authors of this book understand the urgency of this moment in terms of global warming, extreme weather events, and the dislocations that come with them. Just as importantly, they understand that new energy regimes have begun to appear in some locations, and they understand that a dire fate is not inevitable. There are new practices and possibilities. There is ground for hope. There are new ways to think about energy's relation to justice, populism, protest, mining, families, gender, and Indigenous communities.

Historically, energy transitions require at least half a century. The improvement, adoption, and dissemination of steam engines began with Newcomen's machine that pumped water out of mines in 1712, but the transition to steam was not complete until the late nineteenth century. The chronology for the steam engine was different in the United States. It had only two working engines in 1783 and relied mostly on waterpower until after 1830, and yet it also reached full adoption of steam by century's end. After that time came improvements in steam engine efficiency even as they were adapted to produce electricity, which made possible another reorganization of both production and consumption. The electrical transition might be dated from the telegraph's invention in the 1840s, but the decisive moment was the 1870s, with the improvement and adoption of arc lighting, the telephone, the incandescent light, and improved dynamos. In the United States, the development of the electrical network required about seventy years and grew rapidly only after about 1900. Petroculture took just as long to emerge and to be normalized. Energy regimes are not fixed; they are always both emerging and declining, depending on where one looks. Many things now considered normal are recent technological impositions, such as mechanical air conditioning, container shipping, and services available only to those with smartphones.

Power Shift helps make sense of impending transformations and suggests the choices becoming possible. New energy regimes can foster or obstruct equality. They can be used to concentrate power and centralize economies or to disperse production and democratize control. Technologies are not deterministic. They are always culturally shaped. Every new energy system implies possibilities, demands imagination, and resuscitates forgotten solutions, such as passive cooling, recycling, renewables, and retrofitting. In the Anthropocene a new politics of energy is imminent, and the results had best be heterogeneous.

David E. Nye, Professor Emeritus,
University of Southern Denmark
Senior Research Fellow,
Charles Babbage Institute,
University of Minnesota

HOW TO USE THIS BOOK

Our vocabulary of keywords is organized alphabetically. In addition to listing the terms in the table of contents, we offer further aid to navigation with several systems of cross-reference.

Each chapter in the book raises a provocative question, which is listed under its respective term in the table of contents. In addition to a conventional index, an Alternative Index organizes the contents of *Power Shift* by grouping these provocative questions into several categories: Actions, Agreements, Belongings, Communication, Economics, Extractions, Infra/structures, Knowledge, Narratives, Non-Humans, Politics, Technologies, Time, Transitions, and Values. These clusters of questions provide one version of thematic groupings that readers—including teachers and students using *Power Shift* in the classroom—can use to make their way through the book. We know that there are many more threads of connection to trace!

Additional aids to navigation appear within the book's chapters. When one of our contributors uses another *Power Shift* keyword in their discussion, we indicate the link by formatting the term in bold. Also, at the end of each entry is a "see also" list of keywords that intersect or overlap with the chapter's term—sometimes in less obvious ways, and in a few cases ironically and even humorously.

This book will offer insights no matter how you use it. Choose your own path. You decide what comes next.

INTRODUCTION

IMRE SZEMAN AND JENNIFER WENZEL

The end of fossil fuels and their ever-increasing greenhouse gas emissions is closer than you might think. Or at least the beginning of the end. By 2030, demand for fossil fuels will peak, and renewables will begin to eclipse fossil fuels in terms of actual use and new investment, creating what the International Energy Agency (IEA) calls a "considerably different global energy system," with more electric vehicles, heat pumps, and other clean-energy-powered technologies than today.[1] In their World Energy Outlook reports of 2022 and 2023, the IEA projected not only that solar photovoltaics would generate more **electricity** by 2030 than was currently produced by the *entire* U.S. grid but also that global CO_2 emissions in the power sector would peak in 2025.[2] Such imminent milestones in the fight against climate change offered surprisingly good news, and their significance was not lost on the authors of the IEA reports. "Global fossil fuel use has risen alongside GDP since the start of the Industrial Revolution in the 18th century," the 2022 report notes; "Putting this rise into reverse while continuing to expand the global economy will be a pivotal moment in energy history."

In contrast to the IEA's good news about energy transition, other assessments have been less optimistic about how far we've come in shifting to renewables and what this progress (or lack thereof) means for the likelihood of containing global temperature rise even to the levels established by the Paris Agreement.[3] Even the World Energy Outlook reports cited earlier emphasize that this recent "phenomenal rise of clean energy technologies" will not achieve climate goals that nation-states have set themselves for 2050. "Today's policy settings," the 2023 report notes, will bring major shifts toward clean energy in the coming years, while also increasing average global temperatures by approximately 2.4°C, which blows past the 1.5°C Paris Agreement target. The impressive progress toward energy transition, the 2022 report warned, will be "far from enough to avoid severe impacts from a changing climate." Regarding these future prospects, one might say that any reduction in emissions and any steps toward the emergence of a green energy world would be cause for celebration, were it not for the fact that these achievements will not be enough to avert severe climate disruption. Whether or not the IEA correctly forecasted a "pivotal moment" of peak emissions, the coming decades seem more likely to bring climate failure than climate success. Simply bending the curve of ever-increasing fossil fuel use that began with the **Industrial Revolution** will be insufficient to head off the broader derangements of the past centuries. Indeed, the 2023 World Energy Outlook anticipated that the projected 2025 emissions peak would be followed by a plateau rather than rapid decline. As journalist David Wallace-Wells observed in conversation with **Greta** Thunberg, "Few people understand just how far a peak is away from zero."[4]

Scholars, writers, and scientists concerned about climate change have, for the most part, emphasized the latter version of these narratives—a version that emphasizes failure. The story of **action** against climate change is understood, at best, as one of missed opportunities, particularly in retrospect. Consider, for example, Nathaniel Rich's analysis of efforts that nearly resulted in a global framework to reduce emissions in the 1980s.[5] An earlier start in addressing climate change could have made a big difference: in the twenty-five years between 1990 and 2015—the period between the collapse of the agreement Rich examines and the year of the Paris Agreement—global greenhouse emissions increased by 43 percent.[6] Such missed opportunities for action are just one among many failures with which advocates for climate action have had to contend. Most obviously, there is the failure of many segments of the population, even now, to recognize in any meaningful way the full import of global warming. There are the widespread failures of existing political systems to address (and, in many cases, even *begin* to address) the impact of climate change on human communities and the Earth System as a whole. The continued success of industry at keeping fossil fuels flowing—and at reaping unprecedented profits even in the midst of accelerating climate breakdown—constitutes a failure for those who prefer that all of it be kept in the ground.[7] The seemingly impossible challenge of unnerving, undoing, and redirecting global systems of **resource** and value extraction has generated its own sense of failure. To be sure, many critics, scholars, scientists, and artists alarmed about climate change have sought to push fossil-fueled citizenries and political and social systems in better directions. But the slow-moving sludge of policy initiatives and the glacial pace of decision making about the environment offer disconcerting **evidence** of the failure of facts, arguments, and new ideas to make any real difference. The failure to arrest climate change is thus evident all around us. Perhaps the starkest confirmation of this predicament is the failure to have reached *any* of the targets that scientists agree are necessary to avoid the most serious consequences of global warming. Moreover, even the perception of failure can itself create a negative feedback loop, a vicious cycle that deepens and entrenches the tendency toward inaction if such action is perceived to be futile anyway.

Amid such seemingly overwhelming evidence of failures past, present, and future, this introduction to *Power Shift: Keywords for a New Politics of Energy* poses two sets of questions intended to open up alternative ways of understanding the current predicament and to chart possible ways forward. The first set of questions, which we address in the following section, takes seriously the unexpected optimism about energy transition in the IEA projections, in order to ask, *what would happen if, rather than failure, we turned our attention to the remarkable successes of the environmental and climate justice movements?* The second set of questions, addressed in a subsequent section of the introduction, seeks to complicate the very notion of what it would mean for the proverbial glass to be full or empty: *what, exactly, constitutes success or failure with regard to climate action and energy transition?* Taken together, these questions are intended to lay out the need for, and possibilities of, a new politics of energy in the present, which we and our contributors examine in this book. *Power Shift* offers a nuanced and comprehensive map of those emergent politics, drawing together the varied perspectives of over a hundred artists, scholars, and activists, who have contributed brief essays on keywords that are crucial to this politics in the making. Our discussion of failure and success in the following sections aims to explain what we mean by politics and why we understand politics as crucial in forging our future, in order to provide context and motivation for the chapters that follow.

* * *

Turning to the first set of questions, what if, rather than ruminating upon endless and inevitable failure, we focused instead on the myriad achievements of climate action and decarbonization movements? How would our understanding of the challenges of energy transition and climate change be transformed (along with our understanding of *change* itself) if we shaped narratives of climate action around the amazing ecopolitical accomplishments of the past fifty or sixty years—depending on whether one takes E. F. Schumacher's *Small Is Beautiful* or Rachel Carson's *Silent Spring* as a baseline?[8] Without ignoring or minimizing the roadblocks to change still being erected at present, nor the serious hurdles undoubtedly looming ahead, might we gain new critical purpose, conceptual insight, and sheer inspiration by recognizing and building upon the prospective energy success of the coming years—rather than lamenting anticipated future failures to reduce emissions by 2030, 2050, and 2100? And might there also be a change in mood, attitude, and outlook—those intangibles so necessary to mobilize political action and to sustain movements for change—that could further energize these positive developments? Failure, after all, provides its Cassandras with the kinds of comfort that come with certainty. Letting go of the comforting surety of inevitable failure might make space for the discomfort, uncertainty, and unsettlement that, we believe, climate politics in the twenty-first century will demand.

The reasons for this proposed shift of perspective have little to do with a naively optimistic desire to look on the bright side of life. There's no question that without profound changes—not only in our environmental practices but also in many of the sociocultural beliefs, desires, hopes, and dreams that now animate the life of human communities—targets for avoiding the worst effects of climate change will not be achieved. The UNEP's 2022 Emissions Gap Report noted that it is impossible to trace a credible path toward the achievement of net-zero targets.[9] But even so, it is important not to gainsay the fact that profound changes are indeed happening. The past decade alone has witnessed remarkable developments, and not merely in the astonishing technological advancements in **renewable** energy generation and storage that promise to make green electricity cheaper than fossil fuels far sooner than anyone expected. Enormous change has also occurred in the realm of public awareness and sociopolitical mobilization. Energy is now frequently a topic of front-page news, and energy transition is an issue about which broad segments of the public have become conversant. Governments around the world have instituted green transition policies, with some even adopting **keep it in the ground** programs to limit or eliminate the production of fossil fuels, rather than focusing solely on limiting consumption (via carbon taxes, for example).[10] Against all odds, even the U.S. Congress passed climate legislation in 2022 that could transform the renewable energy landscape, to the bewildered consternation of European states left worrying about their capacity to compete with U.S. industry in decarbonization.[11] Fossil fuel companies are increasingly being held to account for the greenhouse gases they emit and the oil they spill.[12] So, too, are wealthier, developed countries—those most culpable for accumulated levels of atmospheric CO_2—as evidenced by the initiation of a process at COP27 (an event deemed by many to be a failure) to address the disproportionate impact of climate change on poorer countries. The affirmation by delegates at COP28 of the need to "transition away" from fossil fuels—the first time the root cause of climate change has been explicitly named, after nearly three decades of UN climate conferences—is a sure sign of where things now stand on the energy future. (This outcome is all the more encouraging given the worries before COP28 about it being held in a petrostate and led by the head of an oil company.)

And there's more. At a subnational level, various communities, organizations, institutions, and governmental entities have become newly attentive to environmental issues. Religious communities, for example, have called for the end of fossil fuel use.[13] Successful efforts to divest university endowments from investments in fossil fuel companies have led to demands for divestment by other kinds of institutions. The Fossil Fuel Non-Proliferation **Treaty** has gained force and momentum by gathering signatures from national governments, cities, Nobel laureates, civil society organizations, and faith institutions.[14] Protests by **Indigenous** communities against new pipelines and fossil fuel developments—including, most prominently, the Standing Rock Sioux Tribe in South Dakota and Wet'suwet'en Nation in British Columbia—have brought attention to the impact of resource extraction on these communities as well as to the broader colonial histories of injustice and dispossession visited upon them for centuries.[15] Similar protests have taken place across the world, with communities mounting resistance to mega-mining projects, new gas plants, waste-to-energy projects, and specious carbon capture initiatives. Writing in the wake of COP27, Mitchell Beer (editor of *Energy Mix*) notes a range of other developments that point to a much faster shift to renewables than previously imagined, including investor interest in net climate-positive projects, the adoption by farmers of solar energy as a second cash crop, and the implementation of energy efficiency measures and development of large renewable projects in European countries, so that, for example, Germany produced more than half of its energy from renewables in 2023.[16]

Somewhat counterintuitively, the aggregate impact of these developments is also legible in desperate attempts by fossil-fuel interests to delegitimize alternatives to the status quo, by pointing to renewables as the cause of infrastructural failure or energy price spikes. Actually-existing energy transition has become enough of a factor to be invoked as a scapegoat at moments of breakdown or crisis in the fossil energy regime. Consider, for example, how Governor Greg Abbott attempted to blame the historic and deadly failure of the Texas electric grid in February 2021 on frozen **wind** turbines. Or consider the campaign by state Republican lawmakers in the United States to punish companies (including investment firms like BlackRock) for taking climate action.[17] Or consider how Gazprom and the German government colluded in late 2020 to create a fake state-sponsored climate foundation that was supposedly supporting offshore wind fields but was actually working to complete the Nord Stream 2 **pipeline** while evading U.S. sanctions on Russia.[18] And progressive petrostates like Canada and Norway are lining up to position themselves as the last, most ethical producers in a "green" near future. What unites these varied examples are the ways that political figures have aligned themselves with the fossil fuel industry against energy transition, but do so by spuriously *invoking* energy transition as either a threat to the status quo or a path to the future, while doubling down on the fossil regime.

In effect, these incumbent interests cling to an earlier, more familiar version of peak oil—not the emissions peak recently heralded by the IEA but instead the mid-twentieth-century peak predicted by American geophysicist M. King Hubbert. This bell-shaped curve of growth and projected decline in petroleum production (dubbed Hubbert's peak) has been repeatedly flattened and extended by the discovery of new reserves and the development of new technologies (such as ultra-deepwater drilling, steam injection, and hydraulic fracturing, or fracking) to exploit previously inaccessible deposits.[19] While it's not surprising that the industry dreams of sustaining some part of the fossil fuel energy regime for as long as possible, their dogged, duplicitous attempts to ensure that the anticipated peak and subsequent decline of fossil fuel use will align with technologically augmented supply, rather than with politically constrained demand, represent a fundamental lack of seriousness about both the existential threats that climate change poses and the industry's culpability in producing these threats. If we understand these tactics as a politics of delay and deferral—pushing the peak further into the future—then we can also recognize that the fossil fuel industry itself acknowledges that energy transition is inevitable. Their public stance is not so much "fossil fuels forever" as "now's not the time; not yet," with every new crisis (coronavirus, Ukraine, Gaza) offering a new reason for putting off the necessary transition. Although the private communications of industry executives may damningly reveal otherwise,[20] their public messaging has created a widespread impression that their heralded future, toward which "clean-burning" **natural gas** is supposed to offer a much-vaunted "bridge," will be a future *not* powered by fossil fuels.

Despite concerted pushback by the fossil fuel industry, these multiple developments in the politics of energy mean that the mid-twenty-first century will look remarkably different than either the mid-twentieth century or its imagined futures, and different even from the beginning of this century. The simple fact of having established climate change as an issue with which the world must contend, and must continue to do so indefinitely, constitutes a remarkable political achievement. Even David Wallace-Wells, a wonted doomsayer who made his name in climate discourse with *The Uninhabitable Earth* (2019), has recently observed that plausible climate futures have veered away from apocalypse, "pull[ing] the future out of the realm of myth and return[ing] it to the plane of history."[21] This shift, Wallace-Wells insists, is because of human actions taken so far, and these futures remain to be written by human actions taken now. This is another way of saying that politics matters, profoundly.

Indeed, letting go of the inevitability of failure brings along with it the imperative of taking up responsibility, which, as Darin Barney (a contributor to this volume) has argued, demands a new kind of "post-masculinist courage: the courage to face the uncertainty of that which we cannot control; the courage to take up the uncomfortable burdens of judgment and action; the courage to be let go into action that begins something truly new and unpredictable; the courage to become political."[22] The new politics of energy that we consider in this book comes precisely out of this uncertainty, emergent from a moment of radical openness in which the need for action seems (newly) urgent, but the terms of struggle are not given. If, as we have argued, the certainty of failure offers a comforting yet unhelpful sense of surety, so too would the certainty of success, as, for example, promised by techno-utopian assurances that as-yet undeveloped technologies will come along to fix the fix that earlier technologies have put us in.[23]

To be "encouraged" in Barney's sense requires a reckoning with the formidable forces of discouragement at work in the present. Among these are the facts that capitalism hasn't ended, that climate denialist and fossil-fuel-loving populist governments continue to sprout around the world, and that narratives about the necessity of economic growth and technological progress at all costs remain compelling to many. Even as it celebrates the anticipated decline of emissions, the IEA continues to blithely imagine that what is most remarkable in its forecasts is the continued expansion of the global **economy** in a world after fossil fuels. And yet discussions about energy transition have moved from *if* to *when* these transformations will take place. Moreover, talk of energy transition very often acknowledges the need for broader changes to existing systems of patriarchy, property, power, and privilege that were, in direct and indirect ways, made possible by fossil fuels.[24] From this indispensable perspective, energy transition is about much more than finding new fuels and retrofitting present systems to run on them; it demands profound transformations of society writ large. It's why technocratic approaches to energy transition are being challenged by demands for energy justice and just **transitions**, and not only by critics and activists, but by governments, intergovernmental agencies, and even industry.[25] Indeed, the very rapidity with which activists' language of climate and energy **justice** has been adopted by institutions might raise new questions about whether they all mean the same thing by *just*.

This book offers an account of how these recent positive developments might provide a platform from which to spur the work of envisioning and making futures in which the upward race of global temperatures slows to a crawl or comes to a decisive halt. *Power Shift: Keywords for a New Politics of Energy* aims to capture the newfound sense of possibility emerging from the multiple challenges to the energy status quo described here. The book features essays on key terms, arranged alphabetically, that together offer a vocabulary for understanding—and forging—the politics of energy in the present. While deeply engaged with questions of political economy, policy, **science**, and technology, *Power Shift* features contributions from a diverse range of academics, activists, and artists whose disciplinary perspectives are grounded in the arts, humanities, and humanistic social sciences. The book extends recent energy humanities scholarship that draws connections between new energy sources and new modes of social relation and cultural imagining. What do we want the post-fossil era to look and feel like, and how might we get there? Our contributors answer these questions by tracing an emergent post-carbon consensus and the role of publics, communities, and individuals in shaping our energy future, while remaining attentive to the all-too-real challenges mounted by industry and government. *Power Shift* is addressed broadly, not only to scholars and specialists but also to students and citizen-readers concerned about how energy and climate will shape the world we'll inhabit in the coming years.

Power Shift is a sequel to our previous book, *Fueling Culture: 101 Words for Energy and Environment*, where we assembled a similarly diverse group of contributors to examine the myriad, complex relationships between culture and energy by writing on keywords ranging from *aboriginal* to *whaling*, *wood*, and *work*. [26] With both books, our aim was not to create an energy dictionary or encyclopedia that would summarize established definitions or narrate simplified histories. Rather, we asked our contributors to propose a new lexicon, in order to stretch existing concepts and open up new ways of thinking. In this, we take inspiration from Raymond Williams's seminal book, *Keywords: A Vocabulary of Culture and Society*, where he distinguishes between a dictionary—neutral, static, and book-bound—and "an active vocabulary," which is dynamic and continually shaped by its users.[27] Like Williams, we understand the vocabulary proposed here to be social, historical, and political, and we share his interest in interconnections among keywords.[28] As with *Fueling Culture*, we intend *Power Shift* to offer what Williams calls "a vocabulary to use, to find our own ways in, to change as we find it necessary to change it, as we go on making our own language and history."[29] Both projects are handbooks of provocations that readers can use to navigate to and through the places where energy and culture (writ large) come together. But if the focus of *Fueling Culture* was on how to think about (and with) energy, our central question in *Power Shift* is more pointed: what can we *do* about energy, now?

While there is an emerging consensus that an energy transition is necessary and inevitable, what is less certain in the public imagination is *how* transition should actually happen, and what a future after oil will look like. How do we decide upon the world that will come after what Stephanie LeMenager (a contributor to this volume) has called "Tough Oil World"?[30] Who is involved in bringing this new world into being, and what forces are pushing back? These are questions of politics. *Power Shift* examines the politics of energy in the present moment—with both *politics* and *energy* construed broadly, and with an eye toward practical and political transformation as elements of an energy transition that moves toward climate and energy justice. Our contributors speak to the ongoing struggles of publics, communities, and individuals to inform, engage with, and act against decisions made about energy and the environment by companies, governments, nongovernmental actors, and international organizations.

Taken as a whole, this collection outlines the current state of energy and environmental action around the world today, in order to identify gaps in current knowledge and strategy regarding what kinds of action are possible in the present, point toward possible new directions, and find practical ways to anticipate and defuse the myriad obstacles to change. With the help of the lexicon offered here, we hope to further energize emerging conversations and movements for energy and environmental change, wherever they exist and whatever critical and political form they take. The book's title, *Power Shift*, speaks to our sense of the profound changes in the politics of energy detailed in this introduction; these changes have put the status quo on its back foot and generated hope even in the face of the overlapping, cascading crises of the present. The title is also intended as a wry rebuke to the vision of the future offered in Alvin Toffler's infamous *Powershift* (1990), where environmentalism is repeatedly named as a problem or threat—unhinged, fundamentalist, and too radical—and energy transition not mentioned at all.[31] In all of his speculations, Toffler could grasp neither how impactful and mainstream the environmental movement would become, nor that fossil fuels were on their way to becoming a thing of the past instead of the fuel of the future.

We know that energy transition is inevitable: the hegemony of fossil fuels—with that status accrued as much by their impact on social life as by their long-time dominance among competing power sources—is at an end (even the IEA thinks so!). The questions that remain radically open are the forms that this transition will take, whether it will succeed in mitigating emissions or fail to do so, and what other social transformations—intentional or otherwise—it will entail. These are questions raised by the contributors to *Power Shift*. Who, for example, will own and manage renewable energy systems? Who will benefit, and who has to endure the consequences of new forms of resource extraction or **waste**? Put most broadly, how will new modes of generating power for industry and everyday life intersect with dominant or emergent workings of sociopolitical power? Clean energy populisms and solar liberal capitalisms might well make the putative successes of energy transition feel like the environmental failures with which we are all too familiar.

* * *

Success versus failure? *Positive versus negative*? *Yes versus no*? We recognize that such simple oppositions might read as conceptually and politically specious. There are good reasons, after all, for the despair felt by some climate activists, for the nihilism of some climate theorists, and for the apocalyptic dystopias imagined in much climate fiction. The problems we face are too big. The speed at which action must happen is too fast. We just don't have enough time. The greenhouse gases that have already accumulated in the atmosphere and oceans are, in effect, there for good, and will continue to increase long after emissions peak. To really work, action on climate change needs to be global, but there are no political entities with sovereignty over the **planet** and no prospect of such a political body on the horizon. While the United Nations—the only significant climate actor one might describe as global—can help muster agreements by which nations might measure their climate progress, it has no real power to enforce them. This limitation applies to the recent major COP breakthroughs, including the agreement on the need to "transition away" from fossil fuels and the major emitters' contributions to a loss and damage fund for poor nations.

Frustration at the lack of significant climate action and at the limitations and impasses of existing political and economic systems has led to radical proposals to overturn the apple cart of capitalist modernity. Some scholars call for **degrowth** and rewilding the planet;[32] others argue that only "war communism" can impose the rapid, extreme political changes necessitated by the climate crisis, or that eco-terrorism is necessary to eliminate the material and political infrastructures that enable fossil-fueled life.[33] Most left-wing writing about climate calls for the end of capitalism in addition to a transition away from fossil fuels—or suggests that the former is a precondition of the latter. In comparison with these radical programs, our proposal to foreground the accomplishments of the climate movement and the steps already taken toward energy transition might seem meek and accommodationist. At best, the idea of emphasizing success can't help but seem to affirm a standard, liberal position that is apparently content for climate action to emerge out of long-standing forms of interchange among business, government, and civil society. On the whole, these modes of thinking seem to add up to little more than the same old politics of energy.

We began this introduction with a first set of questions about what would happen if we were to flip our approach to climate politics by emphasizing possibilities rather than limits, successes rather than failures. But this flip can itself come across as a kind of failure, taking us back toward energy business as usual. Part of the problem is the very inadequacy of these terms, *success* and *failure*, for assessing the present circumstance, as we explain in this section of the introduction by considering a second set of questions. Can we even think in terms of success or failure with regard to climate action and energy transition? How do demands for climate and energy justice complicate the kinds of goals and targets that would count as success? And how does the current planetary predicament put pressure on what we mean by *politics*? Here too we follow Raymond Williams's example in recognizing how "the problems of [a word's] meaning [are often] inextricably bound up with the problems it was being used to discuss."[34] Williams advocated a relationship toward language that was "at once conscious and critical"; in the face of sociopolitical contestation, he acknowledged, this approach could admittedly yield "not resolution but perhaps . . . just that extra edge of consciousness."[35] This edgy, reflexive stance is, we believe, one of the most indispensable modes of humanistic thought: a critical examination of the given, including received terminologies and frameworks of analysis. It characterizes not only the essays in *Power Shift* but also this introduction, in which we advance arguments about success and failure while also subjecting those terms to critical reflection.

In the context of climate action, failure and success can be difficult to discern. At the planetary **scale**, the fundamental target and measure of success has been established as the achievement of global net-zero emissions, ideally in a way that attends to the political and ethical implications of vast differences in historical energy use and greenhouse gas emissions in different parts of the globe. The year 2050 is taken to be the finish line toward which we are all racing. Between now and then, nation-states (and millions of corporations and other institutions) will have to traverse a lot of ground in a big hurry: the challenge is something like running a marathon while sustaining the pace of a hundred-meter sprint. In the dominant social imaginary, the mid-twenty-first century is often assumed to be the moment when problems with the climate are finally fixed and the deleterious effects of greenhouse gas emissions put to rest. The **net-zero** target is sometimes understood as something like a zero-sum game: to reach it is to win against climate change. The more intractable implications of accumulated atmospheric emissions and their persistent effects on the Earth System have only recently begun to be discussed more widely in the popular media. While the 2050 finish line remains important to reach, many observers are beginning to understand that crossing it—if we manage to do so—won't constitute the victory we might have hoped for. In the big scheme of things, what might look like climate success could actually obscure and even entrench climate failure.

Indeed, even beyond the persistent effects of accumulated emissions that will persist long into the future, net zero is a highly imperfect measure of success. One might instead envision the finish line as a return to preindustrial levels of greenhouse gas emissions or to the "safe" level of 350 parts per million (ppm) of atmospheric CO_2. Or the return of the planet to pre-anthropogenic populations of **animals**, particularly insects and birds. Or the elimination of any or all of the many causes behind the symptom known as climate change—most obviously, capitalism and colonialisms of various kinds. After all, the vast majority of the planet's inhabitants need *more* access to energy, not less: a foreboding imperative in the face of the desperate need to reduce planetary levels of CO_2. Counterhistories like these (with their attendant alternative futures, some impossible, some utopian) play a crucial role in identifying the depth and complexities of the overlapping challenges of the present circumstance. They also provide a powerful rationale for contemporary climate politics and constitute an ethical reproach to all those apparently comfortable with the current state of the planet. If 350 ppm is the limit of safety, then those who continue to add CO_2 to an atmosphere with 426 ppm (as of July 2024) are making decisions and taking actions that must be forcefully challenged.

The language of failure or success, it seems to us, has become constitutive of climate politics. The limits and confusions of this language are, at core, consequences of how it frames what counts as climate politics, which includes a commitment to failure as the ultimate guarantor of criticism and discounts anything resembling success as always already too little, inadequate, or a ruse. Given this predicament, our first step in a new climate politics may well be to reimagine what we mean by *politics* itself. There is, we think, something distinctive and emergent about the politics of energy at the present moment. In an earlier era, one might have traced energy politics primarily in terms of geopolitical contests between Cold War superpowers or other hegemons over the control of energy, particularly oil reserves located in volatile client states in the Global South. A further, secondary gesture in this kind of analysis would consider how these geopolitical rivalries played out in various national contexts, attending to their deformation of political structures and the consequences for citizens of oil states.[36] To be sure, this kind of analytic reemerged with the Russian invasion of Ukraine; new alliances and divides formed around the question of who would buy the Russian oil and natural gas widely understood to sustain the Putin regime and to fuel its expansionist war efforts.[37]

Even with the conflict in Ukraine and the consequent fuel crunch that challenged consumers and governments around the globe, this analytic inherited from the twentieth century seems inadequate to explain the present for several reasons, some of which have to do with major shifts in our understanding of *geopolitics* itself. In suggesting that *geopolitics* is taking on new meanings, we refer not only to the shifting strategic polarities of a post–Cold War world but also to the emergence of a multiscalar *politics of the earth (and the atmosphere and oceans)*, rather than the old geopolitical rivalries whose terrain was arguably less the earth or the planet than the globe.[38] European colonialism and Cold War brinksmanship were certainly premised upon control over the extraction of innumerable material substances, but the perceived effects of those contests upon the planet itself tended to be framed merely in terms of the apocalyptic future prospect of mutually assured nuclear destruction.

By contrast, the earthly environmental politics of the twenty-first century involves the emergence of and **solidarity** among localized social movements at sites of extraction and production as well as broader movements for climate justice that would demand a transition away from fossil fuels. The new politics of energy in the present certainly includes, but is not limited to, forces that are either aligned with or opposed to the imperative of energy transition as well as the modes of power and forms of authority that will shape the terms of that transition. Whereas one can read mid-twentieth-century politics at a global scale as having been framed as and organized around a bipolar choice between two superpowers, one might read the fundamental dilemma of the present in increasingly stark terms: as a choice between life and capital. Of course, one could counter that the spectacular hot and cold wars organized around bipolar superpower conflict in the mid- to late twentieth century also boiled down to a not-dissimilar dilemma. But in this emerging planetary politics, nature and Earth itself are explicitly becoming a new *kind* of pole, around which ties of alignment, affiliation, identification, solidarity, and survival are being forged.

Cold War geopolitics saw the emergence of conflicts among the First, Second, Third, and Fourth Worlds; geopolitics in this century will have to reckon with the planet.

This new politics of the Earth is inherently messy. It's messy not merely because it must reckon with multiple, vast forms of agency not easily accommodated within political forms envisioned around human individuals or relatively small human collectivities. Additionally, the temporality of this politics confounds conventional understandings of how change is made, a predicament with profound implications for the meaning of failure or success. Though often described as such, the politics of energy and climate is not necessarily the politics of an interregnum, an in-between moment that promises the emergence of new forms of socioeconomic and political organization as old forms pass away. To be sure, there will be a transition to a world in which renewable energies are used more widely and intensively than they are today and in which fossil fuel use will come to an end. There likely will not be a moment, however, when climate change as we currently understand it is resolved or fixed or transmuted into some other form.

Within the timescales of politics as conventionally understood, this struggle over the future history of Earth will continue indefinitely, and that fact must bear upon how we understand the present. Perhaps the messy ongoingness of this challenge is unhelpfully obscured by the metaphor of the finish line at the end of the marathon sprint. (Put another way, the *finish line* metaphor is simply one version of an ur-narrative of progress that understands historical change as linear and unidirectional, aiming ever upward.) This ever-receding horizon of success is the corollary to Wallace-Wells's argument that technological and sociopolitical changes have recently returned the climate crisis from the supra-human realm of apocalypse to the temporal realm of human action and politics. From now on, the task of *not* destroying the biophysical conditions that have supported human life will involve a permanent struggle and ongoing deliberation among possible presents and futures.

This new politics of the Earth, we insist, is multiscalar, involving billions of human individuals in varied relationships to the more-than-human and the planetary. *Power Shift* aims to understand the new forms that old geopolitical rivalries are taking in the present, the new gestures toward multilateral governance that aim to address shared challenges at a planetary scale, as well as a new emphasis on the region, on the extraction site, and on the varied sites of politics that now constitute daily life.[39] There's no question that what still stands in the way of transition are the entrenched financial and political interests of petrostates and fossil energy companies. But change is also limited by the embodied, everyday attachments of ordinary people to the worlds that fossil fuels have made, and which in turn have made them, from their bodily affects to their sense of achievement and success in life. Where, then, does one find a new politics of energy in the making? The multilateral treaty negotiating table, the corporate boardroom, the path of a proposed pipeline, the voting booth, the classroom, the soccer field, the wetland or forest, and the kitchen table—in our view, these are all important sites of politics, and there are many more types of sites that demand the attention of anyone wanting to understand the possibilities for making change in the present. This capacious understanding of politics has shaped the terms we chose to include in our vocabulary, which includes some words that readers would likely expect to find in a collection of this sort (**blockade**, **Green New Deal**) and others that they probably wouldn't (**animals**, **sport**). In the introduction to *Living Oil: Petroleum in the American Century*, Stephanie LeMenager exhorts, "The global movements of these fossil fuels and their imbrication in all aspects of social and economic endeavor, let alone the personal nature of their effects in our bodies, demand the collaborative efforts of academics, artists, scientists, industry, and everyone—which means *everyone*—living in Tough Oil World."[40]

This inclusive gesture of building **community** and gathering publics inspires our work in *Power Shift*. We invite you, as part of this *everyone*, to recognize your own varied imbrications within the politics of energy, and to think with our contributors about sites and **scales** of politics that lie within your everyday experience, as well as beyond it.

* * *

The "extra edge of consciousness" that we value in Williams's work has implications for our own critical practice. The question of what constitutes success or failure with regard to energy and climate action poses a particular challenge for writers and thinkers not directly involved in scientific discovery, technological development, or public policy, for whom the direct link between their efforts and the putative finish line of net zero can be difficult to trace. The sense of commitment and desire to *do something* that justifiably provokes critics to confront urgent questions of energy and environment in their writing are difficult to calculate in terms of their measurable effects in the world. What would it mean, for example, for *Power Shift* to *succeed*, or to have *accomplished something?*

There are some additional comforting certainties worth naming here, whose seemingly opposed tendencies might actually point in the same direction. The first tendency can be understood (at least tacitly) as a certainty of success, in which literary and cultural critics overestimate the effects of their work in the world, in a kind of category mistake. In a recent essay, the two of us trace this predicament with regard to the overstated ways that extractivism has been taken up as a topic in recent scholarship and public discourse, where critics seem to assume that they're doing something against material practices of extraction by finding metaphorical analogues of it everywhere.[41]

The second tendency amounts to a not-so-tacit certainty (or perhaps necessity or valorization) of failure, by which we mean that critical seriousness tends to be equated with negativity. Narratives of success are, from this vantage, necessarily suspect. Rather than deconstructing an argument or idea—taking it apart to see what it's made of, what it does, for whom, how, and why—the critical task is to dismantle and destroy. (The latter mode is often what beginning graduate students assume that graduate school is *for*.) The work of criticism is made simple when one knows from the outset that one will find the present to be empty—lacking either the planetary richness of the bygone past (to be lamented, or invoked as something that must be restored) or the promise of a future on the other side of a crisis that is deemed inevitable (for example, capitalism can't help but falter because there is nowhere left for it to generate profit from capital invested).

The fundamental role of criticism is to reveal the fictions of the given: to refuse the seeming inevitability, naturalness, and unchangeability of the status quo. What joins these particular comforting certainties of success and failure is our sense that some modes of actually existing critical thinking—as well as some of the ways that criticism articulates (explicitly or implicitly) the legitimacy of what, why, and how it does its work—can themselves begin to skew toward the conventional, the always-already expected, and the status quo. In thinking through political questions, or in understanding the putatively matter of fact *as* political, the various traditions and modes of critical analysis at work these days don't necessarily foster nuance, contradiction, and ambivalence, let alone recognition of progress or success. The interpretive equivalent of fine motor skills, analytic subtlety, is often absent. The reflections on success and failure in this introduction seek to provoke a metacritical reflection on the counterintuitive comforts of climate criticism. They are also a demand for critical nuance and for close attention to an energy politics that is not foreclosed and impossible. They are a plea for the kind of courage that Barney describes, born out of uncertainty.

The reflective and critical dispositions associated with the humanities can feel sharply at odds with the technocratic race for solutions in the face of climate collapse, and this contradiction is only deepening. One unlikely similarity between CO_2 and critical work is that their effects are fundamentally accretive:

building up over time, often nonlinearly. The environmental and energy humanities have grown exponentially over the past two decades, at the same time that the ongoing accumulation of CO_2 increasingly poses a challenge for thinkers trained to emphasize the importance of careful, patient critical reflection over the urgency of unconsidered action. Raising fundamental questions about the histories, interests, and values that underwrite the wicked problems of the present may become even more of a hard sell than it already is—even to the humanities scholars asking the questions. Who wouldn't prefer, at least at first glance, a quick solution? What is the role of such critical questioning in the present (and future) circumstance of an already-warming world?

The threat of climate change induces us to think about immediate action. But the blending of scientific knowledge with immediacy has proven to be a poor way of mobilizing change; it has served instead to mark what hasn't been done, what won't be done, and what can't be done. What this predicament means is that we need to set aside the comforts of failure in order to shoulder the work of what needs to be done. We now have myriad examples of what can be done, of what works. To foreground these successes is not to quit the race or give up the fight, nor to abandon demands for radical change in favor of the concessionary approaches named by climate mitigation or adaptation. It is, rather, to reframe how we imagine our critical work and to allow ourselves to be energized by what has been accomplished and by the promise of accomplishments yet to come. It's also to confront the contradictions among the multiple scales, multifarious actors, and very short windows and very long horizons of struggles present and future, for the future.

"For a while now I've been getting used to imagining the future without flinching," says the narrator of Italo Calvino's "The Petrol Pump," a short story written amid the oil shock of the early 1970s. Running on empty, then filling his tank with gasoline whose price has quadrupled almost overnight, the narrator imagines backward in time to the prehistoric fossilization of ancient creatures into modern fuels, and forward in time to an apocalyptic moment of "the ultimate cataclysm, the simultaneous drying up perhaps of oilwells pipelines tanks pumps carburettors oil sumps," "as if all engines everywhere had ceased their firing and the wheeling life of the human race had stopped." Whereas Calvino's narrator associates the end of oil with the end of the world, we who inhabit the energy crises of the twenty-first century are tasked with facing a formidable future in which it is not scarcity but instead the ongoing surplus of fossil fuels that threatens the world and Earth as we have known them. Facing this future, many activists have cried out against and worked to undo a lack of action. Scholars in energy humanities have observed and attempted to remedy the absence or disregard of energy in our narratives. Like LeMenager, we believe that living in Tough Oil World and building the world to come demands the work of *everyone*. We also believe that that work entails not only insisting that things must be done but also recognizing and building upon what has, in fact, been done. This vocabulary offers a partial inventory of the struggles and obstacles so far and an archive of critical knowledge and experience intended to help answer the question, what comes next?

January 2024

Notes

1 International Energy Agency, "World Outlook Report" (2023), https://www.iea.org/reports/world-energy-outlook-2023.

2 International Energy Agency, "World Outlook Report" (2022), https://www.iea.org/reports/world-energy-outlook-2022.

3 See, for example, the alarming projections in a 2022 report by the United Nations Environment Programme (UNEP), which insists that current levels of greenhouse gas (GHG) reductions are inadequate to achieve the goals set for 2050. UNFCCC Secretariat, "Nationally Determined Contributions under the Paris Agreement: Synthesis Report by the Secretariat" (October 26, 2022), https://unfccc.int/documents/619180.

4 David Wallace-Wells, "Greta Thunberg: 'The World Is Getting Grimmer by the Day,'" *New York Times*, February 8, 2023, https://www.nytimes.com/2023/02/08/opinion/greta-thunberg-climate-change.html.

5 Nathaniel Rich, *Losing Earth: A Recent History* (New York: Picador, 2020).

6 U.S. Environmental Protection Agency, "Climate Change Indicators: Global Greenhouse Gas Emissions" (July 23, 2024), https://www.epa.gov/climate-indicators/climate-change-indicators-global-greenhouse-gas-emissions.

7 UNEP, "2021 Production Gap Report" (2021), https://www.unep.org/news-and-stories/press-release/governments-fossil-fuel-production-plans-dangerously-out-sync-paris. UNEP estimates that global fossil fuel production will continue to increase until 2040.

8 See E. F. Schumacher, *Small Is Beautiful: Economics as if People Mattered* (1973; repr., New York: Harper, 2010) and Rachel Carson's *Silent Spring* (1962; repr., Boston: Mariner Books, 2022 [1962]).

9 UNEP, "Emissions Gap Report" (2022), https://www.unep.org/resources/emissions-gap-report-2022.

10 The Beyond Oil and Gas Alliance is composed of countries that have made the decision to forgo fossil fuel production. See https://beyondoilandgasalliance.com.

11 For an overview of the Inflation Reduction Act (IRA) from the perspective of advocates for climate change, see Ciaran Clayton, "A 'New Day for Climate Action in the United States' as U.S. Congress Passes Historic Clean Energy and Climate Investments" (Nature Conservancy, August 11, 2022), https://www.nature.org/en-us/newsroom/us-house-passes-landmark-climate-bill/. See also Paul Krugman, "Did Democrats Just Save Civilization?," *New York Times*, August 8, 2022, https://www.nytimes.com/2022/08/08/opinion/climate-inflation-bill.html. Critics of the bill have argued either that it does too much, i.e., that it's too expensive and will do little to address inflation, or that it does too little and in the wrong way. Matt Huber has pointed to a number of deficiencies in the act, including huge concessions to the

fossil fuel industry and a tax-credit program that will largely benefit middle-class homeowners looking to electrify their lives. See Huber, "Mish-Mash Ecologism," *New Left Review Sidecar*, August 18, 2022, https://newleftreview.org/sidecar/posts/mish-mash-ecologism.

12 For some examples, see Valerie Volcovici, "Minnesota Sues Exxon, Koch and API for Being 'Deceptive' on Climate Change," Reuters, June 24, 2020, https://www.reuters.com/article/us-usa-climatechange-oil-idUSKBN23V2XY; and Roger Harrabin, "Shell: Netherlands Court Orders Oil Giant to Cut Emissions," *BBC News*, May 26, 2021, https://www.bbc.com/news/world-europe-57257982. Much remains to be done to hold industry culpable for their emissions and oil spills, which have impacted soil and water around the world.

13 Pope Francis has repeated the call he first put out in *Laudato Si'* (2015) for religious communities to attend to the environment and climate change more actively. See Associated Press, "Pope Calls for Courage in Halting Use of Fossil Fuels to Protect Planet," *Guardian*, September 24, 2022, https://www.theguardian.com/world/2022/sep/24/pope-calls-for-courage-in-halting-use-of-fossil-fuels-to-protect-planet?CMP=Share_iOSApp_Other.

14 See Fossil Fuel Non-Proliferation Treaty Initiative, "Fossil Fuel Non-Proliferation Treaty" (n.d.), https://fossilfueltreaty.org. As of December 14, 2022, the treaty has been signed by (for example) 72 cities and subnational governments, 200 health organizations (including the WHO), and faith institutions representing 1.5 billion people.

15 There are a growing number of studies of Indigenous protests against extractive industries. Examples include Kai Bosworth, *Pipeline Populism: Grassroots Environmentalism in the Twenty-First Century* (Minneapolis: University of Minnesota Press, 2022); Nick Estes, *Our History Is the Future: Standing Rock versus the Dakota Access Pipeline, and the Long Tradition of Indigenous Resistance* (New York: Verso, 2019); and Winona LaDuke, *To Be a Water Protector: The Rise of the Wiindigoo Slayers* (Halifax: Fernwood, 2020). The StandingRockSyllabus is an excellent resource for additional texts on resistance and resilience against predatory extraction. See NYC Stands with Standing Rock Collective, "#StandingRockSyllabus" (2016), https://nycstandswithstandingrock.wordpress.com/standingrocksyllabus/.

16 Mitchell Beer, "Don't Let Wreckage of COP27 Distract from Climate Wins," *Energy Mix Weekender*, November 27, 2022, https://energymixweekender.substack.com/p/dont-let-wreckage-of-cop-27-distract. In 2022, renewables became the European Union's top energy source, exceeding nuclear and fossil fuels as a percentage of the whole. See Daisy Dunne, "Wind and Solar Were EU's Top Electricity Source in 2022 for the First Time Ever," *Carbon Brief*, January 31, 2023, https://www.carbonbrief.org/wind-and-solar-were-eus-top-electricity-source-in-2022-for-first-time-ever/. On Germany, see Reuters, "Renewable Energy's Share on German Power Grids Reached 55% in 2023, Regulator Says," *Globe and Mail*, January 3, 2024, https://www.theglobeandmail.com/business/industry-news/energy-and-resources/article-renewable-energys-share-on-german-power-grids-reached-55-in-2023/.

17 Both the scapegoating of wind for the Texas grid failure and legislation against corporate climate action in several U.S. states are the brainchild of the Texas Public Policy Foundation. See David Gelles, "The Texas Group Waging a National Crusade Against Climate Action," *New York Times*, December 4, 2022, https://www.nytimes.com/2022/12/04/climate/texas-public-policy-foundation-climate-change.html.

18 See Katrin Bennhold and Erika Solomon, "Shadowy Arm of a German State Helped Russia Finish Nord Stream 2," *New York Times*, December 2, 2022, https://www.nytimes.com/2022/12/02/world/europe/germany-russia-nord-stream-pipeline.html.

19 As fossil fuels are a nonrenewable resource, Hubbert's peak marked the moment at which global oil production would reach its maximum level, before declining and moving toward zero.
20 Hiroko Tabuchi, "Oil Executives Privately Contradicted Public Statements on Climate, Files Show," *New York Times*, September 14, 2022, https://www.nytimes.com/2022/09/14/climate/oil-industry-documents-disinformation.html.
21 David Wallace-Wells, "Beyond Catastrophe: A New Climate Reality Is Coming into View," *New York Times Magazine*, October 26, 2022, https://www.nytimes.com/interactive/2022/10/26/magazine/climate-change-warming-world.html.
22 Darin Barney, "Eat Your Vegetables: Courage and the Possibility of Politics," *Theory & Event* 14, no. 2 (2011): n.p.
23 Imre Szeman, "System Failure: Oil, Futurity, and the Anticipation of Disaster," *South Atlantic Quarterly* 106, no. 4 (2007): 805–23.
24 On patriarchy and oil, see Cara Daggett, "Petro-masculinity: Fossil Fuels and Authoritarian Desire," *Millennium: Journal of International Studies* 47, no. 1 (2018): 25–44. See also Andreas Malm and the Zetkin Collective, *White Skin, Black Fuel: On the Danger of Fossil Fascism* (New York: Verso, 2021), which offers a critical overview of ongoing developments in fossil capitalism and political power.
25 Unlikely advocates for just transition include the World Bank (see World Bank, "Just Transition for All: The World Bank Group's Support to Countries Transitioning Away from Coal" [2024], https://www.worldbank.org/en/topic/extractiveindustries/justtransition), the Government of Canada (see Natural Resources Canada, "Public Consultations and Engagements" [May 21, 2024], https://www.rncanengagenrcan.ca/en/collections/just-transition), and the Brookings Institution (see Michaël Aklin and Johannes Urpelainen, "Enable a Just Transition for American Fossil Fuel Workers through Federal Action" [Brookings Institution, August 2, 2022], https://www.brookings.edu/research/enable-a-just-transition-for-american-fossil-fuel-workers-through-federal-action/).
26 Imre Szeman, Jennifer Wenzel, and Patricia Yaeger, eds., *Fueling Culture: 101 Words for Energy and Environment* (New York: Fordham University Press, 2017).
27 Raymond Williams, *Keywords: A Vocabulary of Culture and Society* (Oxford: Oxford University Press, 2015), xxvii.
28 Admittedly, most of our contributors are not as interested as Williams in tracing shifts in a word's meaning across the centuries, nor in considering close cognates or synonyms, but they often do emphasize, as he does, social contestation over a keyword.
29 Williams, xxxv–xxxvi. In his introduction, Williams emphasizes this open-endedness by noting that his publisher has included some blank pages at the end of the book "as a sign that the inquiry remains open" in the ongoing "shaping and reshaping" of "our common language" (xxxvii, xxxv). While we have never actually seen an edition of *Keywords* that included such blank pages, we offer *Power Shift* in the same spirit of ongoing, active *use* of a language to imagine and create a new history of energy.
30 Stephanie LeMenager, *Living Oil: Petroleum in the American Century* (Oxford: Oxford University Press, 2014), 18.
31 Alvin Toffler, *Powershift: Knowledge, Wealth, and Violence at the Edge of the 21st Century* (New York: Bantam Books, 1990).
32 See Giacomo D'Alisa, Federico Demaria, and Giorgos Kallis, eds., *Degrowth: A Vocabulary of a New Era* (London: Routledge, 2014); Kohei Saito, *Marx in the Anthropocene: Towards the Idea of Degrowth* (Cambridge: Cambridge University Press, 2023); Matthias Schmelzer, Andrea Vetter, and Aaron Vansintjan, *The Future Is Degrowth: A Guide to a World Beyond Capitalism* (New York: Verso, 2022); Troy Vettese and Drew Pendergrass,

Half-Earth Socialism: A Plan to Save the Future from Extinction, Climate Change and Pandemics (New York: Verso, 2022).

33 Andrea Malm, *Corona, Climate, Chronic Emergency: War Communism in the Twenty-First Century* (New York: Verso, 2020); and Malm, *How to Blow Up a Pipeline* (New York: Verso, 2021).

34 Williams, *Keywords*, xxvii.

35 Williams, xxxv.

36 For perspectives on the refiguration of geopolitics at the end of oil, see Aymeric Bricout, Raphaeul Slade, Iain Staffell, and Krista Halttunen, "From the Geopolitics of Oil and Gas to the Geopolitics of Energy Transition: Is There a Role of European Supermajors?," *Energy Research & Social Science* 88 (2022): 1–8; Indra Overland, "The Geopolitics of Renewable Energy: Debunking Four Emerging Myths," *Energy Research & Social Science* 49 (2019): 36–40; and Daniel Yergin, *The New Map: Energy, Climate, and the Clash of Nations* (New York: Penguin, 2021).

37 Helen Thompson, "Europe's Unsolved Energy Puzzle: How the Quest for Resources Has Shaped the Continent," *Foreign Affairs*, September 27, 2022, https://www.foreignaffairs.com/europe/europe-unsolved-energy-puzzle; and Jessica Lovering and Håvard Halland, "Russia's Nuclear Power Hegemony: The West Is Dependent on Moscow for More Than Just Gas and Oil," *Foreign Affairs*, June 8, 2022, https://www.foreignaffairs.com/articles/russian-federation/2022-06-08/russias-nuclear-power-hegemony.

38 For expanded discussions of this distinction, see Jennifer Wenzel, "Planet vs. Globe," *English Language Notes* 51, no. 1 (2014): 19–30; and Wenzel, *The Disposition of Nature: Environmental Crisis and World Literature* (New York: Fordham University Press, 2019).

39 On region, see Darin Barney, "Energysheds," *Energy Humanities*, December 8, 2022, https://www.energyhumanities.ca/news/energysheds; and Imre Szeman, "On the Politics of Region," in *Dimensions of Citizenship: Architecture and Belonging from the Body to the Cosmos*, ed. Nick Axel, Nikolaus Hirsch, Ann Lui, and Mimi Zeiger (Los Angeles: Inventory Press, 2018), 90–101.

40 LeMenager, *Living Oil*, 18.

41 Imre Szeman and Jennifer Wenzel, "What Do We Talk about When We Talk about Extractivism?," *Textual Practice* 35, no. 3 (2021): 505–23.

2040

What genre will the future be?

PHILOMENA POLEFRONE

A number of ominous graphs intersect at 2040. A year when oil production may plateau and begin to fall. A year by which it will be clear whether warming can be limited to 1.5°C or even 2°C. A year of rising temperatures, water scarcity, extreme weather—a year that is *near* chronologically but distant in almost every other way.

Between the present and these possible futures lies a space for speculation and politics. Speculation *as politics*: the political work of imagining and shaping futures involves **storytelling** about what is possible and what needs to be done. The year 2040 thus also sits at an intersection of political-environmental narratives, narratives that must be understood as forms of speculative fiction as much as political platforms. Understanding the genres of these fictions is critical to understanding how they function politically.

Consider *peak oil*. Oil executives and OPEC have recently projected oil production to peak and begin declining around 2040.[1] The date keeps moving back and, with it, the end of "life as we know it" that will supposedly come with the end of this era-defining fuel source. Conveniently for oil executives, the end is always nigh, but never quite here. The threat of a peak—the end of endless growth—never arrives as scheduled but stays close enough to allow for the continued expansion of oil and gas infrastructure. For others who imagine peak oil, it is less a negotiating strategy than an eschatology, an end-of-days narrative beneath a broader system of beliefs. Matthew Schneider-Mayerson argues that between 2004 and 2011 this "peakism" became an ideology unto itself, with belief in the "apocalyptic consequences of energy depletion" inspiring adherents across political spectrums to prepare for widespread political, economic, and social collapse and prompting them to stockpile supplies and isolate themselves from society and collective movements, going off-grid literally and politically.[2] Peakism as ideology begins with pessimism, a sense that mitigation is futile and that adaptation by any means necessary is the only remaining option. It is perhaps not merely coincidence that James Howard Kunstler, an influential theorist of peakism, writes postapocalyptic fiction.[3] Postapocalyptic is the genre of peakism, in fiction and in practice.

Much climate change discourse and activism has likewise been structured by apocalyptic narratives, though used strategically. The year 2040 is the far edge of predictions for peak oil, but it is the midway point of two deadlines for global emissions reduction named in the 2018 **IPCC** special report: 2050, when emissions must reach net zero to limit warming to 1.5°C, and 2030, when emissions must be reduced to 45 percent of 2010 levels.[4] These are more ambitious targets compared to the 2014 IPCC report aimed at limiting warming to 2°C, a recommendation received by many as both inadequate and unlikely, and which inspired a wave of strategic, apocalyptic pessimism.[5] Roy Scranton's *Learning to Die in the Anthropocene* (2015) argues for accepting that modern society is as good as dead and for adapting to the afterlife of an unimaginable future by asking the hard philosophical questions that future raises.[6] David Wallace-Wells's "The Uninhabitable Earth" uses pessimism as a corrective for the reckless, even paternalistic optimism of some climate scientists, who he claims are watering down their climate forecasts to make them more palatable and, perhaps, more politically actionable.[7] As with peakists, climate pessimists often view their apocalypticism as adaptive and optimism as naïve.

Optimists, in turn, argue that apocalyptic thinking is irresponsible. Elena Bennett and colleagues warn of the dangers of climate dystopianism: because people act based on futures they deem most likely, pessimism risks becoming self-fulfilling.[8] The EcoModernist Manifesto challenges the perceived misanthropy of apocalyptic fatalism with the utopian idea that humanity's *extraordinary powers* can be used for good.[9] The suggestion is nothing new. Charlotte Perkins Gilman's *Herland* (1915) and Ernest Callenbach's *Ecotopia* (1975) have shown that utopian fiction can structure speculative environmental politics just as powerfully as the postapocalyptic.[10]

Apocalypse and utopia seem at once necessary to think with and woefully simplistic in addressing environmental futures. Perhaps one needs the subtitular "ambiguous utopia" of Ursula K. Le Guin's *The Dispossessed: An Ambiguous Utopia* (1974)—both critical and forward-looking.[11] Such is Kim Stanley Robinson's *New York 2140* (2017),[12] which imagines fifty feet of sea level rise by 2140, drawing on a report by James Hansen et al. that Robinson has described as strategically pessimistic, "nonfiction speculative fiction."[13] From Hansen's careful apocalypse, Robinson sketches an ambiguous utopia in which climate resilience offers an occasion for the redistribution of wealth, rewilding, and community organizing. In Robinson's 2140—and perhaps in 2040, too—utopian and dystopian futures exist simultaneously. So do the acceptance of hard truths and the collective will to thrive.

The year 2040 could be a disaster. It probably will be. It could be a key moment of transition. It certainly must be. The meaning, today, of 2040 (or 2030, 2050) is a matter of its mobilization. Summoning an inevitable decline into survivalism? Delaying emissions reductions even further? Or dealing with the complexity and ambiguity of a future that many of us will live to see but can't predict? The shape of this future may depend on the genres through which it is imagined.

See also: **Abandoned, Communication, Nonlinear, Scales, Transitions**

Notes

1 Claudia Carpenter, "Peak Oil Production Seen by 2040 as IEA Calls 2020 'Turning Point' for Energy," *S&P Global Platts* (blog), August 31, 2020, https://www.spglobal.com/platts/en/market-insights/latest-news/coal/083120-peak-oil-production-seen-by-2040-as-iea-calls-2020-turning-point-for-energy; Rowena Edwards, "Opec Sees Oil Demand Peak around 2040," *Argus*, October 8, 2020, https://www.argusmedia.com/en/news/2148360-opec-sees-oil-demand-peak-around-2040.
2 Matthew Schneider-Mayerson, *Peak Oil: Apocalyptic Environmentalism and Libertarian Political Culture* (Chicago: University of Chicago Press, 2015), 3–6, 18–20.
3 James Howard Kunstler, *World Made by Hand* (New York: Grove Press, 2009).
4 Intergovernmental Panel on Climate Change, "Global Warming of 1.5°C" (2018), https://www.ipcc.ch/sr15/.
5 Intergovernmental Panel on Climate Change, "Climate Change 2014: Synthesis Report," ed. R. K. Pachauri and Leo Mayer (Geneva: Intergovernmental Panel on Climate Change, 2015).
6 Roy Scranton, *Learning to Die in the Anthropocene: Reflections on the End of a Civilization* (San Francisco: City Lights Books, 2015).
7 David Wallace-Wells, "The Uninhabitable Earth," *New York Magazine*, July 9, 2017, https://nymag.com/intelligencer/2017/07/climate-change-earth-too-hot-for-humans.html. Wallace-Wells released a book version in 2019.
8 Elena M Bennett et al., "Bright Spots: Seeds of a Good Anthropocene," *Frontiers in Ecology and the Environment* 14, no. 8 (October 1, 2016): 441.
9 EcoModernism, "An EcoModernist Manifesto" (last modified September 11, 2022), http://www.ecomodernism.org/manifesto-english/.
10 Charlotte Perkins Gilman, *Herland and Related Writings*, ed. Beth Sutton-Ramspeck (Peterborough, ON: Broadview Press, 2013); Ernest Callenbach, *Ecotopia: The Notebooks and Reports of William Weston* (Berkeley, CA: Banyan Tree Books, 1975).
11 Ursula K. Le Guin, *The Dispossessed: An Ambiguous Utopia* (New York: HarperCollins, 1974).
12 Kim Stanley Robinson, *New York 2140* (New York: Orbit, 2017).
13 Kim Stanley Robinson, "Adapting to Sea Level Rise: The Science of 'New York 2140,'" YouTube, May 9, 2017, https://youtu.be/tC7Cr8g-ru0.

ABANDONED

Who is tending to the present?

LIZ HARMER

In my novel *The Amateurs*, I depict a character standing on a bridge over the 403 in Hamilton, Ontario, a many-laned highway over which I used to ride my bike when traversing the city from one end to another.[1] There is something sublime about an enormous freeway, perhaps especially one free of cars. During one of the early **coronavirus** lockdowns, this empty 403 of my imagination came to reality. Residents of my former hometown sent me a photo of the very thing I had envisaged: all the lanes of the highway empty, the road stretching out into a gray, spooky horizon.

When I was writing this postapocalyptic novel of depopulation that occurs because of Big Tech, one of my mentors called it a novel of technological rapture. Even when I was a devout Christian, I had never belonged to any tradition that believed in the Rapture, the event during which God would vanish away the saints and after which the rest of humanity would enter a time of tribulation. But even *I* had read the first book of the *Left Behind* series, novels meant to thrill Christians and fill them with a sense of mission.[2] Writing a speculative novel during the Anthropocene, I remembered how many times an abandoned Earth had piqued my imagination. I was frightened by these desolate worlds, abandoned by God, in which people were abandoned to their baser instincts. The idea of abandonment—abandoned spaces, abandoned principles, and even an abandoned Earth—figures prominently in our visions of climate disaster and in our responses to our environment more broadly.

The COVID-19 pandemic caused a temporary abandonment of public spaces, and many people across North America have also abandoned any idea of a public will or a public good. In *History of the Plague in England* (1722), Daniel Defoe's narrator repeats several times that the people had "abandoned themselves to despair."[3] One can be raptured or rapturous; one can abandon people or places, or one can be a person of reckless abandon, abandoned to delight, evil, and despair. Despair about climate change is an excellent psychological mechanism for abandoning the cause of preventing its worst effects.

In my novel of an almost entirely abandoned Earth, people were choosing to leave the present for rosier pasts or futures via a ubiquitous portal technology. When I invented these machines, I first imagined them as similar to our own technological traps: as with the smartphone, you chose to be in the portal instead of in the real present of your life. My mentor was right: the portals were a technological means of taking people away from an increasingly lonely, uninhabited Earth. These invented portals were a way of thinking through the problem of *presence*, of *being present*. You cannot go back in time, and the future is imaginary, dissolving, when it arrives, into the past. So, who is a steward to our present reality, not lost to fantasy or memory or addiction? Is anyone here? Had we already abandoned the Earth?

The environmental propaganda of my childhood—tree-saving fairies and polluting monsters—led me to believe that there were adults in the world who had all this *in hand*. Now, however, adults with the most power are instead building portals to elsewhere. As I write this, Elon Musk is fantasizing about Planet B, a Mars to be terraformed as a kind of ark.[4] Rather than solve the political problems he's created, Mark Zuckerberg made a VR world to distract us from the distressing present. In Southern California, where I live, a constant influx of truck and train traffic delivering cheap consumer goods to warehouses (such as those belonging to Amazon) causes a deadly amount of pollution.[5] Meanwhile Jeff Bezos, without irony, has claimed that space exploration is necessary because "we can move all polluting industry off Earth and operate it in space."[6] Bezos can more easily imagine space colonies filled with pollution-belching factories than imagine a more sustainable version of his company here. The fantasy of abandoning the mess on Earth appears to be a significant temptation to the companies most responsible for the state it's now in.

We cannot just leave and hope that our problems will go away. Even the attempt to transition to clean energy has led to another problem: abandoned fossil fuel infrastructure. At the lowest estimate, hundreds of thousands of inoperative—"abandoned and orphaned"—oil and gas wells and coal mines continue to pollute. They "contaminate ground water, release dangerous air pollutants and, in some cases, lead to explosions."[7] They affect the usability of this land and emit enormous amounts of methane. There are solutions to this problem that, like all climate initiatives, require money and political will. All of this makes my dystopian novel, which pokes fun at the absurdity of the imagination of tech megalomaniacs—absurdities that nevertheless have catastrophic effects on the **planet**—seem more like realism than science fiction. A new politics of energy will require less reliance on sexy ideas set a hundred or a thousand years in the future and more on a stewardship of the present, a terraforming of our own planet, a cleanup that requires a communal effort and for which we can expect no glory.

See also: **Documentary, Nonlinear, Scales, Science, Storytelling, Trans-**

Notes

1 Liz Harmer, *The Amateurs* (Toronto: Vintage Canada, 2019).
2 Tim LaHaye and Jerry B. Jenkins, *Left Behind: A Novel of the Earth's Last Days* (Carol Stream, IL: Tyndale House, 1995).
3 Daniel Defoe, *History of the Plague in England* (1722), sec. 13, http://www.telelib.com/authors/D/DefoeDaniel/prose/plagueengland/plagueengland013.html.
4 Jamie Seidel, "Elon 'Ark' is Getting Ready to Fly: Eccentric Billionaire Races Ahead with His Mars Colony Dream," *Advertiser*, May 31, 2019, https://www.adelaidenow.com.au/technology/science/elon-ark-is-getting-ready-to-fly-eccentric-billionaire-races-ahead-with-his-mars-colony-dream/news-story/47e44d18f1c8ba3116b6dd218cdb4e63.
5 Stats on health impacts in the Jurupa Valley region of Southern California are cited in an unpublished article by Liz Harmer and include information provided by the Center for Community Activism and Environmental Justice (CCAEJ) and the Southern California Air Quality Management District (SCAQMD).
6 Caitlin Yilek, "Jeff Bezos on the Future of Spaceflight: 'We Can Move All Heavy Industry and All Polluting Industry Off of Earth,'" *CBS News*, July 21, 2021, https://www.cbsnews.com/news/jeff-bezos-space-heavy-industry-polluting-industry/.
7 Jillian Neuberger, Tom Cyrs, and Devashree Saha, "How the US Can Address Legacy Fossil Fuel Sites for a Clean Energy Future" (World Resources Institute, September 27, 2021), https://www.wri.org/insights/addressing-us-legacy-fossil-fuel-infrastructure.

ACTION

What becomes possible when climate action fails?

STEPHEN COLLIS

Conversations *about* action—calls to and deployments of the language of action—litter the terrain of the climate struggle. The UN has a "climate action" webpage—replete with references to "mobilizing action" and its "global call to action," to which "everyone can contribute"—and Goal 13 of the UN's Sustainable Development Goals is to "take urgent action to combat climate change."[1] In the face of all this talk *about* (and arguably in the place of) action, **Greta** Thunberg aptly intones, "blah blah blah."[2]

Perhaps it should be no surprise that speech about action, rather than action itself, has dominated political debates about energy and the climate. Politics in the Aristotelian tradition is, after all, a matter of speech acts. In Canada, since 2015 the Trudeau government has given a masterclass in duplicity, paying lip service to climate action while its *actual* actions, such as purchasing the Trans Mountain Pipeline, clearly put the lie to their words.[3] Nonetheless, many still legitimately ask why meaningful action against climate change is so hard when the need for action is so widely acknowledged. Perhaps this is not even the right question anymore: has *action* perhaps become an empty political signifier, suggesting an approach to social change that is currently ineffectual, if not out of reach?

The explanations for our inaction are varied and perhaps predictable. Naomi Klein has speculated that "something deep in our human nature . . . keeps us from acting in the face of seemingly remote threats" (the evolutionary argument).[4] Shoshana Zuboff points the finger at digital platforms and social media, instruments with unprecedented powers "to nudge, coax, tune, and herd behavior toward profitable outcomes" (the techno argument).[5] Klein herself came around to what she calls "market fundamentalism" (the structural argument), whereby "we are stuck because the actions that would give us the best chance of averting catastrophe . . . are extremely threatening to an elite minority that has a stranglehold over our economy, our political process, and most of our major media outlets."[6] That conclusion is hard to argue against. However, both Klein and Zuboff—among many others—assume an unexamined sense of self-willed action as something that is simply being held in check, whether by human nature, technology, or the corporate-state-media nexus. This idea of (in)action assumes an "inward space of lived experience," in Zuboff's words, where "*the will to will* is the inner act that secures us as autonomous beings."[7]

Klein and Zuboff's apparent faith in a self-willing and autonomous individual aside, ours is an age in which agency—that key political ingredient of self-motivated action—has, in the thinking of a number of philosophers, been broadly expanded: to our fellow **animals**, plants, minerals, and all kinds of objects that are seen to network together to collaborate on the genesis of the notable acts of the world.[8] The wide distribution of agency might offer a more accurate, relational ontology; nonetheless, certain actors (investors, the legal persons that are **corporations**, lobbying groups, etc.) still leverage outsized impacts across the distributed network, while others are unequally on the receiving end of actions in which they never had a say.

In the context of politics, action is inevitably a question of power, and yet, as Zygmunt Bauman argues, power "(that is, the ability to do things) has been separated from politics (that is, the ability to decide which things need to be done). . . . The sole agencies of collective purposive action bequeathed to us by our parents and grandparents, confined as they are to the boundaries of nation-states, are clearly inadequate, considering the global reach of our problems."[9]

Here Bauman makes what I might term the "governance" argument about political inaction. Action—in the face of the rift he identifies between the governmental functions of deciding and doing (including how this plays out both within and *between* nations)—too often feels like frantic windmill tilting, a sort of doing that knows it has no access to the decision process. But if doing-without-deciding leads to a reactive and anxious form of "activism," then deciding-without-doing is self-annihilating business as usual. Perhaps counterintuitively, it may be that what is needed is not "action," or even an expanded sense of networked "agency," but instead sustained, methodical struggle—**community** self-governance and capacity building, and the forging of relationships that alone can lead to what we really need: the production of new forms of social life. In this context, referring, for instance, to Indigenous land defense as "action" belittles what is actually going on: the (re)production of relations and forms of life that run counter to sclerotic colonial extractivism.

The rub, of course, is that the dominant sensibility in the climate movement is, legitimately, one of white-knuckled urgency, while the necessary action, running counter to this urgency, is likely a form of inaction—a slower doing that is in fact an undoing: a radical diminishment of the networks of production, consumption, and circulation; a stranding of assets; and a walking away from many large resource investments. It seems likely that we will only *decide* and *do* these large-scale systemic changes through the slower relational processes noted in the previous paragraph. In the meantime, the urgent desire for action will not soon go away. It seems hard to imagine anything other than a dual form of political action, one that, on the one hand, acts to put pressure on government "to institute reforms" (thus *doing* to propel *deciding*) while, on the other hand, and simultaneously, "acts" in a diametrically opposite way, "as if the state does not exist," thus *deciding* to *do* otherwise.[10]

See also: **Civil Disobedience, Evidence, Indigenous Activism, Keep It in the Ground, Protest**

Notes

1 United Nations, "Climate Action: Home" (n.d.), https://www.un.org/en/climatechange; United Nations, "Sustainable Development Goals: Goal 13: Take Urgent Action to Combat Climate Change and Its Impacts" (n.d.), https://www.un.org/sustainable development/climate-change/.

2 Speech delivered at the Youth4Climate Summit, Milan, Italy, September 28, 2021. After her repeated "blah blah blahs," Thunberg asserts, "Our hopes and dreams drown in their empty words."
3 Mia Rabson, "It's 'Not Nothing': Trudeau Defends Climate Change Action," *Global News*, September 11, 2021, https://globalnews.ca/news/8183335/trudeau-climate-change-push-back/.
4 Naomi Klein, *This Changes Everything: Capitalism vs. the Climate* (Toronto: Simon & Schuster, 2014), 16.
5 Shoshana Zuboff, *The Age of Surveillance Capitalism* (New York: Public Affairs, 2019), 8.
6 Klein, *This Changes Everything*, 18–19.
7 Zuboff, *Age of Surveillance Capitalism*, 291.
8 See the works of Jane Bennett and Bruno Latour.
9 Zygmunt Bauman, *Liquid Modernity* (Cambridge: Polity, 2000), viii.
10 David Graeber, *Direct Action: An Ethnography* (Oakland, CA: AK Press, 2009), 203.

AFRICA

Energy for whom and at what cost?

ERIN DEAN AND KRISTIN D. PHILLIPS

Rose Kiliba leaves her office job in urban Arusha, Tanzania, where electricity rationing has cut short the productivity of her workday. She catches a bus powered by imported diesel to her home in the peri-urban outskirts of the city. She cooks dinner over a charcoal jiko, using charcoal purchased from a tradesman transporting it from a rural region. At sunset, power returns and her children study at the table using electric light from Tanzania's hydropowered electrical system, joined by two children from a neighboring home as yet unconnected to the grid. Later, with their prepaid **electricity** balance running low, Rose switches to the system powered by a rooftop solar panel as she washes dishes by hand while listening to a battery-powered radio.

Like the lives of other residents across Africa, Rose's daily life involves an energy bricolage of partial infrastructures, regular and unplanned interruptions, evolving technologies, gendered expectations, and disparate resources. Her experience is framed by the intersecting but at times paradoxical political projects taking place within and across national borders on the African continent: the expansion of electrical access to previously unserved populations, the accompanying push for increased national energy production, and the transition to sustainable and healthy energy sources. Across these diverse agendas, a host of actors pursue and negotiate their interests, tastes, and values, provoking the vital questions that frame this piece: *Energy for whom? Toward what social ends? In which forms? And at what cost?*

The intertwined projects of expanded access, increased production, and energy transition frame an African energy story that is both complex and contradictory, haunted by the long and dark history of violent extractions: the horrific capture of human energy through transatlantic and Indian Ocean slave trades, the exploitation of labor and resources by colonial powers, and the siphoning of fossil fuels and minerals by authoritarian states and international corporations. These projects are also shadowed by the specter of climate change, as the value and flow of water, land, capital, and fossil fuels shift in response to changing climatic conditions and aspirations to decarbonize global energy systems. Yet Africans are not simply providers of energy for global consumption; they are innovators, consumers, and social theorists of energy in their own right, generating technologies and knowledge to make meaningful lives and livelihoods. The African energy landscapes encompass not only the corporate petropolitics of fossil fuel extraction but also state-led megaprojects for rural electrification and hydropower, the green capitalism of **renewable** energy entrepreneurs, the politics of global and African climate change activists, and the innovation and claims making of communities and households.[1]

The expansion of energy access, particularly through electrification, aims to address numerous inequalities. Roughly half of the people residing in sub-Saharan Africa live without household electrification, and the International Energy Agency estimates that by 2040, 90 percent of people in the world living without electricity will live in Africa. However, these statistics do not capture how electricity shapes daily life. Binary approaches to measuring electricity (access or lack) obscure how Africans live *with* electricity, in multiple forms and across multiple spaces.[2] In the absence of household electrification, access points such as mobile phone charging stations, internet kiosks, churches, mills, mosques, and schools create itinerant, flexible, and uneven geographies of electrification, inspiring new mobilities and power sharing socialities.

Ambitious rural electrification projects and regional power pools are state-sponsored attempts to smooth out this uneven landscape, but African governments want more than equitable distribution of existing supplies; they want more energy. As one Tanzanian university student told us in 2021, "We need megawatts." Her country is aiming for six times its current capacity—enough to say goodbye to power rationing and the economic uncertainty it wreaks. To fuel these networks, African governments have doubled down on oil, gas, and hydropower initiatives that promise increased production and rent-seeking opportunities but too often result in appalling violence against environments and communities. Yet even within these sacrificial energy zones, individuals and communities carve out spaces of autonomy. Omolade Adunbi describes how **youth** in Niger Delta creeks operate artisanal oil distilleries, tapping corporate pipelines and drawing on local expertise and cultural claims to oil.[3] Petro and hydropower megaprojects address economic goals in ways that meet the needs of some at the expense of others, but they also generate new forms of collective identity and resistance.

Coupled uncomfortably with the politics of increasing energy supply is the project of transition, in which governments, NGOs, and community leaders confront climate change by pushing shifts to alternative energy sources. Though Africa accounts for only 2 percent of global energy-related CO_2 emissions, the continent has become a pivotal site of green energy entrepreneurship and experimentation. Yet such projects are often bracketed as provisional or inferior options. At the national and regional levels, alternative energies offer less return in megawattage, and project timelines can extend beyond the electoral cycles that make hydroelectric so attractive to politicians. At the household level, many people take advantage of decentralized energy options, but household users often perceive renewables as technologies of the disconnected poor in relation to grid electricity, which has come to mark status and national inclusion. Moreover, with these alternative energies dominated by expatriate opportunists and foreign manufacturers who are struggling to make them profitable, the promise of autonomy or equality may fall short.

As Rose Kiliba considers whether to top up her home's prepaid electricity balance, when to replace her solar **battery**, where to buy charcoal for tomorrow's cooking, or if her bus fare will increase, she—like each of the roughly 1.3 billion people across Africa—is navigating the entangled energy politics of expanding access, boosting production, and transitioning to sustainable sources. These unfinished projects are enmeshed in an unjust energy landscape that is too often driven by extractivist logics, but also increasingly focused on meeting African energy needs with African energy infrastructures and resources. These energy politics reflect an ongoing negotiation of values and ideas. Attending carefully and creatively to these tensions will be essential to pursuing a more ethical and just African energy future.

See also: **Air Conditioning, Cuba, Development, Gender, India, Solar Farm, Transitions**

Notes

We gratefully acknowledge research support from the National Science Foundation (awards 1853185 and 1853109) as well as research permission from the Tanzanian Commission for Science and Technology.

1 Michael Degani, Brenda Chalfin, and Jamie Cross, "Fueling Capture: Africa's Energy Frontiers," *Cambridge Journal of Anthropology* 38, no. 2 (September 2020): 1–18, https://doi.org/10.3167/cja.2020.380202.
2 Paul G. Munro and Anne Schiffer, "Ethnographies of Electricity Scarcity: Mobile Phone Charging Spaces and the Recrafting of Energy Poverty in Africa," *Energy and Buildings* 188–89 (2019): 175–83, https://doi.org/10.1016/j.enbuild.2019.01.038.
3 Omolade Adunbi, "Crafting Spaces of Value: Infrastructure, Technologies of Extraction and Contested Oil in Nigeria," *Cambridge Journal of Anthropology* 38, no. 2 (September 2020): 38–52, https://doi.org/10.3167/cja.2020.380204.

AIR CONDITIONING

Has air conditioning become a technology of citizenship?

ANUSHREE GUPTA AND AALOK KHANDEKAR

One consequence of global warming is a pressing need for indoor cooling, particularly in tropical regions. While the United States currently leads the world with approximately 40 percent of its residential buildings air-conditioned, countries such as India, Indonesia, and China could account for half of the world's air conditioning (AC) by 2050.[1] **India** has seen steady growth in the demand for AC across residential, commercial, and public spaces. AC is increasingly common in corporate offices, theaters, shopping malls, airports, and even public transit vehicles like buses and metro trains. Approximately 24 percent of households across India own an AC, a figure that reaches 40 percent in urban areas.[2] A 2019 study of high-energy-consuming, high-income neighborhoods in Delhi found that of the 43 percent of households with AC, 87 percent were purchased in the past few years.[3] As summers become increasingly harsh, with rising humidity and temperatures reaching 104 to 114 degrees Fahrenheit for several consecutive days, AC is displacing extant, vernacular cooling practices.

Long before the invention of AC, indoor air temperatures in India were managed through architectural techniques including windows and interior courtyards for ventilation, lattice or perforated structures such as *jaalis*, as well as shaded overhangs or building envelopes that insulate the indoor environment.[4]

Such "passive cooling" measures can reduce the need for mechanical active cooling, making buildings more energy efficient and sustainable in the long run. But these practices have been written out of mainstream architectural sensibility and **design**, so that AC becomes a vital necessity. Urban growth in India increasingly reflects a *petromodern* imagination, materialized in the form of new concrete and cement buildings with all-glass facades and steel mainframes that assume AC as indispensable to the cooling demands of urban residents.[5]

Some challenges associated with widespread AC usage in densely built urban environments are widely known. AC units emit harmful hydrofluorocarbons that exacerbate climate change.[6] AC also drives a vicious cycle in the "urban heat island effect"[7]: dense clusters of AC units emit high amounts of heat; modern construction materials like concrete, cement, and glass increase thermal mass so that buildings absorb more heat; these overheated areas increase demands for cooling that then require heavier AC units that emit even more heat. Less obvious, perhaps, are the ways that AC embodies and configures social and political relations. Even as air conditioners have become seemingly ubiquitous, the need for AC is underwritten by particular trajectories of urbanization that produce segregated, enclaved, and unequally air-conditioned spaces. Therefore, while AC is becoming infrastructural, an almost-necessary support for inhabiting extreme climatic conditions, the unevenly distributed ability to access and afford it sharpens preexisting inequalities and opens up new terrains of contestation, resistance, and adaptation.[8]

The social experience of heat is reconfigured both by the increasing presence of AC and by the increasing stratification of comfort, along axes of social difference including **gender**, **class**, and age. In ongoing research in an informal settlement in Hyderabad, we have observed that women spend more time in poorly ventilated indoor spaces, particularly kitchens. While the elderly and children are commonly understood to be most vulnerable to the impacts of extreme heat, gender also structures thermal vulnerabilities. Cooking on gas stoves exposes women to heat, which causes higher levels of perspiration and severe thermal discomfort in the summer. Similarly, individuals engaged in livelihoods that increase their exposure to heat in outdoor settings also experience higher thermal discomfort. Occupational trajectories are shaped by cultural histories of migration and settlement that limit individuals of marginalized social locations (caste, class, or ethnic backgrounds) to low-skilled or daily wage work such as construction, sanitation, street vending, domestic work, or contractual home-based labor, underwriting thermal experience in specific ways. In the patchwork and makeshift housing structures of informal settlements, installing an air conditioner is rarely possible, given both space and financial constraints. An electric fan or desert cooler, both of which provide limited respite from the furnace-like heat they experience in peak summer, is often the most that residents can afford. Secondhand AC markets also thrive in cities like Hyderabad, with middle-income households able to afford only these inefficient devices to cool indoor spaces. In other words, rising income inequalities also produce hierarchies of thermal comfort and lead to a differentiated domestication of AC in urban contexts.

This stratification creates new sites for technopolitics in shifting urban socionatural environments. The seeming ubiquity of AC across everyday urban spaces maps onto existing social relations. As a potent cultural object, the air conditioner not only signifies comfort and convenience but also indexes luxury, aspiration, and desire.[9] In metropolitan areas across India, platform aggregators such as Uber and Ola offer air-conditioned taxi cabs as a luxury service that is frequently used by middle- and upper-class urban residents for commuting around the city. However, when both temperatures and fuel prices soared in the summer of 2022, platform drivers found themselves struggling to provide this luxury service. Given the insecurity of employment in the gig economy, the increased fuel and maintenance costs of providing AC proved nearly impossible. Drivers launched a "No AC" **protest** campaign: refusing to turn on the AC unless customers paid extra. This campaign was a political act that foregrounded platform workers' lived realities and articulated their struggles to sustain themselves amid the challenging, precarious, and generally un-air-conditioned conditions of life on the margins. The workers' effort to reclaim their agency through this campaign reframes the energy crisis as an overlapping crisis of urban citizenship, sustainability, and livelihoods that circumscribes the everyday socioeconomic lifeworlds of the majority of inhabitants in "southern cities."[10] Turning thermal comfort into a site of contestation, the "No AC" campaign laid bare the question of who pays for luxury.

See also: **Africa, Civil Disobedience, Electricity, Technology**

Notes

1 International Energy Agency, "The Future of Cooling" (2018), https://www.iea.org/reports/the-future-of-cooling.

2 Pratap Vardhan, and the Hindu Data Team, "Data | How Many Indians Own a Fridge, AC or a Washing Machine: A State-Wise Split," *The Hindu*, June 14, 2022, https://www.thehindu.com/data/data-how-many-indians-own-a-fridge-ac-or-a-washing-machine-a-state-wise-split/article65526597.ece.

3 Radhika Khosla, Anna Agarwal, Neelanjan Sircar, and Deepaboli Chatterjee, "The What, Why, and How of Changing Cooling Energy Consumption in India's Urban

Households," *Environmental Research Letters* 16, no. 4 (March 25, 2021): 044035, https://doi.org/10.1088/1748-9326/abecbc.

4 Feza Tabassum Azmi, "How India's Lattice Buildings Cool without Air Con," *BBC*, September 21, 2022, https://www.bbc.com/future/article/20220920-how-indias-lattice-buildings-cool-without-air-con; International Energy Agency, "Future of Cooling."

5 Stephanie LeMenager, "The Aesthetics of Petroleum, after Oil!," *American Literary History* 24, no. 1 (2012): 59–86.

6 Climate & Clean Air Coalition, "Hydrofluorocarbons (HFCs)" (2013), https://www.ccacoalition.org/en/slcps/hydrofluorocarbons-hfcs.

7 C. Ramachandraiah, "Weather and Water in Urban Areas," *Economic and Political Weekly* 32, no. 43 (1997): 2797–800, http://www.jstor.org/stable/4406004.

8 Brian Larkin, "The Politics and Poetics of Infrastructure," *Annual Review of Anthropology* 42 (2013): 327–43.

9 Amita Baviskar, "The Social Experience of Heat: Urban Life in the Indian Anthropocene," *India Forum*, June 13, 2022, https://www.theindiaforum.in/article/social-experience-heat-urban-life-indian-anthropocene.

10 Gautam Bhan, "Notes on a Southern Urban Practice," *Environment and Urbanization* 31, no. 2 (2019): 639–54.

AIRPLANE

Why might we still need to fly?

PARKE WILDE

John Hodgman once asked his audience to choose what superpower they desire, flight or invisibility.[1] If we asked an airplane this question, it seems obvious that its superpower would be flight. But the airplane's actual superpower is invisibility. Like Wonder Woman's invisible jet, airplanes have a magical way of evading our attention, even as they roar overhead.

In a time of climate crisis, aviation emissions should be a major topic of conversation. Hour-for-hour, flying is one of the fastest ways to burn fossil fuels and release greenhouse gases. Aviation emissions increased almost sevenfold from 1960 to 2018 and are now responsible for 3.5 percent of the human impact on the climate.[2] Moreover, aviation emissions are greatly unequal, making up more than 40 percent of carbon dioxide equivalent emissions for the top 1 percent of European Union households, even while most humans never fly.[3]

Despite their importance, aviation emissions are commonly ignored. In the annual greenhouse gas inventory for Boston (near my home in eastern Massachusetts), the official statistic is zero, but only because aviation emissions are omitted: "The inventory does not currently include emissions from airplane travel at Logan Airport."[4] In Seattle, by contrast, the official greenhouse gas inventory shows that air travel accounts for 1.2 million metric tons of CO_2 equivalent per year, equivalent to 21 percent of total emissions.[5]

Aviation is even more invisible on the international stage. Emissions from international flights account for 65 percent of total aviation emissions, but they are not addressed within the commitments that individual nation-states made as part of the 2015 **Paris Agreement** on climate change because they were deemed separate from any one country's jurisdiction. Instead, cross-border goal setting was relegated to the International Civil Aviation Organization (ICAO), a UN agency with long-standing responsibility for aviation infrastructure and navigation standards. In 2017, the ICAO announced a new Carbon Offsetting and Reduction Scheme for International Aviation (CORSIA). Environmental organizations judged CORSIA as inadequate in substance, overreliant on dubious offsets, and lacking transparency.[6] Evading the spirit of its task from the Paris Agreement, this scheme set no aviation emissions targets but instead offered an unspecified mix of limits and offsets that would apply to emissions above a baseline to be set in 2019–20. Airlines anticipated two more years of rapid growth and planned to limit or purchase offsets for subsequent growth. Although the **coronavirus** pandemic suppressed the volume of flights in 2020, the aviation industry was able to persuade ICAO to use 2019 as the CORSIA emissions baseline, thereby avoiding what would have been an unexpectedly lower and stricter target.[7] Taking stock of the international climate regime as a whole, Gössling and Humpe conclude, "a large share of emissions is unaccounted for in global mitigation plans for aviation."[8]

Even in university and academic settings, the airplane can be overlooked. The Association for the Advancement of Sustainability in Higher Education provides a rubric for universities to be awarded *stars* for meritorious effort on environmental causes, but it does not require universities to include aviation in a greenhouse gas inventory. In some annual sustainability reports, flying may be excluded as scope 3 emissions, outside the control of the institution, or counted only in part, for example, by including only those flights booked through an official university travel agent. The #flyingless initiative (flyingless.org), an environmental project started in 2015, promotes a petition asking universities and research institutions, academic associations, and research funders to set explicit goals and transparently report their progress in reducing flying. Measurement is the first step in accountability.

Academics and researchers sometimes perceive frequent flying as essential for field research, conferencing, and a high-quality lifestyle. But there are alternatives. With planning, we can make effective use of work time on trains and buses, conduct field work in fewer and longer trips, save more time for writing by skipping conferences less crucial to our research, and maintain our professional networks in novel ways. In the field of music and psychology, Richard Parncutt and Annemarie Seither-Preisler organized an innovative global conference in 2019 that replaced a traditional in-person global conference with a hub-based model offering sites on four continents, preserving the delights of a real in-person experience while drastically reducing the jet fuel.[9] The pandemic has accelerated changes in how academic and research communities approach travel. "Before the pandemic grounded travel, academics were often hypermobile," write Tullia Jack and Andrew Glover, but the pandemic has induced a rapid increase in familiarity with and expectations for lower-carbon alternatives to in-person conferencing.[10]

If we imagine the airplane as transparent, perhaps we can peer through its aluminum surface to recognize the underlying source of the good things it offers: transportation services, obviously, but also connection with friends and colleagues, cross-cultural contact, direct experience of new places, and an escape from our ordinary lives at home. These good things are more fundamental than the airplane itself. We can rise to the challenge of our times, preserving the good we do through our work and the joy we find in our leisure, with less frequent flying.

See also: **Expert, Globalization, Natural Gas, Net Zero, Online, Shipping, Wind**

Notes

1. John Hodgman, "Invisible Man vs. Hawkman. Superpowers," *This American Life* (National Public Radio, 2006), https://www.thisamericanlife.org/178/superpowers /act-one-4.
2. Stefan Gössling and Andreas Humpe, "The Global Scale, Distribution and Growth of Aviation: Implications for Climate Change," *Global Environmental Change* 65 (2020): 102–94. See also David S. Lee, D. W. Fahey, Agnieszka Skowron, M. R. Allen, Ulrike Burkhardt, Q. Chen, S. J. Doherty, et al., "The Contribution of Global Aviation to Anthropogenic Climate Forcing for 2000 to 2018," *Atmospheric Environment* 244 (January 2021): 117834.
3. Diana Ivanova and Richard Wood, "The Unequal Distribution of Household Carbon Footprints in Europe and Its Link to Sustainability," *Global Sustainability* 3 (July 2020).
4. City of Boston, "Boston's Carbon Emissions: Explore Boston's Community and Municipal Greenhouse Gas Inventories" (2021), https://www.boston.gov /departments/environment/bostons-carbon-emissions.
5. Seattle Office of Sustainability and the Environment, "2016 Seattle Community Greenhouse Gas Emissions Inventory" (2018), https://www.seattle.gov/documents /Departments/OSE/ClimateDocs/2016_SEA_GHG_Inventory_FINAL.pdf.
6. Jocelyn Timperley, "Airlines Around the World Have Recently Begun to Monitor Their CO2 Emissions as Part of a UN Climate Deal," *Carbon Brief*, April 2, 2019,

https://www.carbonbrief.org/corsia-un-plan-to-offset-growth-in-aviation-emissions-after-2020/.

7 International Civil Aviation Organization, "ICAO Council Agrees to the Safeguard Adjustment for CORSIA in Light of COVID-19 Pandemic" (June 30, 2020), https://www.icao.int/Newsroom/Pages/ICAO-Council-agrees-to-the-safeguard-adjustment-for-CORSIA-in-light-of-COVID19-pandemic.aspx.

8 Gössling and Humpe, "Global Scale, Distribution and Growth of Aviation."

9 Richard Parncutt, Annemarie Seither-Preisler, and Alastair Iles, "Live Streaming at International Academic Conferences: Ethical Considerations," *Elementa: Science of the Anthropocene* 7, no. 1 (2019), https://doi.org/10.1525/elementa.393.

10 Tullia Jack and Andrew Glover, "Online Conferencing in the Midst of COVID-19: An 'Already Existing Experiment' in Academic Internationalization without Air Travel," *Sustainability: Science, Practice and Policy* 17, no. 1 (2021): 292–304.

ALBERTA

How many nuclear bombs does it take to develop an oil field?

JEREMY J. SCHMIDT

On two occasions, decades apart, engineers and government officials contemplated nuking northern Alberta's oil sands. Each time, the goal was to find an economical way to separate bitumen from sand deep underground, *in situ* (literally "in place"). In the 1950s, using nuclear heat to melt geological formations was part of promoting "peaceful nuclear explosions" under the U.S. program known as Project Plowshare.[1] The second effort was spearheaded in the 1970s by the wealthy U.K. financier Edmund de Rothschild, most (in)famous in Canada for his role in the Churchill Falls hydroelectric plant in Labrador that flooded the territories of **Indigenous** peoples. Objections by Canada's federal government derailed both plans to use nuclear technology to get at the 167 billion barrels of recoverable bitumen in Alberta, of which 80 percent is accessible only through in situ methods of extraction because it lies too deep for surface **mining** and is too viscous for conventional drilling.

Atomic oil extraction is a jarring idea. It also focuses attention on the immense payload that other technologies have delivered to Alberta's subsurface. These are not onetime detonations. Rather, technological force is distributed through operations that unite the planetary history of Alberta with the economic geology of oil extraction. To explain how this feat is organized, a peculiar Venn diagram is needed: one sphere includes the scientific practices that disclose planetary history and geological properties worth speculating on; another sphere corrals an *economic* understanding oriented toward extracting value from land; and a third assembles the technologies used to mobilize bitumen, including the know-how that converts carbon into commodity. This diagram doesn't float abstractly in mathematical space. Instead, the political structure through which Alberta organizes state-led science, economic geology, and extractive technology forms a political geography written against Indigenous sovereignty. Diagrammed together: settler geology.

Formed in 1842, the Geological Survey of Canada underpinned virtually all development of Alberta's oil and gas in the late nineteenth and early twentieth centuries. State science coincided with state making such that geology became constitutive of settler-colonial claims to territory: after an 1888 Senate report declared Alberta's oil sands part of Canada's "great reserve," the Geological Survey of Canada's follow-up report in 1893 asserted that history in the region "commences in 1778" with the "Old Establishment": the first white settlement by Peter Pond.[2] This assertion was flat erasure; Indigenous peoples had been sovereign there for millennia, which the government knew when it negotiated Treaty 8 in 1899. By assembling its own claims in reference to geology, however, the state began anchoring itself to planetary history rather than to relations with Indigenous polities.

Nobody was more active in mapping Alberta's oil sands than Sydney Ells. Ells produced the first comprehensive map of the oil sands in 1914. He went on to survey the region repeatedly over subsequent decades. By 1962 he bragged about having cajoled Indigenous children into carrying equipment for him and complained when the government refused to reimburse him for moldy tobacco that he used to pay Indigenous men for their labor.[3] State officials and potential investors understood the importance of scientific work undertaken by Ells and others, which all but marked *x* on the map for industry. In this process, state science mapped planetary history to economic geology and produced a political geography that declared Indigenous peoples outside of history.[4]

Oil sands technologies today are more sophisticated than maps sketched from canoes and more subtle than subterranean bomb threats. In situ technologies superheat steam and force it underground to liquefy geologic formations. The force is tremendous. During 2013–14, one set of wells raised the ground by thirty centimeters before planetary history erupted into the present with flows of bitumen emulsion (steam-saturated bitumen) to the surface in four sites miles apart.[5] Oil carpeted the forest floor. Over the winter, the snow lifted the oil higher, leaving rings on trees two feet high. In one case, one-third of a lake had to be drained to expose and remediate cracks of seeping oil more than

120 meters long. The explanation of these events by industry and the state emphasized technological factors and contrasted them with a geological map of deep time. Technological forces were deemed knowable, while the uncertainties of prehistoric environments were suspected as the real causes of oil seeping to the surface; ancient river channels, old seabeds, and glacial rills made by retreating ice ages were all to blame. This was no neutral use of science but instead one that used the Venn diagram of settler-colonial geology to enlist the uncertainties of planetary history for prevailing structures of governance.

The distribution of force in Alberta's oil sands is not a one-off explosion. It is regulated through a social structure that organizes time itself, such that links of state science, technology, and the mining economics and labor relations of Alberta's fossil fuel workforce appear in unmediated relation to geology. State connection to planetary history is asserted at the direct expense of Indigenous peoples. Their rights of self-determination, treaty, and relations suffer violence linked to state structures that begin history with white settlement and reference time immemorial to geologic praxis.

Extractive industries put Earth's history up for sale while sidestepping relations of Indigenous peoples to land and **law**. In this context, any energy transition for Alberta that is not also anticolonial is merely an exercise in redistributing forces that dispossess Indigenous peoples while facilitating planet-altering practices linked to climate change. Many Indigenous peoples are more direct: an anticolonial transition means meeting Indigenous demands for returning their land.

See also: **Keep It in the Ground, Neoextractivism, Settler Colonialism**

Notes

1 M. L. Natland, "Project Oilsand," in *The K. A. Clark Volume: A Collection of Papers on the Athabasca Oil Sands*, ed. M. A. Carrigy (Edmonton: Research Council of Alberta, 1963), 143–56. Cf. Scott Kirsch, *Proving Grounds: Project Plowshare and the Unrealized Dream of Nuclear Earthmoving* (New Brunswick, NJ: Rutgers University Press, 2005).

2 Richard G. McConnell, *Report on a Portion of the District of Athabasca Comprising: The Country between Peace River and Athabasca River North of Lesser Slave Lake* (Ottawa: S. E. Dawson, 1893); John Schultz, "Select Committee on the Resources of the Great Mackenzie Basin," *Journals of the Senate of Canada* 22 (1888): 157–72.

3 Sidney Ells, *Preliminary Report on the Bituminous Sands of Northern Alberta* (Ottawa: Department of Mines, 1914); Sidney Clarke Ells, *Recollections of the Development of the Athabasca Oil Sands* (Ottawa: Department of Mines and Technical Surveys, 1962).

4 Tara Joly, "Growing (With) Muskeg: Oil Sands Reclamation and Healing in Northern Alberta," *Anthropologica* 63, no. 1 (2021): 1–26.

5 Jeremy Schmidt, "Settler Geology: Earth's Deep History and the Governance of *in Situ* Oil Spills in Alberta," *Political Geography* 78 (2020): 102–32.

ANIMALS

How do animals organize?

JAIMEY HAMILTON FARIS

In 2017, I spent time at a protest camp in the Henoko coastal district of Nago, Okinawa, where a group of local residents had staged a decades-long resistance to the relocation of a U.S. Marine Corps base to Henoko-Ōura Bay, near Camp Schwab. The U.S. military expropriated land from Okinawan families long ago; in the next phase of coastal land reclamation, it was set to dredge and infill parts of the bay for new runways.

As I walked along the chain-link fence that separated the **community** harbor from the base, activists pointed out dozens of banners that celebrated the Okinawan dugong, a relative of the sea manatee. One banner read, "Save the dugong, Save Ōura." Another featured a dugong riding a medieval horse. The hilarity of the latter image fed on local Okinawan mythology and English Arthurian legend alike. In Okinawan folklore, spirits were said to have come to the Ryūkyū archipelago (as Okinawa was known prior to Japanese conquest) riding on the backs of dugongs, bringing assurance of an abundant harvest from both land and sea.[1] The story relates **Indigenous** environmental knowledge about the important function of dugongs in fertilizing coral beds and maintaining healthy sea grasses. In the updated activist myth, the dugong was the knight in shining armor riding in to help the activists win their battle.

While not quite as mediagenic as the polar bear or whale, the dugong has become the iconic animal face of resistance to the U.S. military in Okinawa. Most Okinawans have long protested occupation by both **Japan** (since the nineteenth century) and the United States (with the cooperation of Japan, after World War II), which has led to the displacement of many Okinawans and the decline of farming and fishing subsistence practices.[2] Protestors have successfully delayed dredging on the U.S. base by establishing that land reclamation would destroy one of the last remaining habitats of the critically endangered dugong as well as those of the loggerhead sea turtle and blue coral.[3] Though activists have not fully succeeded in shutting down the project, they have broadened their collaborative and organizing skills, not only by emphasizing animal rights and building on the image politics of iconic mammals but also by returning to the wisdom of animals and what it might mean for sustaining peace. As one young activist, Hideaki Gushiken, poetically put it, "How can we create a society that doesn't spawn bases?"

Gushiken challenged the naturalization of military-industrial reproduction of defense and death, while also celebrating coral spawning—a wondrous multispecies life-perpetuating event. Corals are some of our most ancient animal ancestors. They are also holobiomes, multispecies assemblages that form a distinctive ecological unit.[4] Each microscopic coral polyp hosts algal mates in its tissue in order to eat, secrete, and grow. Coral polyps live in massive reef communities that spawn simultaneously once a year to ensure fertilization across vast distances and in strong currents. Spawning depends not only on algal symbionts but also on whales, dugongs, and other large sea mammals whose excrement provides the right chemical matrix for the coral gametes. Corals also need reef fish to clear new landing sites, as well as multitudes of viruses, microbes, crabs, and more.

Gushiken's provocation *energizes* transnational, transcultural, transspecies politics and societal arrangements in general. The question *how do animals organize?* invites renewed curiosity about the entangled cultural, biological, neurological, and environmental structures of animals that upset human exceptionalism.[5] As Anja Kanngieser and Zoe Todd suggest, learning how animals **organize** themselves with other animals, plants, and elements can shape human responsibility and positionality in these relations.[6] To do this well requires a relational method that prioritizes dispositions of listening to (rather than speaking for) animals. Their methodology proposes to turn again to place-based environmental wisdom such as Okinawan folklore, which expresses the nonlinear ecologically abundant relationships among dugong pods, coral reefs, seaweed, fisheries, and human dwellers.

Thinking about transspecies relations overturns the question that has pervaded the Western scientific hierarchization of life: how can humans organize animals? These organizations have conveniently objectified and exploited *lesser* species for food, labor, and energy.[7] Even Darwin's 1859 *Origin of Species*, in which humans were linked to all other animals, invited a taxonomic enterprise subordinated to human (especially European colonial and entrepreneurial) purpose.

Sea creatures in particular have been the basis of modern energy extraction. The frenzy for whale oil led to the demise of multiple whale species by the early twentieth century. In a matter of two centuries, petroleum- and natural-gas-based cultures have consumed much of the ancient plankton that took millions of years to compress and transform into fuel.[8] Even the so-called renewable technologies of solar and wind energy depend on new deepwater mining methods that will vacuum up the benthic biome in pursuit of rare minerals for batteries, thereby muddying the ocean's **water** column, disturbing plankton communities that need consistent sunlight, and affecting fisheries and atmospheric oxygenation.[9] Across these energy transitions, numerous animal species have been organized for extraction.

The continued devaluation of animal life will haunt whatever energy source is next identified. Placing kin-centered approaches at the forefront can supplant notions of individualism and competition with lessons learned from polyps, pods, and other trans-species communities. While species **solidarity** may start with animal rights in antimilitary protests, it can also demand that human societies organize around energy-element-plant-animal co-becoming.

See also: **Extinction, Farm, Nonlinear, Solidarity, Trans-**

Notes

I'd like to thank artist James Jack, who organized the visit to the protest site, and Azusa Takahashi for her research and translation assistance.

1 Ōi Kōtarō, *Okinawa Kodai No Mizu No shinkō* (Okinawa: Bunkyō Shuppansha, 1973).
2 Douglas Lummis, "On the Political Culture of Okinawa," *Asia-Pacific Journal* 17, no. 2 (December 1, 2019): 1–8.
3 Mission Blue, "Japan's First Hope Spot Honors Rare Coral Reefs and Dugong Habitats" (last modified October 24, 2019), https://mission-blue.org/2019/10/japans-first-hope-spot-honors-rare-coral-reefs-and-dugong-habitats/.
4 Donna Haraway, "Symbiogenesis, Sympoiesis, and Art Science Activisms for Staying with the Trouble," in *Arts of Living on a Damaged Planet*, ed. Anna Lowenhaupt Tsing, Nils Bubandt, Elaine Gan, and Heather Anne Swanson (Minneapolis: University of Minnesota Press, 2017), M25–M50.
5 Kari Weil, *Thinking Animals: Why Animal Studies Now?* (New York: Columbia University Press, 2012).
6 A. M. Kanngieser and Zoe Todd, "From Environmental Case Study to Environmental Kin Study," *History and Theory* 59, no. 3 (September 2020): 385–93.
7 Mel Y. Chen, *Animacies: Biopolitics, Racial Mattering and Queer Affect* (Durham, N.C.: Duke University Press, 2012).
8 Vaclav Smil, *Energy: A Beginner's Guide* (London: Oneworld, 2006), 85–88.
9 Jeff Drazen, Craig Smith, Kristina Gjerde, et al., "Opinion: Midwater Ecosystems Must Be Considered When Evaluating Environmental Risks of Deep-Sea Mining," *PNAS* 117, no. 30 (July 28, 2020): 17455–60.

ART

How are art and energy related?

CAROLYN FORNOFF

How does art animate energy as a structuring societal force? An answer to this question might be found in Marcela Armas's 2009 intervention "Exhaust." Armas set out to show how Mexico City's car culture was responsible for the city's famously poor air quality. To do so, she set up six cars in a circular formation underneath the city's elevated Periférico highway. Their exhaust inflated a massive plastic wrap that emulated the shape of a highway support beam. In a video of the intervention, passersby gaze up at the towering inflatable, through which gasoline combustion was made tangible as an entity that structures the city. While Armas's ephemeral installation drew the attention of the public that walked or drove by that day, its positioning within the realm of fine art also targeted wealthier museumgoing audiences who tend to be car owners. Studies have shown that high-income households are overwhelmingly responsible for the dramatic increase in vehicles on the road in Mexico City, which doubled between 1980 and 2010, degrading air quality for all city dwellers.[1] Armas's inflatable support endows form to the individual and collective contributions to the haze that hugs the city, animating emissions as unevenly engendered and yet experienced by all.

Another recent project that echoes Armas's double critique of energy regimes and the place of art within them is Gonzalo Lebrija's *Breve historia del tiempo* (A brief history of time), a sculpture installed from 2020 to 2021 in the plaza of Museo Jumex, a contemporary art museum in Mexico City's upscale Polanco neighborhood. Updating Futurism's obsession with the speeding automobile, Lebrija suspended a shiny car above a reflecting pool so that it hovered vertically in the air as if it were photographed just before plunging in. The inverted black vehicle makes a striking showpiece, particularly when viewed from the Jumex's second story. From above, the car's profile is brilliantly duplicated in the pool below; from the side, it is offset by the adjacent Soumaya art museum, a glittering swell of a building financed by billionaire Carlos Slim, who (among other industries) owns the third largest domestic stake in Mexican mining.

Surrounded by these architectural jewels, Lebrija's suspended animation awakens desire, reinforcing the values of petroculture.[2] Read through this lens, Lebrija's work exemplifies how art cultivates attachments to car culture, just as Armas points out how the emissions of those attachments shape a world that privileges individual mobility and speed over collective health.

Another way of getting at how energy animates art in Mexico is by tracing its sources of funding. Ever since the petroleum industry was nationalized in 1938 by President Lázaro Cárdenas—a radical move that expropriated subsoil resources from U.S. oil companies and was in keeping with the Mexican Revolution's dreams of energy sovereignty—oil revenue has been the condition of possibility for countless **development** initiatives, including Mexico's historically robust public arts programming.[3] The discovery of massive oil fields in the 1970s briefly transformed Mexico into a petrostate, until the global crash of oil prices in the 1980s catalyzed a decade-long depression, which was accompanied by austerity measures that included the opening up of the arts to private funders like Slim and the FEMSA Biennial, whose fortunes are made through extractive industries (**mining** and **water**, respectively). Although Mexico is no longer considered a petrostate (a definition that emphasizes oil exports), oil continues to be a socioeconomic linchpin. Until recently, revenue from state-owned Pemex (Petróleos Mexicanos) has constituted 40 percent of the federal government's operating revenue. The fiscal reliance on oil means that any state-funded enterprise, including cultural industries, is enmeshed, however obliquely, in the waxing and waning of global energy prices and supply. In recent years, this reliance has been even more tumultuous. The steady decrease of oil reserves and global oil price fluctuations have translated to a series of increasingly severe budget cuts to state institutions, social programming, environmental agencies, and arts initiatives, at the same time that the state has upped investment in extractive technologies and exploratory fracking.

What to make of oil as the animating force behind Mexico's public arts funding? Argentine critic Verónica Gago observes that throughout Latin America, extractivism has been embraced across the ideological spectrum on the grounds that it can fund much-needed social and cultural programming. The state delegitimizes potential critics by aligning "anyone who opposes the extractive model [as also] opposed to a form of financing poor populations," or, we might add, the arts.[4] Particularly in the case of Big Oil producers like Mexico, Venezuela, and Nigeria, as Cajetan Iheka writes, oil has operated as "a medium of world building" that is central to national "claims for modernity."[5] As we think about the past and future intersections of art and energy, it is crucial to acknowledge how energy mediates artistic production—even when it is not the object of artistic representation. The task put before artists, scholars, and activists who aspire to imagine beyond oil and move toward a post-carbon world is to reckon with these structural ties, in addition to accounting for how art makes the social costs of oil visible and reinforces petroculture. In this way, we might not just think about how art mobilizes us to see current energy regimes but mobilize around art to interrogate its role in sustaining an unsustainable status quo.

See also: **Air Conditioning, Finance, India, Music, Neoextractivism, Scales, Storytelling**

Notes

1 Erick Guerra, "The Geography of Car Ownership in Mexico City: A Joint Model of Households' Residential Location and Car Ownership Decisions," *Journal of Transport Geography* 43 (2015): 171–80.
2 Stephanie LeMenager, *Living Oil: Petroleum Culture in the American Century* (Oxford: Oxford University Press, 2014); Sheena Wilson, Adam Carlson, and Imre Szeman, *Petroculture: Oil, Politics, Culture* (Montreal: McGill-Queen's University Press, 2017).
3 Germán Vergara, *Fueling Mexico: Energy and Environment, 1850–1950* (Cambridge: Cambridge University Press, 2021).
4 Verónica Gago, *Neoliberalism from Below: Popular Pragmatics and Baroque Economies*, trans. Liz Mason-Deese (Durham, NC: Duke University Press, 2017), 171.
5 Cajetan Iheka, *African Ecomedia: Network Forms, Planetary Politics* (Durham, NC: Duke University Press, 2021), 5.

AUTONOMY

Why do we want energy autonomy?

DARIN BARNEY

In her essay on the welcome intrusion of Gaia into contemporary Anthropocene discourse, Isabelle Stengers warns, "*Autonomy* is a delicate word to use; that is, it is a word easy to misuse."[1] Among the misuses of the word are reducing its meaning to something like individual freedom, independence, self-determination, or even sovereignty. Such reductions are common, and predictable, in cultures where the (white, masculine, colonial, capitalist) fantasy of being an unencumbered, carefree, self-sufficient maker and master of one's fate continues to exert a powerful hold on the social imaginary. This fantasy does not comport well with the environmental principles of Gaia, which posit the planetary interdependence and co-constitution of all things.

The reduction of autonomy to nondependence upon others—whether human, animal, vegetal, mineral, or technological—is commonplace. However, as Kennan Ferguson has argued, "Such autonomy is a lie. An absurdity. A pipe-dream so unrealistic, so fanciful, that believing in it boggles the mind. The presumptions inherent to it prove so strange that it invites a series of questions as to why it has such a hold on so many. . . . But before asking such questions, it is first strangely necessary to argue the obvious: freedom is impossible."[2] Freedom (as autonomy) is impossible because we actually depend on human and more-than-human others to live, just as we depend on variously configured elements and the infrastructures that mediate our relations with them. Scholars in intersectional disability studies explain why this point can be difficult to grasp: for normative, enabled subjects, the material facts of their dependency are erased by the liberal ideology of autonomy.[3] If being an independent agent is all that autonomy means, then it is indeed a lie. We are necessarily indebted, or, as Ferguson remarks, "All are beholden," even if some bear the burdens of indebtedness and dependency disproportionately, such that others are thereby relieved of them.[4] This unequal distribution of indebtedness probably accounts for why some who reject the individualist pretensions of autonomy and freedom are reluctant to let go of them as ideals that inform struggles against racialized, gendered, and colonial oppression.[5]

Even if autonomy is a lie, establishing control over sources of energy has always been a means by which some have sought to reduce their dependence on others and increase the dependence of others upon them. Perhaps for this reason, people oppressed by extractive energy regimes have articulated energy autonomy within their political programs for self-determination.[6] In discourses of environmental sustainability, *energy autonomy* is defined as "an energy system [that is] fully functional through its own **local** production, storage, and distribution systems while simultaneously fostering local environmental and social goals . . . as a potential way of creating a sustainable, low-carbon energy system."[7] *Autonomy* names a range of desires, including "independence from energy markets and utilities, stability in the face of fluctuating energy prices, environmental concerns, and better energy security."[8] This is why energy autonomy resonates strongly with efforts to establish sustainable, decentralized energy systems based on renewable fuel sources.[9]

Does energy autonomy adequately express desires for environmental **justice**? Drawing on the meaning of *autonomia* in Italian workerist thought and politics, Karen Pinkus speculates whether a conception of *autonomy for* (not *from*) might support "an openness to different futures and revolutionary aspirations" in relation to energy systems, that might "embrace differential temporal and economic modes that are not easily assimilable to the repetitious and exhausting labor of factory work, or to the regular work hours of the office, or the odd hours of the temp, or to the hyperenergized flows of international capital."[10] Feminist theorists have similarly argued for rejecting the liberal, individualist, and masculinist conception of autonomy in favor of a more relational account in which "relatedness is not, as our tradition teaches, the antithesis of autonomy, but a literal precondition of autonomy, and interdependence a constant component of autonomy."[11] From this perspective, energy autonomy would be conceived as an ongoing, mutual relationship with whatever produces the energy we use, and with what is produced by it.

Whether such uses of (energy) autonomy are "delicate" (in Stengers's words) enough to recuperate the concept for a politics of environmental **justice** is hard to say. As with any concept, the meaning of *autonomy* depends upon the position and context of those who speak it, and the discourses and material practices with which it is articulated. For some, it will be hard to shake the traces of its origins and deployment in European ideologies and practices in which presumed capacity for autonomy marked the difference between people bearing the rights of citizens and the human and nonhuman others who could be exploited, silenced, dispossessed, and killed. As Elizabeth Povinelli writes, autonomy in settler-colonial contexts is "not merely an illusionary construct but a weapon of the enlightened liberal state in its constant maneuvering against **Indigenous** people."[12] Thinking with Aileen Moreton-Robinson, a feminist theorist and elder of the Goenpul tribe and Quandamooka nation, Povinelli points out that, by contrast, "Indigenous relations to land are not defined by autonomy as signified by a settler notion of sovereign possession of or over oneself, another, or a place."[13] This tension suggests that, for some, achieving environmental justice will require something other, or better, than energy autonomy.

See also: **Class, Decolonization, Design, Settler Colonialism, Trans-**

Notes

1 Isabelle Stengers, "Autonomy and the Intrusion of Gaia," *South Atlantic Quarterly* 116, no. 2 (April 2017): 381–400, 381.
2 Kennan Ferguson, "Beholden: From Freedom to Debt," *Theory & Event* 24, no. 2 (April 2021): 574–91, 579.
3 Christina Crosby and Janet R. Jakobsen, "Disability, Debility and Caring Queerly," *Social Text 145* 38, no. 4 (December 2020): 77–103.
4 Ferguson, "Beholden," 587.

5 Kouslaa Kessler-Mata, "Rescuing Freedom from Autonomy," *Theory & Event* 24, no. 2 (April 2021): 592–97. See also Achille Mbembe, *Out of the Dark Night: Essays on Decolonization* (New York: Columbia University Press, 2021).
6 See Thea Riofrancos, *Resource Radicals: From Petro-Nationalism to Post-Extractivism in Ecuador* (Durham, NC: Duke University Press, 2019); Robert D. Stefanelli et al., "Renewable Energy and Energy Autonomy: How Indigenous Peoples in Canada Are Shaping an Energy Future," *Environmental Review* 27 (2019): 95–105; Dominic Boyer, *Energopolitics: Wind and Power in the Anthropocene* (Durham, NC: Duke University Press, 2019).
7 Jouni K. Juntunen and Mari Martiskainen, "Improving Understanding of Energy Autonomy: A Systematic Review," *Renewable and Sustainable Energy Reviews* 141 (May 2021): 1–10, 1.
8 Juntunen and Martiskainen, 5.
9 Fanny Lopez, Margot Pellegrino, and Olivier Coutard, eds., *Local Energy Autonomy Spaces, Scales, Politics* (London: Wiley, 2019). See also Hermann Scheer, *Energy Autonomy: The Economic, Social and Technological Case for Renewable Energy* (New York: Earthscan, 2007).
10 Karen Pinkus, "Intermittent Grids," *South Atlantic Quarterly* 116, no. 2 (April 2017): 327–43, 329, 339.
11 Jennifer Nedelsky, "Reconceiving Autonomy: Sources, Thoughts and Possibilities," *Yale Journal of Law and Feminism* 1, no. 7 (1989): 7–36, 12.
12 Elizabeth A. Povinelli, "The Ends of Humans: Anthropocene, Autonomism, Antagonism, and the Illusions of Our Epoch," *South Atlantic Quarterly* 116, no. 2 (April 2017): 293–310, 301–2.
13 Povinelli, 302. She refers to Aileen Moreton-Robinson, *The White Possessive: Property, Power and Indigenous Sovereignty* (Minneapolis: University of Minnesota Press, 2015).

BANKRUPT

Is bankruptcy the crisis we need?

CALEB WELLUM

On July 18, 2013, the City of Detroit filed for chapter 9 federal bankruptcy protection, the largest filing of its kind in U.S. history.[1] Five months later, America's once prosperous "Motor City" was officially declared bankrupt, owing an estimated $18.5 billion.[2] A twenty-first-century time traveler arriving in 1950s Detroit with a prophecy of such ruin would have been dismissed as delusional. A beneficiary of American oil hegemony, Detroit was the epicenter of the world's most powerful and dynamic automobile industry and a key engine of American power. Yet over only a few decades of deindustrialization, white flight, and global competition, Detroit became a symbol of urban crisis. The 2007–8 financial crisis wreaked even more havoc: home foreclosures; wage, job, and pension cuts; and further depopulation. Bankruptcy was merely the latest in a long line of crises for a long-beleaguered city.[3]

Although Detroit has become a "paradigmatic city of ruins," the city's bankruptcy has often been framed as a starting point for change and renewal.[4] Dozens of stories have been told about Detroit's "rebirth" as a potential path to a better future, whether in the form of **community** solar infrastructure projects or in the transformation of vacant land into vibrant urban farms.[5] Indeed, the story of Detroit, "a place broke enough to survive without [the broken promises of U.S. capitalism]," inspired Raphaëlle Guidée to theorize the post-bankrupt "precarious utopia" as a place where new experiments in living can be carried out, though without any promise of success.[6] It's a provocative notion that raises the question of whether bankruptcy is the crisis we need in climate and energy politics. How might bankruptcy aid a sustainable and just energy future struggling to be born?

The trouble with bankruptcy as it is commonly understood is that it revolves around failure. As such, it entails relying on crisis—the decisive moment of failure to repay debts that have come due—to somehow reset relations and create conditions for change. But crises are impossible to direct. Those awaiting a final fiscal crisis to cripple fossil-fueled capitalism could end up waiting longer than we can afford to wait. Think back to the 1970s energy crisis, one of the previous century's most significant opportunities for energy transition, which exposed the depth of U.S. dependence on cheap fossil fuels and ignited fears of insolvency for ordinary Americans, the U.S. state, and the Western oil-consuming nations. By raising oil prices (one of the causes of the energy crisis) the Organization of the Petroleum Exporting Countries (OPEC) triggered a massive transfer of wealth in the form of "petrodollars" that threatened the financial stability and hegemony of the United States and its allies. Rather than precipitate energy transitions, that decade's energy and petrodollar crises inspired economic, diplomatic, and policy innovations through which the United States restructured and retained its hegemony while reinvigorating carbon capitalism in a neoliberal form.[7] OPEC helped to unleash a potentially devastating financial crisis that, in the end, revealed the adaptive power and flexibility of carbon capitalism. A crisis unleashed is an unpredictable thing.

Post-bankrupt Detroit may have shown that "crises can lead to progressive alternatives to neoliberalism," as Seth Schindler argues, but the suffering that accompanied the city's downfall and which continues to fester further illustrates the trouble with bankruptcy as a strategy for a new politics of energy.[8] In an age of populist rage, vicious inequality, and epistemic polarization, moreover, how often will crisis translate into progressive—or even moderate—alternatives to neoliberalism, rather than deeper immiseration and more suffering? As Randolph Starn once warned historians, we ought to be "knowing and careful" with crisis, a malleable term we would do well to think seriously about, and beyond.[9] Indeed, we might, with Janet Roitman, ask how to "narrate the future without crisis."[10] This question may seem ironic in the context of the climate crisis and energy transition—not to mention the **coronavirus** pandemic—but it is one we need to ask if we are serious about making a viable way forward.

Rather than embracing bankruptcy as the crisis we need, perhaps there is a way forward in understanding it as a source of **solidarity** and a starting point for a politics of obligation, of repaying debts. With global debt nearing $300 trillion in 2021, living on the brink of bankruptcy is an increasingly common experience that may well precipitate a decisive crisis.[11] For two centuries, we have been accumulating environmental debts—some polities much more than others—as we withdraw resources without adequate repayment to the earth and to those who are exploited by systems of excessive extraction and who suffer the uneven effects of climate change. The climate crisis is the result of social and economic systems out of balance. It threatens to make bankrupts of us all. What might be possible if we responded to this looming ecological insolvency by embracing debts to the earth, to each other, and more besides as the foundation for a new politics of obligation and repayment? It may be tempting to imagine bankruptcy as liberation from obligation and the rejection of failure. But as the American farmer and critic Wendell Berry notes, a healthy relation to the earth and to ourselves involves "production, consumption, *and return.*" It demands obligation, responsibility, and care for the Other.[12] In this sense, at least, bankruptcy may be the opportunity we need to form new ways of living after crisis.

See also: **Blockade, Commons, Decolonization, Degrowth, Economy, Sabotage**

Notes

1 Max Ehrenfreund, "Detroit Files for Bankruptcy Despite Automakers' Success," *Washington Post*, July 19, 2013, https://www.washingtonpost.com/business/economy/detroit-files-for-bankruptcy-despite-automakers-success/2013/07/19/64be8182-f09f-11e2-bed3-b9b6fe264871_story.html.

2 Raphaëlle Guidée, "The Utopia of Bankruptcy," *differences* 31, no. 3 (2020): 77.

3 Seth Schindler, "Detroit after Bankruptcy: A Case of *Degrowth Machine Politics*," *Urban Studies* 53, no. 4 (2016): 822.

4 Dora Apel, *Beautiful Terrible Ruins: Detroit and the Anxiety of Decline* (New Brunswick, NJ: Rutgers University Press, 2015), 10.

5 Shane Brennan, "Visionary Infrastructure: Community Solar Streetlights in Highland Park," *Journal of Visual Culture* 16, no. 2 (2017): 167–89; Quinn Klinefelter,

"Detroit's Big Comeback: Out of Bankruptcy, a Rebirth," *NPR*, December 28, 2018, https://www.npr.org/2018/12/28/680629749/out-of-bankruptcy-detroit-reaches-financial-milestone.
6 Guidée, "Utopia of Bankruptcy," 88.
7 David M. Wright, *Oil Money: Middle East Petrodollars and the Transformation of US Empire, 1967–1988* (Ithaca, NY: Cornell University Press, 2021).
8 Schindler, "Detroit after Bankruptcy," 832.
9 Randolph Starn, "Historians and 'Crisis,'" *Past and Present* 52 (August 1971): 22.
10 Janet Roitman, *Anti-Crisis* (Durham, NC: Duke University Press, 2013), 89.
11 Dhara Ranasinghe, "Global Debt Is Fast Approaching Record $300 Trillion—IIF," Reuters, September 14, 2021, https://www.reuters.com/business/global-debt-is-fast-approaching-record-300-trillion-iif-2021-09-14/.
12 Wendell Berry, "The Uses of Energy," in *The Unsettling of America: Culture and Agriculture* (1977; repr., Berkeley, CA: Counterpoint, 1996), 89–90.

BATTERY

Can batteries foster a radically just energy transition?

ISAAC THORNLEY

If batteries are imagined as the key to unlocking a **renewable** energy future, it is because they promise an alluring combination of continuity and change. By filling the intermittency gaps of renewable energy sources like wind and solar, they embody the desire to sustain the current scale of material and energy throughput without the negative social and ecological effects of fossil fuels. As the capacity for battery storage becomes crucial to building out renewable energy systems and electrifying sectors currently dominated by fossil fuels (including transportation), batteries are forcing a confrontation with the realities of what energy transitions actually entail. What must change in the face of the ecological degradation caused by fossil fuel dependence? What will remain the same in an economic system oriented around limitless accumulation and territorial expansion? How renewable, rechargeable, and recyclable will the next generation of batteries be? Can they foster a radically just energy transition?

A battery is a technology for storing energy to power applications at various scales, ranging from consumer electronics to electric vehicles (EVs) and **electricity** grid storage. Batteries function through the movement of atoms in a cell, each of which contains an anode (a negative electrode) and a cathode (a positive electrode); the movement of charged atoms between anodes and cathodes enables the storage and discharge of electrical energy. Batteries express a logic of modularity and substitutability. We swap out used-up batteries for ones that are fully charged, and we electrify and decarbonize by swapping out fossil fuels with batteries that store energy from renewable sources.

Technological change and market mechanisms will play a role in enabling and constraining energy transitions, but so will political struggles over the shape of supply chains for battery manufacture and the industrial systems in which they are embedded. In the fossil fuel era, energy-related extraction involved mining for the fuel itself, but the era of renewables involves a pivot toward extraction for the purpose of energy storage. The contemporary technological landscape is rapidly shifting, with various emerging battery types and chemistries (such as the sodium-ion battery) vying for dominance.[1] Lithium-ion batteries are the most common type of battery in use today, though there are multiple lithium-ion "cathode chemistries," such as nickel manganese cobalt and nickel cobalt aluminum.[2] While the uptick in battery use is part of a global process of decarbonization and electrification, it coincides with a surge in demand for these "critical minerals" and the emergence of new extractive frontiers that generate colonial dispossession and environmental injustice.[3]

Critical mineral is a political category rather than a scientific one; these minerals are defined as critical to their economy and national security, where criticality relates to economic importance and vulnerability to supply chain disruptions.[4] A 2020 World Bank report concluded that massive increases in the extraction of critical minerals would be necessary to meet growing global demand for clean energy storage, including an increase of more than 450 percent for lithium, cobalt, and graphite and 200 percent for nickel by 2050.[5] Some analyses doubt that there is "enough energy and minerals to *replace* combustion engines with electric ones and fossil fuels with other forms of *green* power."[6]

Yet the mineral intensity of the energy transition is not locked in place, and understanding batteries merely as a replacement or substitute for fossil energy defers the more pressing need to qualitatively transform industrial processes of energy production as well as social assumptions about energy use.[7] The supply chains—from the extraction, processing, and refining of minerals, to the production of batteries, their primary application, and ultimate recycling—are emergent, in flux, replete with gaps, and subject to political contestation.[8] While transition hinges on batteries, not all battery technologies or applications are equal in terms of social necessity or **resource** efficiency. The politics of EVs are particularly fraught; batteries enable the electrification of passenger cars yet thereby entrench a mode of mobility that is resource-intensive and inaccessible to many people. As Riofrancos and colleagues argue, policies that incentivize mass transit, reduce the size of EVs and their batteries, and responsibly source their minerals will foster a rapid and equitable transition.[9]

In North America, energy transition understood as an imperative for industrial and social transformation could entail more public investment in clean energy infrastructure and public transit, the creation of high-quality unionized jobs, and increased coordination between organized labor and the Indigenous and climate justice movements to force such concessions from capital and the state. In an era marked by supply chain disruptions, giving deliberate shape to these emerging industrial processes and networks will involve coordination between anti-extractive resistance movements, affordable housing and transit advocates, and organized labor in the mining, refining, recycling, and battery manufacturing sectors. If the mineral intensity of transition remains as yet undetermined, such uncertainty and the political contestation that it entails provide an opening for labor, social and environmental movements, and **Indigenous** communities to shape the future of these supply chains and energy systems—particularly the question of who owns and controls them—in a way that maximizes energy democracy and access to mobility while minimizing environmental injustice and degradation.

See also: **Africa, Digital, Mining, Transitions**

Notes

1 Keith Bradsher, "Why China Could Dominate the Next Big Advance in Batteries," *New York Times*, April 12, 2023, https://www.nytimes.com/2023/04/12/business/china-sodium-batteries.html.

2 Thea Riofrancos et al., "Achieving Zero Emissions with More Mobility and Less Mining" (Climate + Community Project, January 2023), 36, https://www.climateandcommunity.org/_files/ugd/d6378b_b03de6e6b0e14eb0a2f6b608abe9f93d.pdf; Govind Bhutada, "The Key Minerals in an EV Battery," *Elements*, May 2, 2022, https://elements.visualcapitalist.com/the-key-minerals-in-an-ev-battery/.

3 Dayna Nadine Scott, "How Does the Settler State Prime a New Extractive Frontier," *Midnight Sun*, October 6, 2022, https://www.midnightsunmag.ca/how-does-the-settler-state-prime-a-new-extractive-frontier/.

4 Riofrancos et al., "Achieving Zero Emissions," 5.

5 Kirsten Hund et al., *Minerals for Climate Action: The Mineral Intensity of the Clean Energy Transition* (New York: World Bank Group, 2020), https://pubdocs.worldbank.org/en/961711588875536384/Minerals-for-Climate-Action-The-Mineral-Intensity-of-the-Clean-Energy-Transition.pdf; Trillium Network for Advanced Manufacturing, *Developing Canada's Electric Vehicle Battery Supply Chain: Quantifying the Economic Impacts and Opportunities* (September 2022), 22, https://trilliummfg.ca/wp-content/uploads/2022/09/Report_SupplyChainReport_v3_20220705_Publish_TNAM.pdf.

6 Andrew Nikiforuk, "The Rising Chorus of Renewable Energy Skeptics," *The Tyee*, April 7, 2023, https://thetyee.ca/Analysis/2023/04/07/Rising-Chorus-Renewable-Energy-Skeptics/.

7 Matthew T. Huber, *Climate Change as Class War: Building Socialism on a Warming Planet* (London: Verso, 2022), 9–10.
8 Saima Desai and Isaac Thornley, "Greenwashing the Ring of Fire: Indigenous Jurisdiction and Gaps in the EV Battery Supply Chain," *Infrastructure Beyond Extractivism*, February 2024, https://jurisdiction-infrastructure.com/research/greenwashing-the-ring-of-fire-report/.
9 Angele Alook et al., *The End of This World: Climate Justice in So-Called Canada* (Toronto: Between the Lines, 2023), 101–2; Riofrancos et al., "Achieving Zero Emissions," 7.

BEHAVIOR

Why do appeals to values usually fail to change behavior?

STÉPHANE LA BRANCHE

Public policies aimed to change citizens' behavior tend to assume that emphasizing environmental values will inspire changes in daily practices. In order to "save the planet," people are encouraged to adjust the thermostat, become vegetarian, stop driving their cars, or throw away their clothes dryer. But such appeals have largely failed to produce the necessary changes in daily practices, routines, and preferences. If ecological arguments have not been sufficiently effective in catalyzing changes in habits and desires, then what kinds of arguments might work better?

Studies in environmental sociology that I have conducted with the International Panel on Behavior Change have yielded some surprising insights about this problem. We conducted over 450 semistructured face-to-face interviews and several large-scale online and telephone surveys. These studies have helped us to understand why it is so difficult to become an energy-sober person.

The social representation of energy has an especially significant impact on the capacity for energy change. Social representations guide us "in the way we name and define different aspects of our everyday reality, in the way we interpret them."[1] This meaning making is the reason why social representations play a more important role in behavior than information alone, and why they shape judgments about the legitimacy of public policies and awareness campaigns. Social representations are intimately linked to our preexisting practices. In the case of energy, our old habits are connected to representations that frame energy as cheap, abundant, and without negative impacts. These preexisting social networks of belief and practice slow down the adoption of new practices more than they accelerate them because learning new habits demands mental energy, time, and effort. Indeed, research on behavioral change and cognitive science tells us that the brain is, in effect, a "lazy" organ and that **habit** is a way for the brain to pursue efficiency by reducing demands upon it. Individuals who consider changing their behavior have to weigh the reasons for and possible benefits of making a change against the possible costs, including the time, money, and physical and cognitive effort involved.

Our project used empirical methods to understand how social representations shape behavior. From our research, we identified seven socio-energy profiles, which characterize various meanings that individuals give to their daily energy practices and the motivations for their behavior.[2]

Technophiles are passionate about new technologies: they tend to be early adopters who enjoy using tech, including apps related to energy consumption. But they don't necessarily understand or care about energy issues. By contrast, *Energiphiles* are highly knowledgeable about energy itself: they understand things like kilowatt-hour and peak shaving. But this interest in energy does not necessarily translate into ecological concern; their efforts to reduce kilowatt-hour consumption aren't necessarily driven by economic motives either. *Thrifties*, on the other hand, want to save money by lowering their energy bills. *Ecophiles* see energy as primarily an ecological problem; of the seven profiles we identified, only Ecophiles are drawn to programs encouraging behavioral change, which they tend to see as a fun challenge rather than a source of stress. But Ecophiles represent only about 10 percent of our respondents. On the other side of the spectrum, the *Resistant* view all energy and environmental campaigns as illegitimate injunctions to "tighten their belt"; to the extent that they accept the need for change, they believe that efforts should be made by governments and businesses rather than individual citizens. The *Indifferent* do not care about energy or its cost; they do not believe that their actions make a difference, and they're unwilling to test their beliefs about these matters. More susceptible to persuasion are the *Helpless*, for whom energy is a concern, and who are willing to attempt new behaviors if they can be shown to yield demonstrable results; however, they do not know how to act or improve on what they are already doing.

This wide range of motivations should make clear that public policies and campaigns organized around environmental values are likely to appeal only to a small fraction of the population. Moreover, social science research tends to underestimate the importance of cognitive efforts in relation to behavioral change. Such change is never easy; habit is a form of inertia that aims to minimize effort and preserve physical and cognitive comfort. People who are trying to change will aim to preserve (or increase) comfort at less or equal cost and effort, calculated according to the prevailing socio-energic logic at work. For nearly all of the profiles, efforts to reduce energy consumption would disrupt comfort and elicit stress or even anxiety. Becoming sober about energy requires an almost constant process of observation and self-assessment, with many hiccups and setbacks that require significant cognitive work to overcome—all of which discourage the adoption of new practices consistent with ecological goals. If, moreover, these efforts are perceived by the individual as unpleasant or distressing, there is even less chance they will succeed.

Ultimately, most actors (states, environmental associations, and local authorities) and prospective energy scenarios greatly overestimate the capacity and willingness of populations to make changes that would help achieve carbon neutrality by 2050. In effect, much **communication** in this area is premised on the faulty assumptions that the majority of the population are Ecophiles and that the citizenry at large will be persuaded by ecological arguments and believe that behavioral change is a fun challenge to take on.

But the relationship between individuals and energy cannot be reduced to just one of the seven profiles we identified. And as long as this diversity of positions is ignored, the messages that public policies intend to communicate will not be heard by most of those to whom they are addressed. Talking in terms that appeal to the Ecophile will mean little to other profiles. In other words, homogenous and uniform campaigns based on ecological arguments alone cannot reach their goals because they do not speak a capacious energy language. Energy awareness campaigns and energy reduction programs need to take into account a diversity of views on and attitudes toward energy if they wish to bring about real change.

See also: **Economy, Expert, Finance, Lifestyle, Vegan**

Notes

1 Denise Jodelet, *Social Representations* (Paris: University Press of France, 2012), 47–48.

2 Stéphane La Branche, "Eléments de sciences sociales de l'énergie," in *Encyclopédie de l'énergie*, last modified October 14, 2015, http://encyclopedieenergie.org/articles/br%C3%A8ve-introduction-%C3%A0-la-sociologie-del%E2%80%99%C3%A9nergie.

BLACK

What are the Black politics of energy?

WALTER GORDON

Toward the end of early African American filmmaker Oscar Micheaux's 1920 film *The Symbol of the Unconquered*, the camera pans across a broad expanse of rural landscape, dotted with oil wells and overhung by a wide cloud of black smoke. Subtitled *A Story of the Ku Klux Klan*, Micheaux's film is mainly concerned with the interracial conflicts among a number of settlers in an unnamed U.S. state over the ownership of a plot of oil-bearing land. The wide shot of that land and the cloud visualizes the scale of the turn of fortunes undergone by the film's protagonist, a Black prospector turned oil baron, and the possibilities for racial uplift afforded by his proprietorship over land rich in natural resources. Indeed, while the image of a "black cloud over a rural town and otherwise pristine landscape" might be "disturbing to the viewer in today's era of ecological awareness," as J. Ronald Green writes, the cloud was likely understood by the filmmaker and his (entirely Black) audiences as a sign of Black productivity, independence, and tenacity. While the language of the intertitle directly preceding the landscape shot—"[the land] was found to contain abundant oil fields, from which enormous plumes of smoke now *obscure the sky*"—might further excite our sense of ecological worry, historical context, as Green again notes, would suggest we resist that urge and instead read it with the "optimism of the 1910s and 1920s" recognizing that "the sense that rampant exploitation of resources might poison the well did not become a broad public concern until later."[1]

What are the political implications of a critical approach in which we eschew the ambiguity of word and image—here stimulated by our retrospective awareness of ecological risk—in favor of the certainty of historical context? An aversion to the projections of presentism and an uncomplicated reverence for the truth value of context can combine to limit critical insights to the already known and reproduce distorted visions of the past in which complexity is flattened into coherence. By rendering Micheaux's smoke cloud an unbothered picture of progress, we reproduce a traditional historical emplotment of ecological awareness that posits a gradual trajectory of increasingly universal concern emanating from and distributed through networks of whiteness, within which Black people are inserted as peripheral to the expansion of "our" contemporary awareness of global climate emergency. It is more difficult to accept this teleology when faced with transparently ambivalent Black reflections on energy like W. E. B. Du Bois's 1907 poem "The Song of Smoke":

> I am the Smoke King
> I am black!
> I am swinging in the sky,
> I am wringing worlds awry;
> I am the thought of the throbbing mills,
> I am the soul of the Soul toil kills,
> Wraith of the ripple of trading rills;
> Up I'm curling from the sod,
> I am whirling home to God;
> I am the Smoke King
> I am black.
> . . .
> I will be black as blackness can—
> The blacker the mantle, the mightier the man!
> . . .
> Souls unto me are as stars in a night,
> I whiten [my] black men—I blacken my white!
> What's the hue of a hide to a man in his might?
> Hail! great, gritty, grimy hands—
> Sweet Christ, pity toiling lands!
> I am the Smoke King
> I am black.[2]

In addition to noting the use of *black* as a positive term of racial identification in 1907 (a moment in which *Negro* still dominated), a reader today might also respond to the equally startling use of smoke—here described, seemingly disturbingly, as the "soul of the Soul toil kills, / Wraith of the ripple of trading rills"—as a material vehicle for reflections on racial **solidarity** and pride. Indeed, while smoke may appear as "the blood of bloodless crimes"—airborne evidence, against the enclosed architecture of the factory, of the violence within—it remains desirable for the speaker to be made in the image of smoky matter.

Finally, smoke *whitens* Black men even as it *Blackens* white: like coal—frequently depicted coating the skin of miners, creating epidermal and racial confusion—smoke is a political equalizer, an index of modern industry as a force destined to counterbalance the system of U.S.-American racial domination.

If Du Bois could see the blackness of smoke with such undecided eyes in 1907—as at once a signal of modern degradation and racial progress—we might imagine that Micheaux could see the dark particulate (part solid, part liquid, part gas) as a similarly fraught subject. With these complex images of the blackness of smoke, these artists open up space to imagine the *Black politics of energy*. How might one narrate or describe a Black politics of energy? How might conceptualizing a Black politics of energy allow us to better understand, following G. Ugo Nwokeji, the global movement from "slave ships to oil tankers" that links centuries of enslaved African American labor to **Africa**'s modern petrostates?[3]

Approaching questions of energy from the perspective of Black politics requires the initial step of drawing from the rich body of Black thinking on energy (which, as in the cases of Micheaux and Du Bois, has been frequently locked behind layers of metaphor), in addition to foregrounding questions of race in broader discourses of energy extraction, distribution, and consumption.[4] A Black politics of energy emphasizes the materially and ideologically co-constitutive relation between racial Blackness and the Blackness of fossil capital, which is equally reliant on black coal and black oil as it is on the management of Black labor, culture, and life.

See also: **Art, Class, Documentary, Storytelling, Trans-, Wind**

Notes

1 J. Ronald Green, *With a Crooked Stick: The Films of Oscar Micheaux* (Bloomington: Indiana University Press, 2004), 64–65; Oscar Micheaux, *The Symbol of the Unconquered* (1920).

2 W. E. B. Du Bois, "The Song of Smoke," in *Creative Writings by W. E. B. Du Bois: A Pageant, Poems, Short Stories, and Playlets*, ed. Herbert Aptheker (White Plains, NY: Kraus-Thomson Organization, 1985), 10.

3 G. Ugo Nwokeji, "Slave Ships to Oil Tankers," in *Curse of the Black Gold: 50 Years of Oil in the Niger Delta*, ed. Ed Kashi and Michael Watts (Brooklyn, NY: PowerHouse Books, 2008), 63–65.

4 Imani Jacqueline Brown and Samaneh Moafi, "Environmental Racism in Death Alley, Louisiana," *Forensic Architecture*, June 28, 2021, https://forensic-architecture.org/investigation/environmental-racism-in-death-alley-louisiana; Shane Brennan, "Visionary Infrastructure: Community Solar Streetlights in Highland Park," *Journal of Visual Culture* 16, no. 2 (August 1, 2017): 167–89; Nicolas Fiori, "Plantation Energy: From Slave Labor to Machine Discipline," *American Quarterly* 72, no. 3

(2020): 559–79; Ryan Cecil Jobson, "Dead Labor: On Racial Capital and Fossil Capital," in *Histories of Racial Capitalism*, ed. Justin Leroy and Destin Jenkins (New York: Columbia University Press, 2021), 215–30.

BLOCKADE

Why are blockades an essential form of climate action?

JOSHUA CLOVER, DEEDEE CHAO, AND EMILY RICH

The blockade is a fundamental tactic within the "repertoire of collective action."[1] Bracketing interstate conflict, it has generally referred to land-based obstruction of a thoroughfare or development site used by formal or informal social movements to prevent the passage of people, goods, or construction in ways that prevent states from controlling territory and/or inhibit capitalist operations.

Land. The blockade concerns control over, access to, protection of, and disposition regarding land. Because the blockade is a land- rather than labor-based struggle (though it may articulate labor-based concerns), it is often situated as response to dispossession (contra exploitation) and colonialism (sometimes contra capitalism). Within the repertoire of land-based **action**, the blockade exists on a spectrum of scale and complexity between barricade and land occupation and is not easily disarticulated from these tactics. A blockade's infrastructure might include metal barricades across a road; these do not constitute the blockade itself, which presupposes ongoing human presence and material reproduction. Meanwhile, the ongoing human presence is often adjunct to or summons forth a land occupation adjacent to the blockade itself. The French ZAD de Notre-Dame-des-Landes (2007–18) against an airport development exemplifies the commingling of these tactics.

Circulation. Presuming the unity rather than distinction of capitalism and colonialism, the blockade might be understood as a "circulation struggle" within the triad of *production*, *circulation*, and *reproduction*.[2] In conventional usage of the term *circulation*, the blockade frequently interferes with the movement of goods, by blockage either of transport (cf. Gilets Jaunes movement, 2018–20) or of circulatory infrastructure (cf. Wet'suwet'en blockades 2010–20). *Circulation* also retains an elaborated technical sense deriving from political economy, as in Marx's "sphere of circulation."[3] In this "noisy sphere" of exchange and consumption, commodity values are realized (intersecting in this regard with the simple circulation of goods); more broadly, the "sphere of circulation" names the social relations oriented by these acts and thus involves populations that are market-dependent but not necessarily within the formal **economy**. As with the previous distinction, the blockade can be difficult to disarticulate from physically or conceptually adjacent actions. It resembles production struggles such as the strike to the extent that its success is regularly premised on degrading capitalist profitability. A sustained blockade often operates only in conjunction with an encampment or land occupation engaged in social reproduction struggles (a nascent commune) enabling the blockade to persist.

State Violence. Because land rights are guaranteed and colonial dispossession is organized by the state, and because the sphere of circulation is generally understood as public despite subordination to private interests, blockades are immediately subject to state security and state violence via the police or military, even when the immediate object of the blockade is a private entity. Land ownership and usage are further dramatized in the case of native subjects who claim treaty rights, unceded land, or other forms of sovereignty, jurisdiction, or legal status. The blockade is thus regularly in social contest with state and capital in combination. Groups using blockades as a central tactic have had on occasion considerable success against the state (cf. Argentina's *piqueteros* 1995–2005).

Indigeneity and Environment. Because the blockade is a land-based or circulation struggle, and correspondingly available to those outside the formal economy, it is rightly though not absolutely associated with anticolonial struggles and in particular with **Indigenous** struggles against or in the aftermath of land dispossession, given that "territoriality is **settler colonialism**'s specific, irreducible element."[4] Such struggles often feature a sense of relation to the land distinct from Western conventions of ownership. This relation converges with the blockade's use against environmental despoliation. While the blockade is a long-standing tactic of Indigenous struggle, it takes on increasing salience in North America (often under the heading *Land Protectors* or *Water Protectors*) in response to the extraction and transport of carbon, ore, and minerals, which have occasioned original and persisting dispossession. The pivotal moment in the contemporary politics of the blockade was perhaps the resistance to the Mackenzie Valley Pipeline (Canada), first proposed in 1974 but never constructed after organized Indigenous resistance, including the Dene Declaration of 1975.[5] Recent examples of Indigenous-led land struggles in the "Anglo settler colonial countries of Canada, Australia, New Zealand, and the United States" (CANZUS) include, respectively, the Idle No More rail blockades (2013), the

Maules Creek coal mine blockade (2014), the Ihumātao antidevelopment blockade (2019–20), and the Dakota Access Pipeline blockade (2016–17).[6] To this we might add Hawaii's Thirty Meter Telescope blockade (2019) as a distinct scene of **Indigenous activism**.

Frequency. Blockades have become increasingly common since the 1970s, particularly in the capitalist core and in response to global projects organized by core nations. Glen Sean Coulthard writes, "All negotiations over the scope and content of Aboriginal peoples' rights in the last forty years have piggybacked off the assertive direct actions—including the escalated use of blockades—spearheaded by Indigenous women and other grassroots elements of our communities."[7] Kristin Ross notes, "In recent years the rise in the number of occupations and attempts to block what have come to be known as 'large, imposed, and useless' infrastructural projects bears witness to a new political sensibility."[8] There are a number of overlapping and conjoining possible reasons for the increased use of the blockade within the repertoire of collective action. These include intensification of land-based struggles in place of waning labor-based struggles, the availability of circulation struggles to those excluded from production, land protection in response to environmental threat, postindustrial shifts toward logistical and infrastructural development, and increasing pressures on extractivism.

Crisis. All of these proposed reasons for the increased frequency and intensity of blockades are entangled with the multiple crises that emerged in the early 1970s and have since congealed and worsened. These crises might be aggregated under the headings *end of growth* and *climate collapse*, signaled by the steep decline in capitalist profitability around 1973 and the "birth of the modern environmental movement" often dated to 1970. These crises provide the two limits within which political struggle takes place. Among other things, they imply a recolonialization of politics even in the capitalist core, given both the impossibility of expanding wage discipline and the inevitability of land-based politics given existential character for both capital and humans of control over resources and their use. Given that this double crisis cannot be reversed within capitalism, the blockade's political importance is certain to increase.

See also: **Bankrupt, Keep It in the Ground, Mining, North Dakota, Pipeline**

Notes

1 Charles Tilly, "Speaking Your Mind without Elections, Surveys, or Social Movements," *Public Opinion Quarterly* 47, no. 4 (Winter 1983): 464.

2 Joshua Clover, "Riot, Strike, Commune: Gendering a Civil War," *New Global Studies* 14, no. 2 (July 2020): 121–31; Clover, *Riot.Strike.Riot: The New Era of Uprisings* (London: Verso, 2019).
3 Karl Marx, *Capital*, vol. 1 (London: Penguin 1992), 279.
4 Patrick Wolfe, "Settler Colonialism and the Elimination of the Native," *Journal of Genocide Research* 8, no. 4 (December 2006): 388, https://doi.org/10.1080/14623520601056240.
5 Steve Darcy, "The Dene Declaration" (Public Autonomy Project, August 23, 2018), https://publicautonomy.org/2018/08/23/the-dene-declaration/.
6 Robert Nichols, *Theft Is Property! Dispossession and Critical Theory* (Durham, NC: Duke University Press, 2020), 6.
7 Glen Sean Coulthard, *Red Skin, White Masks: Rejecting the Colonial Politics of Recognition* (Minneapolis: University of Minnesota Press, 2014), 166.
8 Collectif Mauvaise Troupe and Kristin Ross, *The Zad and NoTAV: Territorial Struggles and the Making of a New Political Intelligence* (London: Verso, 2018), ix.

CARBON MANAGEMENT

Is carbon management a moral imperative or a moral hazard?

REBECCA TUHUS-DUBROW

Some see the accumulation of carbon dioxide in the atmosphere as a symptom of a sick society and a sign that we need to overhaul that society. But from another perspective, rising concentrations of atmospheric CO_2 might be seen as mere side effects of modernity and prosperity, problems that can be managed. In the latter view, fossil fuel consumption should not be moralized. Instead, as journalist Elizabeth Kolbert has written in her characterization of this ethos, carbon dioxide emissions "should be regarded much the same way we look at sewage."[1] We don't judge the production of excrement; instead, we have systems to manage it, by preventing it from fouling the environment and harming human health.

And so, this line of thinking goes, we must treat CO_2 similarly. We must prevent it from entering the air and oceans, not only by using alternative forms of energy but by trapping carbon dioxide at the source at power plants. We must also remove legacy emissions that are already in the air and oceans—cleaning up the mess that was allowed to accumulate for centuries. Just as bottles can be recycled and manure can serve as fertilizer, CO_2 can be put to use—in root beer, say, or to enhance plant growth in greenhouses, or to produce assorted carbon-based materials. Lululemon, the high-end yoga wear company, is partnering with the startup LanzaTech to convert carbon emissions into feedstock for polyester. Alternatively, carbon might be securely stored: the Swiss company Climeworks permanently binds its captured CO_2 to rock deep underground at a facility in Iceland.

These approaches have given rise to a flurry of overlapping buzzwords and acronyms: carbon capture and storage (CCS); carbon capture, utilization, and storage (CCUS); and carbon dioxide removal (CDR), sometimes known as simply carbon removal and also known as negative emissions technologies (NETs). They are all examples of carbon management.

Carbon management is not yet a household term, but it is becoming more widely used in academic and policy circles, from the International Energy Forum to Columbia University, which recently established a Carbon Management Research Initiative. In July 2021, the Biden administration officially changed the name of the Department of Energy's Office of Fossil Fuels, rechristening it the Office of Fossil Fuels and Carbon Management. "Our new name is also a new vision," the web page reads. "The Biden-Harris Administration recognizes that to meet our climate goals, we have to manage the carbon that comes with the legacy and continued use of fossil fuels."[2]

For some, carbon management may be the optimal answer to climate change. Others might support it more reluctantly—more out of resignation than out of can-do optimism—and endorse it only as a necessary supplement to a rapid transition away from fossil fuels. The math is so punishing that we simply must pull vast amounts of carbon out of the air: the 2018 **IPCC** special report concluded, "All pathways that limit global warming to 1.5°C with limited or no overshoot project the use of CDR on the order of 100–1,000GtCO2 [billion tons] over the 21st century."[3] *Nature-based* solutions such as planting trees and restoring mangroves are important but not sufficient.

The efficacy and necessity of CCS—capturing carbon at the source of power plants and other industrial processes—may seem more debatable, because in theory it might be possible to rapidly shift away from that infrastructure. But to dismiss CCS is to ignore all of the fossil fuel plants that have recently been built or are in construction throughout the world, particularly in poorer countries whose populations have yet to enjoy the benefits of abundant energy. "These facilities have economic lives of decades," noted a 2020 report, "and a large global fleet of coal- and gas-fired power stations are expected to remain in operation well past the middle of this century." The report argued, "This next decade will be central to any successful climate strategy, and respecting the primacy of carbon management is essential for success."[4]

Some environmental groups worry, with good reason, that these approaches could serve as a lifeline to the fossil fuel industry; indeed, the industry has started to embrace them. The term often invoked is "moral hazard." But at this point, rejecting out of hand the various methods of carbon management is no longer viable. One of the most prominent voices to make this case is Olúfẹ́mi Táíwò, a Georgetown University philosopher. He argues that the question is not *whether* to pursue large-scale industrial carbon removal, but *how* to do so in a just way: "Arguments that keep meticulous track of what is in it for ExxonMobil are often uncurious about what will happen to Africans and global Southerners in the scenarios they describe."[5] After all, the Global South is poised to suffer the most from the effects of climate change that these techniques have the potential to mitigate. Carbon management may well be a moral hazard; it's also, at this point, a moral imperative.

The concept of carbon management is at once uninspiring and comforting, suggesting that someone is going to take charge of a situation that is manifestly out of control. The term implies that if, say, you lose your house in a climate-related megafire, or even if you're just distressed about the discombobulated seasons, you could complain to management. It hints that we, as a global community, might cope and muddle through—not necessarily thrive, certainly not solve climate change or save the planet, but manage. It is deeply unsexy. Instead of the hubris of **geoengineering** or the romance of living in harmony with nature, we get the bureaucracy of carbon management.

Carbon management extends the promise, and the threat, that we can continue on essentially the same path, retrofitting some power plants and adding a new sector to the economy—the circular carbon economy, some have called it.[6] How can we pursue the technologies we need while resisting the inertia they encourage? The challenge is to pursue carbon management in ways that disrupt old power relations rather than entrenching them: for example, the companies and countries that have benefited most from fossil fuels could be compelled to remove their legacy emissions from the atmosphere. Another challenge is to deploy such technological solutions while maintaining momentum for deeper changes, from remaking our energy system to reimaging urban planning to reconsidering our material desires. We must manage as best we can. But we must not merely manage.

See also: **Net Zero, Resource, Retrofit, Sustainability, Waste**

Notes

1 Elizabeth Kolbert, *Under a White Sky: The Nature of the Future* (New York: Crown, 2021), 152.
2 Office of Fossil Energy and Carbon Management, "Our New Name Is Also a New Vision" (July 8, 2021), https://www.energy.gov/fecm/articles/our-new-name-also-new-vision.
3 Intergovernmental Panel on Climate Change, "Summary for Policymakers," in "Global Warming of 1.5°C" (2018), https://www.ipcc.ch/sr15/chapter/spm/.
4 Columbia/SIPA Center on Global Energy Policy, "Net Zero and Geospheric Return: Actions Today for 2030 and Beyond" (2020), https://www.energypolicy.columbia.edu/research/report/net-zero-and-geospheric-return-actions-today-2030-and-beyond.
5 Olúfẹ́mi Táíwò, "An African Case for Carbon Removal," *Africa Is a Country*, September 29, 2020, https://africasacountry.com/2020/09/an-african-case-for-carbon-removal.
6 International Energy Forum, "The Circular Carbon Economy" (n.d.), https://www.ief.org/programmes/circular-carbon-economy.

CIVIL DISOBEDIENCE

What rules must be broken to build a better world?

ALICIA MASSIE

Today we find ourselves flattened under a boot darkened by the stain of capitalism and dripping crude oil. The time for talk is finished. It's time for **action**.

Human history is a history of struggle. A history of a few having and the majority having not. It is also a history of those having not coming together and telling those who have: *enough*.

Fifteen hundred years ago, the plebeians, the people of Rome without land, vote, or voice, rose up in the *secessio plebis*, the secession of the plebes. In what is often called the first general strike in history, they abandoned their city in nonviolent action and refused to serve in an army to defend their oppressors. The plebeians' collective action was so effective that it fundamentally transformed the Roman system of law and governance and won representation for them, the plebeian tributes. Thus, alongside law and order, we inherit another civility from Roman predecessors: dissidence, uprising, and disobedience.

Civil disobedience means asking difficult questions. It means standing up and saying something is wrong. It means making people uncomfortable, angry, and awkward. It means being uninvited, disliked, and sidelined. It means getting in the way, breaking things, and forcefully saying *no*. It means putting your happiness, your freedom, your life at risk. Fundamentally, it means rejecting those rules, laws, and conventions put in place to restrict you and your vision for a better world.

Civil disobedience tactics are rarely popular. I've stood in a central intersection in Vancouver, Canada, being screamed at by a port worker as we blocked his route to work. I've sat on a bus silently cursing whoever was causing my transit detour and my resulting lateness. Throwing soup at a Van Gogh masterpiece, chaining yourself to a tree or a fence, blocking traffic, suspending yourself from an important bridge, deflating SUV tires, encasing your limbs in concrete, occupying an office until you are physically dragged from the building—none of these things are likely to get you praise from the public. Blowing up a **pipeline**? Not likely to be popular either. And yet, as Andreas Malm tells us in *How to Blow Up a Pipeline*, "the ruling classes really will not be talked into action. They are not amenable to persuasion; the louder the sirens wail, the more material they rush to the fire, and so it is evident that change will have to be forced upon them."[1]

As Malm's history of revolution reveals, the most successful movements combine nonviolent *and* violent action. Apartheid, civil rights, suffrage, labor rights, political revolution, religious upheaval, **democracy**—pick your historical struggle: you will find those in power staunchly defending their position. And you will also find a political turning point where direct action, and the threat of even greater future action, finally cowed leaders to accept some form of compromise with those challenging them.

Often it is only through historical distance that we come to view disruptive actions in a positive light. It is only in retrospect that we view Martin Luther King Jr.'s actions as almost universally heroic, a model of **protest** and nonviolence. During his lifetime, Dr. King was wildly unpopular: following his "I Have a Dream" speech, 74 percent of Americans believed mass demonstrations hurt his cause.[2]

In our current struggles, we must shake off our chains of **law** and order, and our hegemonic rule of obedience. We must listen to our ancestors, philosophers, and leaders who remind us to think for ourselves. They point us to rebellion and disobedience not to reject society or reason, but to reclaim what is rightly ours: our knowledge and conscience of what is right and just, and our authority to act to demand it.

Civil disobedience is revolutionary not only in action but in thought. It pries from the cold hands of "the way it is" the seed of something greater: a future where the world can be something different. It allows us to realize that we reject the rules that govern us not out of action *against* something, but rather as action *toward* something greater.

To fight a catastrophe on the scale of the climate disaster we face, it will take all kinds of action. It will take backroom political plotting, international summits, academic peer review, and policy reform. But it will also require a revolutionary reimagining and rejection of the status quo that will shake the foundations of what is known and comfortable. It will take arrests, violence, and many people being late for work.

There is unimaginable human and ecological cost to inaction. Civil disobedience, in most of its forms, seeks to disrupt systems that are inherently violent. Climate change has already brought and will continue to bring death and destruction upon us. Starvation, plague, displacement: if climate change is a tsunami, we're already experiencing its first big waves.

You may not always like the ways of others who are disobedient. You may find their means irritating, grating, or "too much." You may cringe at their earnestness, shake your head at their youth, and think you know better. You may hear others (but quietly echo the thoughts to yourself) say that "this isn't the way to get things done," "you are asking for something impossible," or "protesting is fine, but this is not the way."

But if you find yourself, like me, comfortably seated in a position where *you* will not be the first to experience the devastation of climate change, it is time to look uncomfortably at your role in this struggle. Not everyone has the luxury of participating in a **blockade**, taking part in a sit-in, or facing arrest or police violence without the threat of death. I do. Perhaps you do, too.

See also: **Class, Organize, Sabotage, Solidarity**

Notes

1 Andreas Malm, *How to Blow Up a Pipeline: Learning to Fight in a World on Fire* (New York: Verso, 2021).

2 R. J. Reinhart, "Protests Seen as Harming Civil Rights Movements in the '60s," *Gallup*, January 21, 2019, https://news.gallup.com/vault/246167/protests-seen-harming-civil-rights-movement-60s.aspx.

CLASS

Why is climate change a class issue?

VALERIE UHER

Climate change is an epoch-defining symptom of class oppression. Yet energy transition does not play much of a role in today's working-class movements. Understanding class as a historically contingent and malleable social category could help orient more working-class movements toward energy transition. Indeed, insisting on the elastic and responsive politics of working-class formation may be the key for getting us out of the mess we're in.

To understand this crucial aspect of class, it's important to grasp the distinction between class as an economic category and class as a social formation. As an economic category, *class* names objectively determined positions in the process of production; these positions constitute a "class relation" in which the ruling class maintains power through the subjugation of the lower class(es). In a capitalist mode of production, the class relation involves two positions: the capitalist class, who own the means of production and pursue profit, and the working class, who sell their labor power to the capitalists in order to live.

But what of class as a social formation? According to Marx, working classes are not simply the mass of people who sell their labor or various job categories accreted into one larger socioeconomic category (like "potatoes in a sack form a sackful of potatoes")[1] but instead exist because of an act of self-constitution, with individuals choosing to combine into a political unit that acts "for itself"—denaturalizing the economic class relation through this act of **solidarity** predicated on choice, rather than chance. This self-constitution occurs when people occupying a particular class position unite as political actors to threaten the dominant class, and *class formation* refers to these groups who coalesce (often temporarily) as collective social units through class struggle. Working-class formation might occur wherever exploitation enacted by the capitalist class is clearly visible and political collaboration is possible, such as in an industrial workplace or at sites of **resource** extraction. There, groups of working-class people might come together to fight for fair treatment in the face of capitalist power, appearing in such well-known forms as unions, committees, action groups, or political parties. They can also include bands of saboteurs, members of blockades, and protestors blocking circulation routes.

Working classes as social formations are therefore not static entities, nor even cultures, but instead outbursts of collectivity shaped by the particular circumstances of work and life in a given era. And despite recent upticks in unionization efforts in the United States, as well as the continued prevalence of strike tactics on a global **scale**, neoliberal capitalism has created conditions that negatively impact working-class formation. Work continues to become more isolated, dispersed, and precarious, with workers driven punishingly to the limits of their productivity, their performance scrutinized with increasingly exacting technical precision. At the same time, underemployment—a result of both underdevelopment in the Global South and deindustrialization in the Global North—has forced increasing numbers of workers into more risky and piecemeal employment, where it's often impossible to know who the boss is, let alone one's coworkers. The growth of the service sector and its micro-hierarchies of managerial classes has further muddied the waters of the class relation, and austerity attacks have vitiated organized labor and left political parties. These tendencies are like a toxic sludge poured over the (once) fertile ground of working-class formation.

The climate crisis adds a new layer of brutality to working-class experience, intensifying the degradations and misery of wage labor. But the uneven impact of climate catastrophe also makes plain the catastrophic consequences of the class relation and opens up new avenues for cooperation and solidarity, potentially igniting new kinds of working-class formations. A new focus on class as "ecological relation" might invigorate both climate movements and those fighting against class inequality.[2] Destruction of land and environment has always been a rallying point for working-class politics, but the finality and destructive capacity of climate change add a new level of degradation to working-class life, further revealing the capitalist as class enemy.[3] The working-class formation that needs to occur, as many activists and scholars have acknowledged, is a political and social movement engaged in de facto struggle against climate change: class struggle is climate change struggle, and vice versa.

A politics of energy that makes class struggle fundamental to fighting climate change locates itself wherever the ambush of capitalist power occurs and employs working-class labor tactics such as the withdrawal of labor and the disruption of production, supply, and circulation via the logic of the strike—which remain the most effective tactics for stealing power from the capitalist class. Equally important, climate struggles are energized by class struggles that focus on human betterment *here and now*. The past thirty years of environmental movements have made clear that the public is not motivated by either "the image of enslaved ancestors" or the image of "liberated grandchildren," no matter how much information is transmitted about the coming catastrophe and its causes.[4] Of course, bridging climate struggle and class struggle in established labor unions or left political parties is far from easy, and environmentalists might point to numerous examples of groups who might be read as "working-class formations" advocating for continued fossil energy use. Perhaps such groups are better understood not as working-class formations but instead as infrastructures that may have arisen through class struggle but need to be reappropriated by workers and redirected toward the promotion of life—working-class life as well as the lives of **animals**, plants, and all more-than-human species.

A working-class formation predicated on an alignment between class struggle and climate struggle acknowledges that the possibility of climate collapse requires new kinds of comrades and relationships of solidarity across the "web of life" because all life forms are under threat of extinction.[5] I propose the name *energy class* for this class formation, which is brought into existence anywhere the life-seeking interests of those in this energy class are visibly antagonistic to the death-seeking interests of capital, whether in flash point struggles at locations of energy extraction, circulation, and processing, or, utilizing the power and infrastructures of unions, via the withdrawal of labor through strikes, protests, riots, or **sabotage** of energy infrastructures. As a member of the energy class, I view capital as the *death class*, my antagonist. *Energy class* names a position to rally from, a place to look for one's comrades, a praxis and politics to navigate the way forward with, and an occasion for hope, connection, and **action**.

See also: **Blockade, Civil Disobedience, Community, Organize**

Notes

1 Karl Marx, "The Eighteenth Brumaire of Louis Napoleon," in *Marx: Later Political Writing*, ed. Terrell Carver (Cambridge: Cambridge University Press, 1996), 117.
2 Matthew T. Huber, *Climate Change as Class War: Building Socialism on a Warming Planet* (London: Verso, 2022), 187.
3 The Salvage Collective, *The Tragedy of the Worker: Towards the Proletarocene* (London: Verso, 2021), 90.
4 Walter Benjamin, "Theses on the Philosophy of History," in *Illuminations*, ed. Hannah Arendt (New York: Schocken Books, 1986), 260.
5 Jason W. Moore, *Capitalism in the Web of Life: Ecology and the Accumulation of Capital* (New York: Verso, 2015).

CLEAN

What does clean energy really mean?

GÖKÇE GÜNEL

"Here, this is clean energy." A Chinese engineer at a solar power station in coastal Ghana pointed to the young man wiping off photovoltaic solar panels with a bright red mop and chuckled at his own joke. Both men wore high-visibility safety vests that set them apart from the lush but dusty landscape. Unlike the engineer, the solar panel cleaner also donned a safety helmet and knee-high rubber boots, a sign of longer periods of time spent outside in the elements.

Opened in 2018, the Meinergy plant employed three Chinese engineers of different positions and thirty Ghanaian and Togolese staff serving various blue-collar roles including cleaner, security guard, and control room operator. The plant was one of the two 20-megawatt solar power producers in Ghana, both owned and operated by the same Chinese technology firm, Beijing Fuxing Xiaocheng Electronic.

Ghana's investment in solar power stations had followed its deep **electricity** crisis between 2012 and 2015, known as *dumsor*, meaning "off and on" in Twi. Instigated by low water levels in hydroelectric dams due to climate change, disruptions to **natural gas** flows from Nigeria, and alleged mismanagement of the grid infrastructure, dumsor had impacted all areas of life in Ghana. Newspapers reported on the buzzing sound of generators commonly heard in Accra's central neighborhoods, a stopgap measure affordable only to a select few. According to the Ghana Employers' Association, about thirteen thousand people lost their jobs. Businesses collapsed. Sparking political unrest and demonstrations against the incumbent National Democratic Congress government, the electricity crisis earned Ghana's then-president John Mahama the nickname "Mr. Dumsor." Publicized widely in major media outlets around the world, the crisis jeopardized Ghana's newfound position as a lower-middle-income country.

In seeking to contain the crisis, the Electricity Company of Ghana (ECG), the sole electricity distributor servicing the south of Ghana, signed forty-three power purchase agreements with international vendors, which by 2021 had resulted in a surplus of power on the grid and posed novel threats to the economic well-being of the country. New independent power producers from countries such as China, Turkey, and the United Arab Emirates set up plants in Ghana, relying on heavy fuel oil, natural gas, and solar power. Even though their combined electricity output was only a small percentage of the grid's 5,000-megawatt electricity volume, the two Chinese solar power stations carried a symbolic weight among energy experts, who found a unique possibility in them: could Ghana perhaps forgo fossil fuels and leapfrog to a **renewable** energy future?

Like many power plants, the Meinergy solar power station was a world of its own. A high fence surrounded the manicured solar field, marking out acres of space for approximately 65,000 photovoltaic panels and 400 inverters, alongside a house for employees and an administrative building. The three Chinese men, all living on-site with no families, jogged along this fence every morning as part of their exercise routine. In their daily runs, they identified stolen cables and broken solar panels, tended to some papaya trees, and monitored unwanted vegetation that could potentially cast shadows on the panels, before retreating to their office building for the rest of the day. Ghanaian and Togolese employees lived in the nearby town, Winneba. Many of them had grown up there. During the sandy months of Harmattan, a period of particularly low efficiency for solar power stations across West Africa, twelve cleaners took turns in mopping the panels, and said they wished it rained more often. Although the language of operation on this plant was mainly English, several members of staff had picked up some Mandarin in the two years since the founding of the solar power station. All electricity produced on-site went into the national Ghanaian grid, managed by the ECG. The staff's collective, ephemeral labor culminated in what my Chinese guide described as clean energy.

In describing the energy produced at the station as "clean," the engineer was playing upon the word's multiple meanings. *Clean* described a set of tangible solar panels that required constant wiping to retain their efficiency, alongside other kinds of repair and maintenance. But the same Harmattan dust that rendered the panels unclean did not soil the papaya trees. In this context, the word *clean* demonstrated that there was a systematic material order in place in the photovoltaic plant. In complying with this order, staff identified and eliminated the elements that threatened productivity, achieving a kind of temporary purity.

The second meaning of *clean*, which rendered the engineer's comment a witticism, was its national and planetary **scale** abstraction. In its second meaning, clean, more specifically clean energy, referred to a global movement to replace or supplement fossil fuels with renewable energy resources, thereby mitigating climate change and rescuing humanity from present and future environmental catastrophes. Even though the construction of renewable energy power stations did not substitute for fossil fuels in the present but simply added to the energy mix, this movement still strove to safeguard humanity against a time when fossil fuels would be depleted or too difficult and costly to extract.

The clean commodities that facilitated the emergence of clean energy worldwide, ranging from solar panels to electric cars to **wind** turbines, did not challenge global political and economic hierarchies but instead entrenched the existing order. In other words, when used as an adjective to describe these commodities, clean did not indicate equitable or just. For instance, by wiping solar panels, cleaners helped the Ghanaian government achieve its future climate change mitigation goals, while in some ways warding off dumsor. But many energy professionals in Ghana believed that their country did not have to be held accountable for climate change impacts they consistently experienced, as they had not participated in the more than 150 years of industrial activity that resulted in such high concentrations of greenhouse gases in the atmosphere. Their perspectives often followed the arguments in the Kyoto Protocol, the first international climate agreement to acknowledge developed countries' higher levels of emissions to place on them a heavier responsibility for mitigating climate change. Regardless, a solar energy future was appealing, as it would allow for the expansion of the grid in what many energy professionals perceived to be an ethical manner and demonstrate that Ghana could skip a proverbial stage of development. A utopian goal that manifests itself in the form of corporate enterprises of different sizes and shapes, such national- and planetary-scale cleaning depended on the thankless task of dusting solar panels each and every day.

See also: **Africa, Air Conditioning, Autonomy, Decolonization, Solar Farm**

COMMONS

What forms of governance could solve the tragedy of the commons?

ALFONSO GIULIANI

In 1968, a photo taken from space offered a startling view of Earth "rising" from the Moon. This spectacular image appeared amid an ongoing debate about the finitude of the planet's resources, catalyzed in part by the acceleration of economic growth after World War II. That same year, Garrett Hardin, an American biologist, entered this debate with his infamous article, "The Tragedy of the Commons."[1] Hardin dusts off nineteenth-century Malthusian theses by turning to British economist W. F. Lloyd, who argued that the increasing use of natural resources to meet growing population pressures would inevitably bring ecosystems to crisis.[2] To explain this crisis, Hardin introduces the metaphor of the "common pasture." In an open-access pasture, each individual farmer is driven to add another animal to their herd to expand their personal wealth. This increase, in turn, burdens the reproductive capacity of the shared **resource**. For Hardin and his advocates, the result is obvious. Since it is impossible to expand the pasture—Earth—increases in wealth will inevitably lead to overexploitation and destruction: the tragedy of the commons. Humanity, according to Hardin, is therefore faced with an enormous problem at a planetary **scale**: what strategies can be adopted to avoid this tragedy? Hardin saw only two alternatives: the management of these resources could be left to either private interests or the state.

However, Hardin's argument about the tragedy that will befall the commons does not hold up. His metaphor captures only the final phase of the enclosure of common lands, a process linked to the needs of emergent agricultural entrepreneurs to raise sheep for the thriving wool market. The enclosure of the commons meant that any land that remained common was ever more intensely exploited by poor communities who depended on it. Hardin also understood common lands as *res nullius*, that is, as completely unmanaged.[3] But common lands have always been managed through customs, habits, and laws that sustained them for the common good over centuries. Economist Elinor Ostrom, to whom we owe much of the rediscovery of the commons, understood that the natural resources she studied and defined as common-pool resources (CPRs)—lakes, rivers, irrigation systems, and more—put the lie to Hardin's assertions. CPRs were "governed collectively," and their management was efficient and sustainable through time.[4]

In contrast to Hardin, Ostrom also offers a more robust and nuanced understanding of the different forms of consumption, competition, and cooperation that drive access to and use of CPRs. She draws a distinction between *rivalry*, in which one person's consumption reduces the amount of a resource available to others, and *excludability*, in which one person can prevent the consumption of others through their ability or willingness to pay for goods. Different kinds of resources are more or less susceptible to rivalry and excludability. Some resources are less excludable than others because the ability to enforce exclusion is limited or costly; consider, for example, the difficulty of excluding others from drawing water from a lake. As a mode of competition, rivalry is associated with the relative subtractability of a resource. When a user subtracts from a stock of the resource, the stock becomes depleted so that there is less available for others to use. Subtractability is also conditioned by flow—the rate of movement in and out of stock. In the case of a lake, flow would involve how much **water** was being taken out how quickly, as well as whether and how quickly the lake was replenished through precipitation, springs, or creeks. Flow and stock are interdependent, and their balance must be ensured for a resource to remain sustainable.[5]

Conventional discourse on commons, common goods, and common property seeks to understand the management and ownership of different kinds of goods, resources, and services. A standard (and simplistic) view holds that some inherent characteristic of goods determines whether they are held in common. However, there is neither any intrinsic quality nor any specific economic criterion that allows absolute distinctions among public, private, and common. Instead, these distinctions are the result of political choices, agreements dictated by power and political relations in any given historical moment.[6] Consider again water, which is often considered a common good, in part because of the difficulty of excluding its use by everyone. And yet, all over the world, water has been subject to privatization, made exclusive by force or law. For something to be common requires forms of governance that challenge social, cultural, and political mechanisms of excludability and that suspend the forms of rivalry that can emerge when resources are extracted. Any resource, good, or service can potentially be made common, even if it is susceptible to competitive pressures of rivalry, excludability, and subtractability that shape how its use is managed.[7]

Just as water can be common, so, too, can fossil fuels; there's nothing intrinsic about black gold that makes it impossible to hold it in common. And just as collective decisions can be made about any good held in common, all that stands in our way of deciding to no longer use fossil fuels is political forces that have done their best to make rivalry and excludability the way of the world.

See also: **Autonomy, Community, Design, Economy, Waste**

Notes

1 Garrett Hardin, "The Tragedy of the Commons," *Science* 162, no. 3859 (1968): 1243–48.
2 William Forester Lloyd, *Two Lectures on the Checks to Population* (Oxford: Oxford University Press, 1833).
3 *Res nullius* is a Latin term derived from Roman law that means literally "thing that has no owner."

4 Elinor Ostrom, *Governing the Commons: The Evolution of Institutions for Collective Action* (Cambridge: Cambridge University Press, 1990).
5 Ostrom, 30.
6 Alfonso Giuliani and Carlo Vercellone, "From the Neo-institutional Economy of the Commons to the Approach of the Common as a Mode of Production," *South Atlantic Quarterly* 118, no. 4 (2019): 767.
7 Carlo Vercellone, "From the Crisis to the 'Welfare of the Common' as a New Mode of Production," *Theory, Culture & Society* 32, no. 7–8 (2015): 85–99.

COMMUNICATION

What are effective strategies for climate communication?

ANDREW REVKIN

More than half of the several million words I've written on climate change since 1988 were typed under the presumption that I could drive clean-energy progress through communication.

I was convinced that the story of climate change that was inside me, absorbed through countless discussions with scientists across an array of fields, from the Arctic to the Amazon, could spill out into the public sphere and jolt people out of their fossil-fueled comfort zones.

I pursued this strategy in a widening array of media—magazines, books, the *New York Times*, an Arctic **documentary**, and multimedia packages. I piled up awards for conveying the environmental and societal impacts of the accumulating atmospheric blanket of tens of billions of tons of heat-trapping gases, mainly carbon dioxide emitted by the modern industrial surge.

It was in 2006, eighteen years into that journalistic journey, that I first interviewed a social scientist, and then a behavioral scientist, and another, and another. Something of an existential nightmare unfolded as I learned about the deep-rooted biases shaping how different people perceive, misperceive, ignore, embrace, or reject the same data and narratives. It was existential because my entire career was predicated on the idea that information matters, and here I was shown reams of peer-reviewed research illuminating how status-quo bias, confirmation bias, cultural cognition, and the finite basket of worries that we all carry contribute so powerfully to making global warming what some scholars call a "super wicked" problem.[1]

The 2006 catalyst for my exploration of the climate and energy communication challenge had been an interview with Helen Ingram, an expert on environmental public policy, then at the University of California, Irvine. At the time, debate was building around *An Inconvenient Truth*, the 2006 documentary chronicling former vice president Al Gore's climate campaign.[2]

Ingram explained matter-of-factly that the kinds of problems that get attention are "soon, salient and certain." Compare human-driven warming with gasoline price spikes or a recession (let alone the pandemic to come), and it's not hard to understand why I began writing about what I called our "inconvenient mind."[3]

Of course, other powerful forces have impeded a national and global shift to clean energy to avert the worst warming outcomes. Industries extracting or dependent on cheap, plentiful fossil fuels decades ago began to exploit those behavioral and societal patterns by hiring the world's best storytellers—public relations firms like Edelman and Ogilvy & Mather (now Ogilvy). These firms, well aware of the science of persuasion (or confusion), built not just compelling counternarratives but also holistic strategies designed around America's political landscape. Edelman in late 2021 claimed to be pivoting toward promoting climate **action**, but whatever such work it does now would, in effect, aim to undo decades of its own handiwork.[4]

Despite these challenges, I still get up in the morning encouraged and engaged because that same body of social science, while revealing sources of division and delay, also reveals a powerful, but very different, communication path to clean-energy progress—one dependent upon listening as much as (and perhaps more than) **storytelling**. A vivid example emerged in 2015 when journalist John D. Sutter visited Woodward County, Oklahoma, as part of a worldwide CNN Opinion project he was writing and shooting ahead of the climate treaty conference that resulted in the **Paris Agreement**. The county, according to a Yale survey he cited, was the place in America most skeptical about the risks of global warming.[5]

Among a host of residents Sutter interviewed was Randall Gabrel, the fifty-three-year-old owner of a local oil company who paid for a statue of a stegosaurus in town with a plaque reading, "A dinosaur like this roamed the Earth 5,000 years ago." At one point in the interview, Gabrel says, "God controls the environment"—not a promising start. But then he tells Sutter he's spent $30,000 on solar panels, adding, "We're going to try to have a solar system big enough to supply all the needs of our house." And then Gabrel says, "If everyone goes to solar, and that works, and that shuts down the oil and gas industry, I'm good with it."

Of course, Gabrel's motivation is independence, not environmental concern, and that same libertarian streak pretty much guaranteed he didn't vote for Hillary Clinton a year later.

But I'm good with that.

If Sutter or anyone else entered Woodward pressing the case for a climate emergency, that effort would surely fail. But when journalists (or citizens) seek common ground on building a healthy energy future independent of existing centers of power, a very different prospect is revealed. Some nascent efforts at a listening approach to forging climate and energy progress are well worth exploring, including the Rural Climate Dialogues and Rural Energy Dialogues run in Minnesota by the Center for New Democratic Processes and Institute for Agriculture and Trade Policy.[6]

The climate crisis is sufficiently wicked that this approach to communication is no magic bullet. But it's a great start.

See also: **Autonomy, Behavior, Solar Farm**

Notes

1 Kelly Levin, Benjamin Cashore, Steven Bernstein, and Graeme Auld, "Overcoming the Tragedy of Super Wicked Problems: Constraining Our Future Selves to Ameliorate Global Climate Change," *Policy Sciences* 45 (2012): 123–52, https://doi.org/10.1007/s11077-012-9151-0.

2 Andrew C. Revkin, "Yelling Fire on a Hot Planet," *New York Times*, April 23, 2006, http://j.mp/yellingfire.

3 Andrew C. Revkin, "An Inconvenient Mind," *New York Times*, November 18, 2010, https://dotearth.blogs.nytimes.com/2010/11/18/an-inconvenient-mind/.

4 Andrew C. Revkin, "From the Fire Hose—'Building Back' Next Steps and Fossil Fuelers' Favorite Ad Agency Pushes 'Climate Emergency,'" *Sustain What*, November 19, 2021, https://revkin.substack.com/p/from-the-fire-hose-building-back-21-11-19.

5 John D. Sutter, "Woodward County, Oklahoma: Why Do So Many Here Doubt Climate Change?," *CNN*, August 3, 2015, https://www.cnn.com/2015/08/03/opinions/sutter-climate-skeptics-woodward-oklahoma/index.html.

6 Tara Ritter, "The Rural Climate Dialogues" (Institute for Agriculture and Trade Policy, November 17, 2020), https://www.iatp.org/documents/rural-climate-dialogues.

COMMUNITY

Is it still possible to trust other people?

DARREN FLEET

My face is itchy and dry. It's not even ten o'clock in the morning and I have already washed it three times. Nothing seems to clean the film of smoke and dead skin away. The haze arrived suddenly on a Thursday morning in July. It's August now, eighteen days later. The haze stretches from the lake to the horizon, masking the near mountains and the far side of town. All of the surrounding communities are on evacuation alert, some even under evacuation orders. My social media feeds are full of images of wildfire, ashen skies, and pleas for mutual aid. Does anyone have an air purifier? Can someone help me fix my windows to keep the smoke out? Does anyone have a place I can stay? One particular message caught my eye: "Prayers up for our little corner of the world here . . . tension is high in paradise."

I live in a small town in southeastern British Columbia, Canada's westernmost province. It is regarded as a progressive oasis in an otherwise conservative region, with a history of draft dodgers, sixties and seventies counterculture, business cooperatives, cohousing communities, marijuana cultivation, and several decades of continuous flight from population centers. It is beautiful and isolated, with an infrastructure unlike any small town in the country—the remnants of a once thriving early twentieth-century **mining** and supply industry. Many people come here for refuge, for healing, for escape. Others still, progressive and spiritually open, are looking for like-minded souls with whom to spend their days trying to reconnect to nature, or Mother Earth, or self-sufficiency. But even in paradise, the contingent and material realities of utopia are crashing in. The privilege of urban exodus. The resource economy that underpins the surrounding area. The entrenched rural constituencies that align with the politics of climate denial and the far right, as well as new-age anti-vaxism, healthism, and, now, anti-maskers. And, of course, no place on Earth today is beyond the reach of heat and smoke.

Timothy Morton once described weather-themed small talk in the context of climate disaster as an uncanny and discomforting engagement; we talk about the weather without *really* talking about the weather.[1] I appreciate this description, but it does not fully convey how the changing weather, *and its causes*, are understood in the context of petroculture, **nationalism**, and the specter of fossil fascism.[2] Today, at least in Canada, weather is a front line of the culture war. Talking about weather means talking about climate. And to talk about climate means to talk about oil. And to talk about oil means to talk about how we understand social life. These equivalencies include everything from beliefs about the **coronavirus** to fuel, to migration, to wildfire, to what we expect from our communities, our neighbors, and our friends.

While as a scholar I relish debates around environmental and energy politics, in my day-to-day life—the world of making lunches for my two kids, commuting or Zooming to work, and making small talk with neighbors and strangers—I often dread the topic of weather. That is because, inevitably, weather reveals our deepest convictions. As such, I experience significant anxiety in casual conversation when, inevitably, a discussion of smoke, or climate change, or record-breaking 49.6°C temperatures *in Canada* will devolve into an unsettling political reveal.[3] I have had many such encounters lately. These include neighborly talks about how solar radiation and pulsating stars are the real cause of climate change, about how chemtrails alter the daytime sky, about how green energy is worse for the environment than fossil fuels, and even about how activists lie about ecological harms to get attention.

Before moving to this place, I lived in a major coastal city. My former urban community looked like a gerrymandered electoral district, a social world formed more around ideas and shared values than around space. Now that I live in a small town, the ability to associate is more circumscribed by geography, weather, and compromise. So what of my actual neighbors? The folks whom I actually spend most of my physical time with?

At the moment this query has a real practical intention. If the wildfire that is currently burning in the next valley climbs the hillside, crosses the slope, and burns down into this one, these neighbors are the people I will rely upon, and who will rely upon me, to evacuate, to provide mutual care, and to survive. To me, this somewhat stark reality is what all questions about how to address climate change and energy transition ultimately come down to: Do I trust my neighbors to care for me when things get real? When I get sick? When the **economy** takes a dive? Do I trust the types of fulfillments that will emerge without the promise of unlimited mobility and other petroluxuries?

I do not know the answers to these questions, but I do know that getting from here to there, from carbon debt to carbon free, will require new understandings and expectations of the joys of collective and community life. It will require a deep and profound sense of neighborliness, rootedness, mutual aid, appreciation for nonmarket encounters, and humility. We will need to escape toward one another.

As Rebecca Solnit describes in *A Paradise Built in Hell*, a book about the spontaneous communities of care and utopic ideas that arise during disaster, the generative possibility of crisis is that it creates space to imagine the world differently.[4] And so, I wonder, as I stare out at the gray wall of smoke that has replaced the mountainside: how will my neighbors and I talk about weather, about oil, about collective politics, and about community, when the horizon clears? What will this wildfire reveal about paradise, other than its contradictions?

See also: **Airplane**, **Commons**, **Family**, **Fire/Bushfire**, **Populism**, **Science**, **Solidarity**

Notes

1 Timothy Morton, *Hyperobjects: Philosophy and Ecology after the End of the World* (Minnesota: University of Minnesota Press, 2013), 99.
2 Zetkin Collective and Kai Heron, "Fighting Fossil Fascism for an Eco-Communist Future," *Roar Magazine*, July 15, 2021, https://roarmag.org/essays/white-skin-black-fuel-interview/.
3 Eric Holthaus, "How Did a Small Town in Canada Become One of the Hottest Places on Earth?," *Guardian*, June 30, 2021, https://www.theguardian.com/commentisfree/2021/jun/30/lytton-hottest-places-world-climate-emergency.
4 Rebecca Solnit, *A Paradise Built in Hell: The Extraordinary Communities That Arise in Disaster* (New York: Penguin, 2009), 1–12.

CORONAVIRUS

What does the coronavirus reveal about the politics of climate crisis?

HELEN PETROVSKY

Generalizing about the coronavirus pandemic is difficult. Even though its acute phase is officially over, we are still living through it. The coronavirus haunts almost every aspect of daily life, which is why we must come to terms with a new reality. What the pandemic unveiled is not so much personal vulnerability in the face of an unknown deadly disease but instead the importance of collective responsibility: our well-being and physical survival depend on how we interact with others.

Certain images of the corona-crisis became emblematic. Dolphins returning to the Venetian Lagoon, clear skies as a side effect of protracted lockdowns, and oil tankers left stranded in the middle of oceans. Such images have now all but disappeared, but they remain the signature of a new and shared experience. This experience can be described as *mutual aid*, a term coined by Russian scientist Pyotr Alexeyevich Kropotkin.[1] Mutual aid was conceptualized by Kropotkin in the 1890s in opposition to the prevailing interpretation of Charles Darwin's theory of the survival of the fittest. Drawing on the findings of his journeys to Northern Asia, Kropotkin concluded that it is not antagonist struggle but solidarity, or anarchist communism, if translated into political terms, that accounts for the progressive evolution of a species. For Kropotkin, the instinct of mutual support in both **animals** and humans has nothing to do with sympathy or love. Rather, he argues, "it is the unconscious recognition . . . of the close dependency of every one's happiness upon the happiness of all; and of the sense of justice, or equity, which brings the individual to consider the rights of every other individual as equal to his own."[2]

Mutual aid is driven by an instinctive sense of **justice**. A person following his or her instincts has no need to evaluate the deed performed. Rather, whoever acts is enhanced by the act, in the process acquiring a new degree of plenitude and joy.[3] In his obituary of Darwin, written prior to his travel to Northern Asia, Kropotkin drew clear connections among **science**, morality, and social commitment, asking who is best adapted to survive, if one comes to think of survival in terms of social interaction. The one who produces everything, who invents, and is capable of working "with his head and his hands," who provides for their own subsistence and develops themselves—in a word, the worker.[4] And although this conclusion may be prompted by nineteenth-century industrial production as well as by Kropotkin's own revolutionary aspirations, the point he makes is clear: it is only the "sociable species, where all individuals are in solidarity with each other," that prosper, and nothing other than solidarity and "shared work" (*travail solidaire*) consolidate the species in its struggle against the hostile forces of nature.[5]

There is much to gain from Kropotkin's understanding of the organization of life. Action—namely transformative action—is the cornerstone of his doctrine. Such action is, however, never exploitation, whether we think of natural resources or the logic of a socioeconomic system. A geographer who studied land formations, Kropotkin teaches us to respect the vast rhythms of Earth, which are beyond human measure. Lonely champions of industrial growth, we have completely ignored mutual aid as a way of being human in every possible sense. This failure is exacerbated by growing and unmitigated violence, wars simultaneously waged against human beings and nature. We need to break this vicious circle by restoring our balance with nature and realizing that we are just one life form among an infinite variety of others. Which means—for instance—that the overexploitation of natural resources should give way to advanced energy solutions.

At the beginning of the pandemic, one Russian expert aptly remarked that oil had stopped being a mega-important element of world politics and economics, becoming "a mere commodity" instead.[6] This geopolitical decentering of oil is a healthy and encouraging development. But it is still far from the collective work needed to ensure our survival as a species, which can happen only if competition gives way to cooperation and combined social **action** increases the powers and interdependence of life. The biggest impact of the coronavirus isn't the brief dip in CO_2 levels brought about by the lockdown. Rather, it is a reckoning with the imperative of being attentive to each other's basic needs across, and even against, existing social institutions at a global **scale**, reminding us of the power of mutual aid to shape our world into the world we want to live in, socially and environmentally.

See also: **Abandoned, Community, Education, Globalization, IPCC, Scales**

Notes

1 Pëtr [Pyotr] Kropotkin, *Mutual Aid: A Factor of Evolution* (1902), https://theanarchistlibrary.org/library/petr-kropotkin-mutual-aid-a-factor-of-evolution.
2 Kropotkin.
3 P. A. Kropotkin, "Spravedlivost' i nravstvennost,'" Biblioteka Anarkhizma, accessed August 8, 2021, https://ru.anarchistlibraries.net/library/piotr-kropotkin-spravedlivost-i-nravstvennost.
4 "Charles Darwin," *Le Révolté*, April 29, 1882, 1.
5 "Charles Darwin," 1.
6 Maksim Averbukh, "Zabud'te pro neft'. Kak Rossiya poshla na popyatnuyu v voine c OPEK—i pochemu eto ne pomozhet," *Novaya gazeta*, April 13, 2020, https://novayagazeta.ru/articles/2020/04/10/84822-zabudte-pro-neft.

CORPORATION

Will an energy transition help fossil fuel companies?

CAMERON HU

It is summer 2021, and the financial pages are celebrating a peculiar insurgency. A hedge fund has persuaded ExxonMobil shareholders to elect new directors to the board of the oil multinational. The new directors pledge, in turn, to reduce the position of fossil fuels in Exxon's global portfolio of extractive projects. The founder of the hedge fund insists this campaign "isn't really about ideology, it's about economics."[1] BlackRock, the world's largest asset manager, and the largest Exxon shareholder to throw its weight behind the activist investors, agrees: this is about the oil corporation's ability to survive an energy transition. It is about Exxon's "long-term strategy" and "long-term shareholder value."[2] If Exxon redirects its program of permanent growth from petroleum molecules to wind farms and carbon capture, perhaps Exxon can endure indefinitely, proving itself as secure an investment as ever.

The bleak prospect of Exxon outlasting the age of oil turns our attention from petroleum molecules to the petroleum corporation and its form—and in particular, to the way the petroleum corporation's orientation toward indefinite existence will shape any political struggle over the "long term." What is the meaning of transition to a future less destructive than the present, if that transition preserves the customary agent of destruction? What changes, and what persists, if Exxon or Chevron prolongs its reign on Earth by swapping one energy **resource** for another within its resource portfolio, swapping wind for oil the way it swapped Arabian deserts for Texan plains in the twentieth century, or Great Plains fracking for deepwater extraction in the twenty-first? Transition may be business as usual for those energy corporations whose form is decidedly "immortalist."[3] We might begin to wonder how features of the petrocapitalist present could endure, perniciously, even after the waning of hydrocarbon energy—not only in the lingering toxicity of corporate petrochemistry, but equally, in that immortalist mode of enterprise one Exxon CEO celebrated with this maxim: "*Presidents come and go. Exxon doesn't come and go.*"[4]

Fossil fuels conjured the world of immortalist corporations. There were, of course, corporations before coal and oil—most familiarly, the imperial company states exemplified by the British and Dutch East India companies—but the wholesale organization of capitalist **economy** around so many death-averse private business corporations arose hand in glove with the advent of large-scale, fossil-fueled infrastructure: the steam-powered railways of the nineteenth century. In choreography with the vast timelines for constructing and profiting from America's transcontinental railroads, signature machines of continental imperialism, the business corporation was reconstructed as a nonfinite device for continuous management and accumulation.[5] Under pressure from railway industrialists and their associates in steel and **finance**, states progressively removed prior restrictions on the location, capitalization, legal immunity, mergers and acquisitions, and indeed *duration* of corporate operations.[6] The immortalist corporation embodies a reorganization of economic and political time internally related to the steam engine's coal-powered reorganization of space-time. Staring westward down the tracks toward a vanishing point, the business corporation began to contemplate infinity.

And if fossil fuels contoured the corporation as a prospective immortal, corporate immortalism, in turn, gave determinate form to fossil-fueled existence in the twentieth century. The project of endless corporate expansion through extraction sent petroleum multinationals throughout the world in order to replenish what they had elsewhere pumped—or else die—and made them omnipresent forces in global economy and ecology. Still more significantly, the aspiring permanence of petroleum multinationals also determined basic features of North Atlantic *thought*. The oil corporation's project of endless survival would generate many dominant modes of theoretical and instrumental knowledge—among them the very conceptual languages through which Earth is today grasped as a vulnerable object that demands transition. Geological science was progressively elaborated and dignified as a discipline in proportion to its value in the continuous renewal of corporate resource portfolios. Futurological methods of **scenario planning**, so basic to global climate governance, emerged within oil multinationals in response to the threat postcolonial oil revolutions posed to their survival. The Gaia hypothesis, forerunner of Earth System Science and Anthropocene discourse, was articulated by James Lovelock as a contract scientist for Royal Dutch Shell. Explaining that affiliation some years later, Lovelock pointed to Shell's ambition to exist forever. He knew, Lovelock wrote, of "no other human agency that plans so far ahead."[7]

So much of our ability to think critically about Earth and its future is an artifact of corporate immortalism. Should fossil fuels disappear at some future date, it will remain to ask what features of petrocapitalism persist in their wake, not least in the corporate form and the encompassing styles of reason and action that grow from corporate immortalism. The climate crisis might turn out to be just the first episode in a hyper-repetitive serial drama inaugurated, but not exhausted, by fossil fuels. As in the following scenario, as plausible as any other:

Exxon's new directors turn the company from oil and gas to some putatively less destructive energy resource—call it resource *x*. The corporation's drive for indefinite life is thereafter routed through the continuous extraction of *x* and the continuous elaboration of new sciences to enable it. *X* proves harmful to the sites of its extraction/production and their communities. But according to the planetary scenarios contemplated in distant metropoles, *x* is essentially preferable to fossil energy. Decades pass, and it becomes first thinkable, and then indisputable: Exxon's elaboration of resource *x* has degraded some basic mechanism underlying collective life—one hitherto unknown to exist, a mechanism whose existence, like the environment or the Earth System, has come into view as a consequence of its destruction.[8] Now activist investors, fearing that resource *x* too will become valueless and illiquid in the long term, endeavor once again to save Exxon from untimely death. The insurgents elect new directors to its board, promise to swap the promising resource *y* for the formerly promising, now disgraced resource *x*. . . .

See also: **Clean, Electricity, Mining**

Notes

1 Clifford Krauss and Peter Eavis, "Climate Activists Defeat Exxon in Push for Clean Energy," *New York Times*, May 26, 2021.
2 BlackRock Inc., "Vote Bulletin: ExxonMobil Corporation" (May 26, 2021), https://www.blackrock.com/corporate/literature/press-release/blk-vote-bulletin-exxon-may-2021.pdf.
3 Abou Farman, *On Not Dying: Secular Immortality in the Age of Technoscience* (Minneapolis: University of Minnesota Press, 2020).
4 Steven Coll, *Private Empire: ExxonMobil and American Power* (New York: Penguin, 2012), 68.
5 Timothy Mitchell, "Infrastructures Work on Time," *E-Flux*, January 2020, https://www.e-flux.com/architecture/new-silk-roads/312596/infrastructures-work-on-time/.
6 Suzana Sawyer, "Corporations," in *The International Encyclopedia of Anthropology*, ed. Hilary Callan (New York: Wiley, 2018), 6.
7 James Lovelock, *Homage to Gaia: The Life of an Independent Scientist* (Oxford: Oxford University Press, 2000), 193.
8 David Bond, "Environment: Critical Reflections on the Concept," Institute for Advanced Study Occasional Paper 64 (2018).

CUBA

What can Cuba teach us about climate action?

ANDREW PENDAKIS

In 2006 (and then again in 2016) the World Wildlife Fund released a report that measured UN human development criteria against the rate at which a nation was overshooting its biocapacity: it concluded (both times) that "Cuba alone" had adequately fed, housed, and educated its population within the limits laid down by strict standards of sustainability.[1] These results, confirmed by further studies, have been almost completely ignored by researchers, journalists, and policymakers. This is no real surprise—to take them seriously is to turn everything we think we know about the world upside down. Ideas about the legitimacy of **democracy**, the rationality of markets, the correct proportion of society to state, as well as the moral, political, and economic valence of communism are all detached from the usual conclusions and allowed to float free in a disturbing zone of undecidedness. Cuba, a nation often framed as historically belated, suddenly finds its apparent obsolescence converted into a kind of rare genius or luck. Its surreally preserved cars and dilapidated colonial facades now appear in a new light—still a glint of backwardness, perhaps, but one that arrives *from the future*.

There is nothing sustainable about Cuba's reliance on nickel mining and exports of cigars and citrus, nor its joint ventures with foreign capitalists in tourism and fossil fuel extraction. For reasons that are as serendipitous as they are intentional, Cuba is nevertheless a uniquely green place. Its beaches, marine ecosystems, and forests are among the best preserved in the region: no country in the world has more legally protected land as a percentage of its total area. The majority of the fresh vegetables consumed in its cities are grown in organic, urban gardens—its famed *organopónicos*—that are built on roofs or in open lots nestled between buildings and along busy streets. These initiatives combine vertical elements (state-subsidized inputs like seeds, fertilizer, and tools, as well as hands-on skills training) with horizontal dimensions (cooperative ownership, voluntary **community** labor) that complicate easy distinctions between state and society, as well as the usual lines drawn between the economic and the ecological. In 2005, thousands of young social workers and students were sent into every corner of the country with the goal of replacing every incandescent bulb they could find with an energy-efficient alternative: no other country has managed to do this. Cuba has begun its transition to solar power, outfitting remote rural schools with photovoltaic batteries and aspiring to generate a quarter of its energy needs with solar before 2030.

Furthermore, bicycles remain the most common form of transportation in Cuba. Its Marxist-Leninist government is rigorously secular, closely collaborates with and consults scientists, and has been able to politically orient itself to the risks of climate change and away from the systemic disinformation and entrenched interests that characterize Western capitalist democracies. While the democratically elected Trump was dismantling the EPA, the Cubans were busy initiating a hundred-year climate plan that would allow them to ban development in areas at risk of flooding and to move people en masse from zones likely to be hit by increasingly frequent and destructive storms. Cubans eat relatively little meat, rarely travel internationally, and dwell in small homes with few amenities or private possessions.

Bill McKibben once wrote that "Cuba won't look like Cuba once Cubans have some say in the matter."[2] Though he is perhaps underestimating the degree of support in Cuba for the gains of the revolution, there is no doubt that many aspects of the present system would be radically changed were Cuba to shift to a liberal democratic model. Cuba is sustainable not despite but *because* it is illiberal; it is an ecological success story precisely because its Stalinist economic system has so often failed to reach its targets (here the U.S. embargo, for all its horribleness, has probably unwittingly abetted the greening of Cuba). Take, for example, the *libreta de abastecimiento*. The libreta is a book of coupons provided to individuals each month that allows them to buy heavily subsidized basic foodstuffs like rice and beans, as well as a little meat and some eggs. Combined with the low wages of the average Cuban and the sluggishness of local food production, the libreta places limits on the aggregate amount of meat consumed in the country while at the same time equally distributing this consumption. This environmentally positive (and arguably socially just) outcome would almost certainly be lost by the full liberalization of the **economy**.

By simply existing, Cuba provokes us into considering two very different kinds of future, both of which are two ways of distributing limits and two ways of being poor. In the first scenario, that of capitalist business as usual, techno-utopian projections fail, and we find ourselves around 2070 in a state of unevenly distributed planetary collapse. This version of maldistributed penury allows a small percentage of the **planet**'s population to flourish—perhaps even to thrive behind armed gates and walls—while a majority starve, suffer, and migrate in potentially unprecedented numbers. Alongside this possibility exists the specter of *socialist austerity*, a version of the future that some will find intolerable, but which may in the end be the only credible alternative to the scenario outlined above. Here we can imagine an eco-socialism that redefines our conceptions of freedom by instituting illiberal (though possibly democratic) political and economic constraints—limiting, say, rights to infinite consumption, travel, wealth, etc.—while maintaining alternative paths to political life, public goods, culture, **sport**, **education**, health care, and **science**. A frugal system in which nobody starves, billionaires have ceased to exist, and everyone has the capacity and time to develop themselves within a natural environment that everywhere slowly rewilds. Unlike those inhabitants of the developed world who remain in denial or expect capitalism to effortlessly avert the worst, the average Cuban, having already endured the austerities of the Special Period and who lives emmeshed in a set of social networks and political mechanisms designed for (and forged by) crisis, is in many ways already prepared for a coming age of blackouts, storms, and empty shop shelves. Whether or not Cuba is a model for the West is beside the point: its very existence is a strange provocation, Lacan's stain of the Real, a view of something we've never seen before that casts everything we think we know into doubt.

See also: **Class, Clean, Cycling, Degrowth, Youth**

Notes

1 World Wide Fund for Nature, "The Living Planet Report 2016: Risk and Resilience in a New Era" (Gland, Switzerland: WWF, 2016).
2 Bill McKibben, "The Cuba Diet: What Will You Be Eating when the Revolution Comes?," *Harpers*, April 2005, https://harpers.org/archive/2005/04/the-cuba-diet/.

CYCLING

Is cycling a solution?

STACEY BALKAN

What is a cyclist? This seemingly banal question is complicated by the multiple political valences of cycling—whether as a sport, a symbol of environmental advocacy, a mode of transportation, or (quite often) an index of poverty.

For a small coterie of professional cycling fans in the United States, cycling is a sport that gained in popularity after Greg LeMond's 1986 Tour de France victory. For those concerned about the global climate crisis, cycling represents the putatively low-carbon **lifestyle** required to reduce atmospheric carbon and thereby avoid environmental catastrophe.[1] For those who are transit-dependent but subject to deficient public transportation infrastructures, cycling is a necessity—that is, not a conscious rejection of petroculture but a means of survival.[2]

For the subaltern cyclist living in Los Angeles or a swiftly sinking Miami, cycling is a basic necessity while also a grave danger: LA and South Florida boast the highest mortality rates for cyclists in the United States.[3] Both of these metropolitan areas instantiate the "uneven cyclescapes" that mark most American cities.[4] In neighborhoods inhabited by the well-heeled (or well-wheeled), there are smoothly paved, efficiently sited bicycle lanes. For the working **class** who must rely on cycling as a form of transportation—often communities of color and undocumented laborers—bike lanes are often sited along disintegrating thoroughfares or adjacent to major traffic arteries, if they exist at all.

Perhaps a better question to ask, then: how does the bicycle serve as both a symbol of urban poverty and an index of the failures of green transition initiatives to address systemic injustices, including gentrification? (Urban gentrification factors as a significant contributor to greenhouse gas emissions in a consumption-based calculus.)[5] In cities like Los Angeles, Miami, Oakland, and Detroit, bicycle justice organizations aligned with the broader transportation justice movement have emerged to combat eco-gentrification initiatives that reinforce counterproductive notions of sustainability and environmentalism by avoiding questions of uneven development and systemic racism.[6]

In *Cyclescapes of the Unequal City: Bicycle Infrastructure and Uneven Development*, John G. Stehlin recounts the trajectory of cycle transport; he pinpoints a moment when the bicycle shifted "from a vehicle of last resort (signifying racialized urban poverty) to a symbol of *choosing* a cosmopolitan, less carbon-intensive life (making visible the return of the largely White middle class)."[7] This shift reflects a postmaterialist environmentalism that has long sustained settler myths: from paeans to uninhabited "wild" spaces that would augur the creation of the U.S. National Park System and the requisite removal of Indigenous communities; to urban beautification projects that would off-site toxic waste to historically Black and Latinx neighborhoods in the name of smart growth; to the uneven cyclescapes across the United States wherein BIPOC communities would be excluded from planning initiatives for bicycle infrastructures including the siting of bike lanes. Stehlin notes that in Austin, Texas, local residents saw the bicycle as a harbinger of violent forms of gentrification: "when the bikes came in, the Blacks went out."[8]

In this context, bicycle **justice** activists like Tamika Butler, a cyclist of color and former executive director of the Los Angeles Bicycle Coalition, pose to planners and cyclists alike the ontological question with which I began this essay: what is a cyclist, in the American imagination?[9] For Butler, the term itself is (and ought to be) contested. Does a bicycle inevitably index the homogenous culture of professional cycling? Or the bankrupt environmentalism of organizations like the largely white, middle-class Critical Mass? For cyclists like Butler, or New York City's unofficial "bicycle mayor" Courtney Williams (the "Brown Bike Girl"), who promote the joys of cycling for communities too often framed in racialized discourses of poverty and urban decay, the answer to such questions is a resounding no![10]

In order to grapple with the uneven cyclescapes of the American political imaginary, bicycle justice advocates like Butler and Williams work to create hard infrastructures that accommodate a more racially and economically diverse ridership. These efforts include bikeshare programs that supplement existing light rail and bus systems, the expansion of bike lanes into historically underserved communities, and initiatives that seek to reestablish understandings of public space as no longer tethered to systems of private property sanctioned by the enclosure of the urban **commons**. That is, bicycle justice entails forging infrastructural publics that secure multiple-use pathways for all stakeholders.

The question is how to forge new infrastructural publics while also attending to the deepening inequalities in urban centers, not to mention the renewed investment in electric modes of transportation including E-bikes. Even as they cause lower emissions than other mechanized transport, and offer a promising alternative for persons with ambulatory obstacles, E-bikes are nonetheless largely powered by coal and **natural gas**. Bike equity initiatives focused on this twofold question are thus induced to speculate on possible political futures delinked from the extractivist logics of petroculture.[11]

Many bike equity initiatives deploy such an intersectional framework, which is arguably the only viable approach to a just energy transformation in the face of imminent climate collapse.[12] It is also an increasingly rare means of cultivating **community** in the otherwise alienating petroscapes of places like South Florida.

See also: **Habit, Justice, Organize, Transitions, Vegan**

Notes

1 Jennifer L. Rice, Daniel Aldana Cohen, Joshua Long, and Jason R. Jurjevich, "Contradictions of the Climate-Friendly City: New Perspectives on Eco-Gentrification and Housing Justice," *International Journal of Urban and Regional Research* 44, no. 1 (January 2020): 145–65.
2 Adonia E. Lugo, *Bicycle/Race* (Portland, OR: Microcosm, 2018).
3 Luke Whelan and Maura Fox, "What We Learned from Tracking Cycling Deaths for a Year," *Outside*, January 29, 2021, https://www.outsideonline.com/outdoor-adventure/biking/what-we-learned-tracking-cycling-deaths-year/.
4 John G. Stehlin, *Cyclescapes of the Unequal City: Bicycle Infrastructure and Uneven Development* (Minneapolis: University of Minnesota Press, 2019).
5 Rice et al., "Contradictions of the Climate-Friendly City," 154.
6 Sarah Dooling, "Ecological Gentrification: A Research Agenda Exploring Justice in the City," *International Journal of Urban and Regional Research* 33, no. 3 (September 2009): 621–39; Aaron Golub, *Bicycle Justice and Urban Transformation: Biking for All?* (New York: Routledge, 2012).
7 Stehlin, *Cyclescapes of the Unequal City*, 11.
8 Stehlin, 7.
9 Tamika Butler, "Why We Must Talk about Race When We Talk about Bikes: Systemic Racism Can't Be Fixed without Tackling It within Cycling," *Bicycling*, June 9, 2020, https://www.bicycling.com/culture/a32783551/cycling-talk-fight-racism/.
10 See Courtney Williams's work on bicycle equity and advocacy: https://linktr.ee/thebrownbikegirl.
11 Elly Blue, *Biketopia: Feminist Bicycle Science Fiction Stories in Extreme Futures* (Portland, OR: Microcosm, 2017).
12 For additional resources and to get involved, see Bicicultures, "Resources" (n.d.), https://bicicultures.com/resources/.

DECOLONIZATION

What can OPEC teach us about the link between energy and decolonization?

SANAZ SOHRABI

In August 1980, Petróleos de Venezuela published an audiovisual project to commemorate the twentieth anniversary of OPEC. *Rhymes and Songs for OPEC* (fig. 1) was a musical recording and photo catalog, produced in Caracas in consultation with the diplomatic missions and cultural offices of OPEC member states and recorded at Universidad Central de Venezuela. For each of the thirteen OPEC member states at the time, the project included national folkloric songs representative of their local histories and cultural landscape. These songs

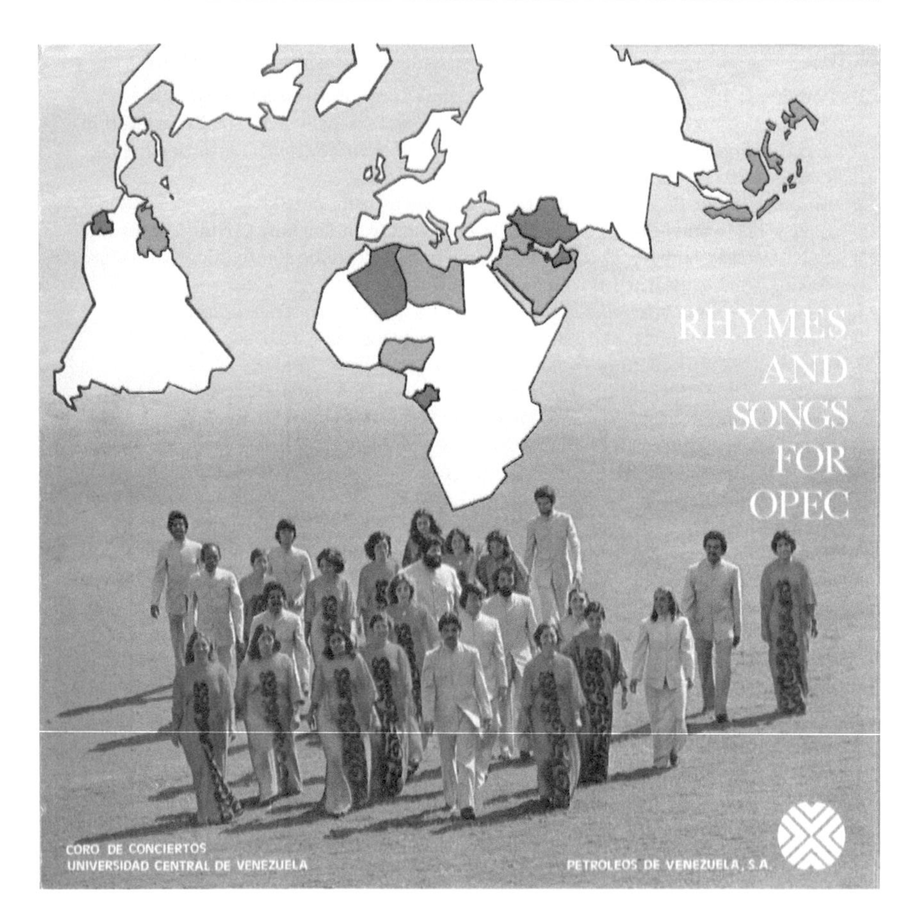

Front cover of *Rhymes and Songs for OPEC*, musical vinyl produced by Petróleos de Venezuela, 1980. Close-up scan from the original archive. Courtesy of the author.

were performed by a Spanish-speaking chorus singing in various dialects of Arabic, Farsi, or Indonesian, even though the singers did not speak the other languages of many member countries; instead, they simply memorized the lyrics. As an audiovisual archive of Third World **solidarity**, *Rhymes and Songs for OPEC* is a monument to and document of OPEC as a project aimed not only at maintaining member states' national control over an important economic resource but also at fostering a transnational movement of cultural and political decolonization.

The worldmaking ambition of this project is legible in the album's cover art. The top half of the frame features a map of the world with North Africa and West

Asia—OPEC's core region—at its center, indicating the emergence of new centers of global power and influence. Territories of OPEC member states are demarcated in red, blue, green, and purple, while the rest of the globe is beige and undifferentiated, suggesting a blank space through which OPEC and the larger project of decolonization might spread. The album cover's bottom half features a photograph of members of the Venezuelan choir walking on a grassy plain under a blue sky, as if heading out of the frame and into a new future. The men wear suits with a beige *achkan* or *sherwani* of a modern, hip-length design reminiscent of the Nehru jacket; the women wear ankle-length flowing gowns in bright red, blue, green, and purple, emblazoned with abstract calligraphic forms evocative of Balinese or Javanese script. The abstracted cartography in the map at the top of the album cover comes to life in these human figures walking on the ground. The sartorial choices in the photograph invoke popular memories of the 1955 Bandung Conference in Indonesia and the early history of the Non-Aligned Movement, upon whose legacy of tricontinental solidarity OPEC is shown to make a symbolic claim.

Despite its goal of promoting transnational economic cooperation and political solidarity among postcolonial states, the OPEC project has not been without disappointment and internal tensions among its members. OPEC's founding in Baghdad in 1960 by Iraq, Iran, Kuwait, Saudi Arabia, and Venezuela paralleled global decolonization movements already underway in the postwar period. But if oil could function as a symbol of transnational solidarity in the Global South, as captured in the *Rhymes and Songs for OPEC*, it has also functioned as a source of geopolitical ambition and international division among OPEC's member states. One month after the release of *Rhymes and Songs for OPEC* in 1980, Iraq's invasion of Iran inaugurated a new era for the organization.

Subsequent changes in OPEC's membership reflect some of these geopolitical pressures. Angola, Equatorial Guinea, and Congo joined in 2007, 2017, and 2018 respectively, while some established members have recently paused or terminated their membership, including Indonesia, Ecuador, Qatar, and Gabon. Three long-standing members, Iran, Libya, and Venezuela, have been under heavy sanctions imposed by the United States and European Union that have undermined their OPEC production agreements as well as their infrastructural development and socioeconomic stability. OPEC's inner divides and evolving South-South dynamics on key issues such as Palestinian liberation, imperially imposed sanctions, and U.S. military presence in the Middle East reveal that the concepts of *Global South* and *Global North* are more inadequate than ever to explain the world that petrocapitalism has created.[1] This discrepancy also highlights, however, the necessity of access to and control over natural resources as a foundation for economic decolonization and political defiance against wealthier, imperialist nations, as well as against broader geopolitical calculations on the part of those seeking to arrest decolonization.

The lessons from OPEC are abundant. Recent assessments of OPEC's remarkable establishment as the first alliance among the oil-producing countries of the Global South have shown how the anticolonial elites of OPEC successfully adopted the framework of international law for self-determination and permanent sovereignty over natural resources, which was also at the heart of the Non-Aligned Movement's "subaltern internationalism."[2] The geopolitical rise of OPEC "unfolded as part of a quarter-century tradition of anticolonial legal and economic thought" inspired by the postcolonial imagination of the Bandung Conference in Indonesia, a country that would join OPEC in 1962.[3] As historian Nelida Fuccaro explains, the establishments of the Arab Petroleum Congress in Cairo in 1959 and OPEC in 1960 were deeply intertwined: both groups aimed to ameliorate the technical deficiencies and lack of national oil knowledge in the Middle East and North Africa, as Arab oil technocrats and government elites sought to build their nations through industrial **development**.[4] Documentaries such as *The Revolution of Machines* (1968) by Egyptian director Madkhour Thabet and *Paykan* (1969) by Iranian director Kamran Shirdel are among the many films commissioned as part of larger educational projects across the Middle East and North Africa. These efforts at cultural and knowledge production framed technological and resource independence as indispensable to liberation from Western economic dominance.[5]

As a filmmaker, I often begin with the image-world of oil. Portrayals of OPEC in the media are shrouded with accumulated layers of post-1973 oil shock discourse. Yet an important counterimage can be found in the album cover for *Rhymes and Songs for OPEC* because it reminds us that, for the diverse national publics of its member countries, OPEC was never imagined or narrativized as a *cartel*. Rather, the map and its embodied equivalent imagine OPEC as a transnational alliance and human community committed to Third World **resource** sovereignty and an expanding global movement for decolonization. When compared with subsequent OPEC maps, it also reveals the fragility of anticolonial solidarity, and the different directions taken by various postcolonial nation-states.

The inclusions and exclusions of the *Rhymes and Songs* map remind us that decolonization requires an economic engine such as energy. They should also remind us that the dreamworlds of decolonization are messy, filled with competing and contradictory political imaginaries, made all the messier when the dream is linked to a resource—oil—that everyone wants, or at least needs. It has been almost half a century since this remarkable map was drawn. What might such a map look like fifty years from now?

See also: **Art, Autonomy, Blockade, Globalization, Music, Nationalism, Neoextractivism**

Notes

1 Vijay Prashad, *The Poorer Nations: A Possible History of the Global South* (London: Verso, 2013).
2 Christopher Dietrich, *Oil Revolution: Anticolonial Elites, Sovereign Rights, and the Economic Culture of Decolonization* (Cambridge: Cambridge University Press, 2017).
3 Dietrich, 64.
4 Nelida Fuccaro, "Oilmen, Petroleum Arabism and OPEC," in *Handbook of OPEC and the Global Energy Order*, ed. Dag Harald Claes and Giuliano Garavini (New York: Routledge, 2020).
5 Dietrich, *Oil Revolution*, 64.

DEGROWTH

Is degrowth inevitable?

DOMINIC BOYER

Degrowth (or *décroissance*) is a concept that has been around since the early seventies. It has been amplified in recent decades as the political struggle against anthropogenic climate change has intensified. This struggle will likely fail unless degrowth can become the new ethos of economic and social activity (including, of course, energy use) in the Global North. With the stakes so high, it is worthwhile to return to degrowth's origins as a concept for possible insights about what might come next.

André Gorz coined the term *décroissance* in 1972.[1] Gorz was a maverick French Marxian economic and ecological thinker whose works, influential in the seventies, deserve reexamination today. The idea of décroissance emerged during the debates that followed the publication of the 1972 Club of Rome report, *The Limits to Growth*.[2] That report famously analyzed global growth trends and **resource** depletion and predicted rapid civilizational collapse unless a state of *global equilibrium* could be achieved. The report was vague on details and solutions and attracted ample criticism as well as praise. With hindsight, for example, the report's warning that the world's oil reserves would be exhausted by 1990 seems charmingly optimistic in its pessimism. Still, the report sold thirty million copies worldwide; its significance for subsequent environmentalist thinking cannot be underestimated—the global equilibrial discourse of **sustainability** would not exist without it.

If the Club of Rome warned of limits, Gorz argued that equilibrium was itself a hopeless aspiration; what was needed was a degrowth trajectory. He asked a pointed question about what was really desired: "a capitalism adapted to ecological constraints; or a social, economic, and cultural revolution that abolishes the constraints of capitalism and, in so doing, establishes a new relationship between the individual and society and between people and nature? Reform or revolution?"[3]

Gorz chose revolution because he viewed the growth logic of capitalist society as fundamental to its operation. On the one hand, paraphrasing Marx, capital simply wishes to propagate itself. But there is a structure to its sprawl: capitalism utilizes scarcity as a means for reproducing social inequality and preserving hierarchy in its **class** structure. New technological achievements and luxuries enjoyed first only by the elite attract the desires of the masses toward them. As the masses gain access to old luxuries, new unattainable luxuries develop to replace them. This treadmill of luxury means that no universal good life will ever be enjoyed in a capitalist society no matter its accumulation of useful things—"the mainspring of growth is this generalized forward flight, stimulated by a deliberately sustained system of inequalities."[4]

Gorz termed this constant process of dispossession through innovation the "poverty of affluence" and argued that to break with its "ideology of growth" a society would first have to affirm a set of values opposed to hierarchy and privilege: "The only things worthy of each are those which are good for all; the only things worthy of being produced are those which neither privilege nor diminish anyone; it is possible to be happier with less affluence, for in a society without privilege no one will be poor."[5] Gorz imagined the world of décroissance as one defined by universally available highly durable goods, beautiful public dwellings and transportation, a twenty-hour work week focused on providing essential needs for all, with the remaining time left over for creative self-realization.

Although much of Gorz's vision is articulated in terms of a true democratization of resources and useful things, energy was a constant consideration for him as well. He was concerned with "energy-wasting private machines" and about the relationship between energy production and environmental pollution.[6] He was also very critical of the nuclearization of France as an alibi for further unchecked growth and predicted the rise of electrofascism.[7]

The world that décroissance seeks is a low-carbon and low-energy modernity oriented toward durable goods and collective infrastructures that minimize individual consumption in the name of mutual thriving. For Gorz, this was not utopian but rather a matter of "ecological realism" that called for active diminishment and slowing down of the pulse of modernity: "The point is not to refrain from consuming more and more, but to consume less and less—there is no other way of conserving the available reserves for future generations."[8]

If this vision sounds remarkably contemporary, it is because fifty years later it could have been voiced by any number of degrowth advocates. Gorzian thinking directly informs the Transition Towns and permaculture movements and their principles of prosperity through degrowing consumption and resource usage.[9] In recent years, degrowth has attracted renewed interest as part of the neoliberal crisis in economics, including important interventions by Kate Raworth and Giorgos Kallis that theorize degrowth as an active program of civilizational transition, one equally committed to combating wasteful energy usage and social inequality.[10]

But, to my mind, where degrowth thinking intersects most generatively with radical energy politics is in the pathbreaking work of Indigenist and ecofeminist critiques of the intertwining extractivist infrastructures of patriarchal economies, capitalist production, and **settler colonialism**.[11] As Bruna Bianchi argues, growth ideology is enabled, fundamentally, by the resilience of women's caring labor that is constantly exploited as an infrastructure for extractive practices. Christine Bauhardt writes that ecofeminism reframes the growth crisis "as the finiteness of natural resources as well as the finiteness of women's caring labor." Ecofeminism has also proposed a solidarity economy as the core of a postcapitalist politics.[12] This solidarity model clearly intersects with Red Deal politics, which promote direct Indigenous **action** toward a "caretaking economy" in place of either capitalist growth or the vague ecopolitics of a sustainability economy.[13] No one knows the teeth of growth better than those **Indigenous** peoples who have been in its jaws for half a millennium. In their view, degrowth is simply a way that the North discovers what the South has known for centuries: capitalism is predatory and creates its luxuries for the few from the miseries of the many.

Returning to Gorz's interest in revolution, I would argue that degrowth thinking represents a "revolutionary infrastructure" for projects of worlding that do not repeat the ecocidal and genocidal trajectories of Northern colonialism and capitalism.[14] Degrowth is an ethical praxis that contests not only the many luxuries of petroculture but also the very rationale for a high-energy modernity. The question today is not whether degrowth is coming. It is coming whether in the form of the widespread collapse of the predatory, extractivist capitalist system or some kind of managed transition to an intersectional ecosocialism. The question is how we can reverse growth in the least destructive way possible.

See also: **Bankrupt, Cuba, Development, Renewable, Transitions**

Notes

1 André Gorz, *Ecology as Politics* (Boston: South End, 1980).
2 Donella H. Meadows et al., *The Limits to Growth: A Report for the Club of Rome on the Predicament of Mankind* (New York: Universe Books, 1972).
3 Gorz, *Ecology as Politics*, 4.
4 Gorz, 7.
5 Gorz, 8.
6 Gorz, 8.
7 Gorz, 102–3.
8 Gorz, 13.
9 See, e.g., Rob Hopkins, *The Transition Handbook: From Oil Dependency to Local Resilience* (White River Junction, VT: Chelsea Green, 2008); Tim Jackson, *Prosperity without Growth: Economics for a Finite Planet* (London: Earthscan, 2009).
10 Kate Raworth, *Doughnut Economics: Seven Ways to Think Like a 21st-Century Economist* (New York: Random House, 2018); Giorgos Kallis, *Degrowth* (New York: Columbia University Press, 2018).
11 See, e.g., Christine Bauhardt, "Degrowth and Ecofeminism: Perspectives for Economic Analysis and Political Engagement" (paper, 4th International Conference on Degrowth for Ecological and Sustainability and Social Equity, Leipzig, 2014); Bruna Bianchi, "Ecofeminist Thought and Practice" (paper, 3rd International Conference on Degrowth for Ecological and Sustainability and Social Equity, Venice, 2012).
12 See, e.g., J. K. Gibson-Graham, *A Postcapitalist Politics* (Minneapolis: University of Minnesota Press, 2006); Val Plumwood, "The Ecopolitics Debate and the Politics of Nature," in *Ecological Feminism*, ed. K. J. Warren (New York: Routledge, 2004), 64–87.
13 Red Nation, *The Red Deal: Indigenous Action to Save Our Earth* (Brooklyn: Common Notions, 2021).
14 Dominic Boyer, "Revolutionary Infrastructure," in *Infrastructures and Social Complexity: A Companion*, ed. Penelope Harvey, Casper Bruun-Jensen, and Atsuro Morita (New York: Routledge, 2016), 174–86.

DEMOCRACY

To save the planet, do we need more democracy, or less?

ROBERT DANISCH

Democracy is a method of organizing social life in ways that emphasize equality among participants in collective decision making. For most of the past century, Western liberal societies assumed the value of democracy as both a political system and a way of life conducive to securing happiness and freedom. But those assumptions no longer necessarily hold. Are enough people still committed to the idea of democracy for it to survive challenges by fascists and authoritarians (including the Republican Party in the United States)? Will more democracy or less help us address the climate crisis and so save the **planet**?

At its best, democracy leverages the capacity of citizens to make decisions that benefit the public. But this outcome is possible only when systems of deliberation have not been co-opted or contaminated by the excesses of capitalism *and* when citizens see the virtue of deliberation through which evidence is considered and options weighed thoughtfully and collectively. Inequality threatens democracy because it constrains possibilities for **communication**. Violence, when construed as legitimate political action, also threatens democracy because it obviates the obligations of deliberation. In the current moment, many citizens publicly reject some of democracy's basic procedures and assumptions as well as the basic scientific facts of rising sea levels and temperatures. In the face of these challenges Andreas Malm recommends escalating forms of climate protest to include intentional **sabotage** and destruction of property.[1] Others, like earth scientist James Lovelock, endorse "eco-authoritarianism" and recommend that we put democracy on hold because of the continued global failure to forge policy responses to ecological disaster.[2]

Perhaps our moment lacks enough democracy to solve the climate problem. The ancient origins of democracy emphasize public conversation and persuasion; all citizens were assumed to be able to deliberate upon moral and political questions. The decision making practiced by ancient Athenians required the effective use of language and rules for orderly debate. This commitment remains with us today. But the Athenians could not have imagined how deeply inequality impedes deliberation, nor how money trumps speech, nor how emergencies can be so dire that thoughtful deliberation lacks sufficient urgency. Do we have enough time to debate the best response to the climate crisis? What would be gained from such deliberation?

American philosopher John Dewey argued that democracy was a way of life: "Democracy is not an alternative to other principles of associated life. It is the idea of community life itself."[3] Dewey championed participatory democracy as a response to Walter Lippmann's skepticism about the average citizen's ability to make good decisions about public issues.[4] According to Lippmann, because of an inevitable discrepancy between the world's complexities and our understandings of it, we are vulnerable to propaganda. Other commentators (including C. Wright Mills and the Frankfurt School) saw the rise of mass communication as a challenge to democratic culture. These critiques may explain public skepticism of scientific knowledge today. Even if we could deliberate as well as proponents of democracy hope, would we have the necessary information to make good decisions? If we are to respond to the climate crisis with deliberative democracy, we also need to worry about the role that media plays in filtering and framing our view of the danger.

Jürgen Habermas, responding to the clash between Lippmann and Dewey and the subsequent critiques of mass communication, explored the status of public opinion within practices of representative government claiming to be democratic.[5] Through assembly and dialogue, the "public sphere" generates opinions and attitudes that guide sociopolitical thought and action. Habermas's emphasis on argumentation and reasoning in the public sphere overlaps significantly with the contemporary commitment to deliberation. Deliberative democracy claims that legitimate decisions can be made only by authentic forms of inquiry that are free from the distortions of unequal political power.[6] Deliberative discussions must be characterized by informed, balanced, conscientious, and substantive arguments.

Climate assemblies are increasingly used by national and local governments to guide decision making. Promoted by civil society organizations, these assemblies engage everyday people to learn about, deliberate, and make recommendations on aspects of the climate crisis—a form of climate **action** almost diametrically opposed to Malm's approach. But is deliberation a panacea for what ails contemporary large-scale democracies? Much political change in the twentieth century was driven by social movements, informal groupings of individuals focused on specific political issues. Social movements often include contentious performances and sustained, organized campaigns that make public claims on targeted audiences; sometimes they involve violence. Social movement scholars have shown how networks and social media can frame political issues, how individuals are mobilized for political ends, and how media spectacles set agendas and influence democratic culture. Climate movements have long been attuned to media spectacle, dating back to Greenpeace's confrontations with oil companies. Will social media hurt or help the fostering of a deliberative culture characterized by equality and broad participation?

Answers to these questions tend to fall on a spectrum from dystopian to utopian. Are our mediated modes of interaction the beginning of a new age of democracy, or the end of the face-to-face deliberative inquiry celebrated by Dewey, Habermas, and the ancient Athenians? Communication practices, both mediated and unmediated, have been the driving force and the dangerous poison within our democratic systems from the beginning. Can we talk our way to effective responses to the climate crisis, or should we abandon our faith in democracy and blow up a few pipelines to create the kind of change that might save the planet?

See also: **Blockade, Civil Disobedience, Cuba, Justice, Nationalism, Populism, Solidarity**

Notes

1 Andreas Malm, *How to Blow Up a Pipeline* (New York: Verso, 2021). The title of Malm's book is a provocation; he does not actually advocate violence against people.
2 Jonathan Watts, "James Lovelock: The Biosphere and I Are Both in the Last 1% of Our Lives," *Guardian*, July 18, 2020, https://theguardian.com/environment/2020/jul/18/james-lovelock-the-biosphere-and-i-are-both-in-the-last-1-per-cent-of-our-lives.
3 John Dewey, *The Public and Its Problems* (Chicago: Swallow Press, 1954), 148.
4 Walter Lippman, *Public Opinion* (New York: Free Press, 1997).
5 Jürgen Habermas, *The Structural Transformation of the Public Sphere: An Inquiry into the Category of Bourgeois Society* (Cambridge, MA: MIT Press, 1991).
6 James Fishkin, *When the People Speak: Deliberative Democracy and Public Consultation* (New York: Oxford University Press, 2011).

DESIGN

What does design make possible?

KELLER EASTERLING

Even at a moment of climate crisis, society is mysteriously unable to design.

The search for singular solutions and singular evils is among the crude but deeply ingrained habits of the modern Enlightenment mind that tend to obstruct design thinking. This is the mind that looks for the righteous one and only and the Manichean struggle. It is attracted to ideational monotheism, monocultures, utopias, dystopias, ultimates, and bounded worlds parsed by an elementary particle. Favoring successive rather than coexistent knowledge, it looks for one idea to kill another in order to exist.

The nation-state embodies these modern Enlightenment logics that, braided with whiteness and imperialism, foster supremacy and hierarchy within its laws and diplomacy. It is a form of governance that, having evolved in relation to war, favors binary thinking and sees even peace as a corollary of war.

These modern Enlightenment logics underwrite a culture that tends to envisage design change in terms of a new technology believed to be a magic bullet or an innovation within a progression. Like the shift from train to automobile, the new technologies need not represent an improvement; they need only to galvanize sufficient power to become the dominant platform. And the same logics use computing power to make social media that, with its ranking and herding of followers, is the whiteness machine par excellence.

Even the activism designed to counter these dominant politics sometimes favors singular or comprehensive solutions. Counterpolitical strategies ranging from uprising to sabotage are just as essential, but a preference for singularity can foster tragic category mistakes or the mistaking of part for whole. An insistence upon the coexistence of multiple evils—capitalism, racism, fascism, xenophobia, or imperialism—can seem like a compromising betrayal of what is regarded to be the supreme struggle. And political superbugs who scramble ideologies run rings around these solutionist approaches.

Dreams of having the right answer are central to what is (so far) regarded to be human. People reward each other for manifestoes and moments when organizations become streamlined, homogenized, and pasteurized into clarified and unified fields. The math is self-reflexively responsive. The world is Turing complete. Unable to conduct trials on speculative scenarios, some of the smartest people in the world can only more precisely measure climate doom with data-heavy quantified proof.

No one stands by waiting for an organization to curdle, get lumpy, or go from solution to colloid—to become patchy, partial, and impossible to parse with any one term. But maybe this lumpy moment is the moment of innovation—the moment of design. Innovations that counter climate change will not be new technologies but instead new protocols for relationships among emergent and existing technologies and practices. These protocols of interplay will not take the form of a shiny new device, software, or vaccine. Instead, like the various **coronavirus** protocols, they will mix many old and new things like vaccines, behavior modification, spatial adjustments, and hygiene practices.

Designing is entangling. Design innovations orchestrate protocols for the way things combine. These interplays do not try to eliminate problems with solutions but rather put problems together—combining their otherwise neglected potentials as resourceful reagents in a productive chemistry. An interplay has a long temporal dimension that allows it to respond to changing conditions and moments when it is politically outmaneuvered.

Applied to the energy economies surrounding mobility, these interplays do not look for fuel and emissions reductions only from electric cars or automated vehicle technology. They do not regard digital technologies as more sophisticated than physical spatial technologies, but instead mix the heavy and the **digital** to make the most information-rich organizations. The only effective way to reduce cars on the road, vehicle miles traveled, emissions, and sprawl is the interplay among transportation modes of different capacities. Designing the interplay is about designing a space for upshifting and downshifting between transit and fleets of shared cars and bicycles. Rather than mobility narratives about speed and autonomy that have been associated with fossil-fueled vehicles, entanglement offers relief, safety, and diverse **community**.

Applied to the energy economies surrounding **development**, spatial interplays overwhelm the financialized abstractions and geometries of capitalized property with multiple modes of exchange or what J. K. Gibson-Graham calls "community economies": all the labors of care and maintenance and all the arrangements of solids relative to each other generate values that are more durable and tangible than puny monetary values.[1] They are often even assembled from the bankrupt spaces that capital regards to be a failure. These interplays offer a variety of alternative organs for collectively cultivating land or spaces that have fallen off the financial ledger. Community land trusts, cooperatives, agrarian trusts, and other forms of **commons** do not offer a simple investment and real estate gamble with the hope of a small return. These economies redouble values beyond those that capital recognizes. They are the forms of mutualism at the heart of the abolitionist thinking that **Black** feminist thinkers have long advocated.[2] Interplays can deploy regenerative techniques and renewable energy sources on agricultural land that has been exhausted by monocultures. And they can be designed to reverse-engineer sprawl and invasive development.

Applied to issues of migration in a wetter, hotter world, can interplays of mobility and knowledge exchange reverse the insular habits of nation-states in order to foster planetary political solidarity? There is no one and only fight and no single replacement for either abusive markets or obstructing governments. There are as many forms of mitigation as there are particular situations. Design can become more planetary as it becomes less universal and more partial or patchy. It can pursue the nonnational, nonmilitary, nonlogistical practices of war-not—addressing other forms of lethality, evacuating rather than invading, and organizing migrations that have less to do with citizenship and more to do with planetary cooperation.

Solutions are mistakes, and it is a mistake to wait for them, when it is possible to begin even now to *design.*

See also: **Airplane, Animals, Cycling, Economy, Law**

Notes

1 J. K. Gibson-Graham, Jenny Cameron, and Stephen Healy, *Take Back the Economy: An Ethical Guide for Transforming Our Communities* (Minneapolis: University of Minnesota Press, 2013).

2 Ruth Wilson Gilmore, *Golden Gulag: Prisons, Surplus, Crisis, and Opposition in Globalizing California* (Berkeley: University of California Press, 2007); Miriama Kaba, *We Do This 'Til We Free Us: Abolitionist Organizing and Transforming Justice* (Chicago: Haymarket Books, 2021).

DEVELOPMENT

Can development be green?

SIDDHARTH SAREEN

This question—inconceivable in the 2000s when **renewable** energy sources struggled to compete with fossil fuels—gathered force in the 2010s and has become pressing to consider in the 2020s. In 2020, solar energy became the fastest growing energy source globally. Major countries have established ambitious climate mitigation goals, notably European Union member states that have committed to a 55 percent reduction of greenhouse gas emissions by 2030 from a 1990 baseline. Floods, wildfires, and other extreme events are increasingly frequent worldwide and routinely cause devastating loss of life and habitat. Even as the interests of incumbents—actors who benefit from the status quo and tend to reproduce it—remain influential in prolonging reliance upon fossil fuels, public appetite for and the financial feasibility of clean energy futures have burgeoned in a rapidly changing global political economy.

One may say that the (unfinished) task of development in the twentieth century was to address rampant inequity in order to lift generations out of poverty toward well-being. The new political economy of affordable renewable energy and the intensifying effects of climate change create both opportunities and barriers toward accomplishing those aims. In many contexts—with war-stricken zones as notable exceptions—it is becoming increasingly untenable to claim a lack of capacity to provision populations with basic energy services, which bring with them improved access to cooking, cooling and heating, and lighting and public transport. The off-grid solar plants of Kenya and the electric rickshaws of India mark significant advances, as do the fully digitalized electric grid of Norway and the renewable energy system of Costa Rica. Yet the impacts of rising greenhouse gas emissions are unevenly distributed, not least in terms of cause and effect: driven by elite consumption, but most impactful on vulnerable populations with low direct responsibility. Climate change exacerbates inequities that the renewable energy boom cannot fully address.

A revised understanding of development to fit twenty-first-century challenges is taking shape in many quarters. A major disagreement among advocates of ecological action has taken shape between the green growth and **degrowth** camps. Proponents of the former emphasize enhanced efficiency in resource use to decouple material consumption from greenhouse gas emissions, whereas degrowth advocates call for a green shift that simultaneously eschews the economic growth paradigm to ensure more modest consumption. The latter group is better positioned to provide disciplined methods "for accommodating the partiality of scientific knowledge and for acting under the inevitable uncertainty it holds," methods that Sheila Jasanoff calls "technologies of humility."[1] Hence, a degrowth approach entails acting in recognition of necessary limits to both consumption and knowledge. Scholars of sustainability transitions argue that the requisite pace of low-carbon shifts to meet the climate challenge necessitates action within contemporary infrastructural configurations, even as those configurations are challenged and transformed.[2] Development from this perspective thus implies both a rapid shift to low-carbon energy futures and an intertwined, more abstract shift to organizing logics that address the societal conditions that engender and perpetuate inequity. This approach transcends the technological domain to constitute a sociotechnical puzzle, one that can be solved only by addressing underlying societal values alongside energy systems.

Influential versions of each of these shifts have come to the fore in the form of "doughnut economics," which emphasizes the safe operating space wherein basic needs are met within planetary boundaries and within the "entrepreneurial state," which points out the state's crucial role in enabling and orientating innovation.[3] The latter has influenced agendas such as the European Green Deal, which envisages a just transition premised on mission-oriented innovation, where advances are channeled to wider societal benefit rather than accruing as private returns to a narrow pool of entrepreneurs who share risks but not benefits with society.

These envisaged shifts and their undergirding principles share a common goal: namely, for societal forces to ensure that renewable energy rollouts enrich the global **commons** and enhance collective well-being. Yet universal energy access often remains the prerogative of an assumed benevolent state, to be secured as part of its social inclusion commitments. By contrast, most rollouts are implemented and owned at large scales by corporate entities concerned with profit maximization. This market construction of energy provision is reflected in elite consumption when billionaires enlist unfathomable quantities of limited and polluting fossil fuels for spectacular space travel, and governments buy fleets of jet-fuel-guzzling military machines with little public scrutiny or outcry. Rather than outliers, these examples are mainstream phenomena routinized in societal protocols that wed growth and development. Such equivalence, premised on commodification, gives the lie to prominent discursive commitments to **net-zero** emission targets by 2050; these grandiose promises rely on scaling up carbon removal technologies that are as-yet unproven, even as emissions continue to rise.

An empirically founded definition of development in a duplicitous world must be wrested away from long-standing normative preoccupations with **sustainability** in a holistic sense. Normativity is co-opted and abused by the very incumbents (including states) who prolong the fossil fuel era; they not only perpetuate inequity but also systematically reproduce it in current shifts to renewable energy infrastructure. Etymologically, *development* combines *change* and *progress*; it connotes a process of directed advancement. In the twenty-first century, the most crucial form of directed advancement at stake is that toward equitable low-carbon energy futures. Such futures entail greater societal control and collective forms of ownership over the production and usage of energy as the ubiquitous stuff that powers all activity. Collective control would enable decision making that steers us away from the high greenhouse gas emission levels that are currently driving unprecedented climate crises. Incumbents cannot be allowed to set the terms for this crucial task, as they benefit from prolonging the status quo to entrench their own privilege and position themselves at low climate risk.

A fit-for-purpose definition of development in a reenergizing world implies not a lesser but a greater role for the state, as the facilitator of equitable low-carbon energy futures. Such a goal requires renewable energy infrastructures to be governed at lower scales, and a flow of benefits from these to wider society, both directly in the form of energy services and revenue flows and indirectly through improved **commons** and climate change mitigation. Importantly, no public support to fossil fuel sources can be tenable under this definition of development, as such support rewards privileged incumbents and prolongs a fossil fuel era inextricably linked with inequity; correspondingly, actors with vested interests must not set the terms of transitions from self-serving, privileged positions. Fundamentally, this view of development befits an era of climate urgency and renewable energy abundance in a manner that resists co-optation: as *directed advancement to equitable low-carbon energy futures*.

See also: **Africa, Air Conditioning, Corporation, Democracy, Design, Electricity, Green New Deal**

Notes

The author gratefully acknowledges support from the Accountable Solar Energy Transitions (ASSET) project funded by the Research Council of Norway, grant 314022.

1 Sheila Jasanoff, "Technologies of Humility," *Nature* 450, no. 7166 (2007): 33.
2 Benjamin K. Sovacool, "How Long Will It Take? Conceptualizing the Temporal Dynamics of Energy Transitions," *Energy Research & Social Science* 13 (2016): 202–15; Gavin Bridge, Stefan Bouzarovski, Michael Bradshaw, and Nick Eyre, "Geographies of Energy Transition: Space, Place and the Low-Carbon Economy," *Energy Policy* 53 (2013): 331–40; Andreas Malm, *How to Blow Up a Pipeline* (London: Verso, 2021).
3 Kate Raworth, *Doughnut Economics: Seven Ways to Think Like a 21st-Century Economist* (London: Random House, 2017); Mariana Mazzucato, *The Entrepreneurial State: Debunking Public vs. Private Sector Myths* (London: Anthem Books, 2013).

DIGITAL

How does the digital reshape the material realm?

SHANE DENSON

How is the digital implicated—and how does it implicate us—in a redistribution of energy and agency within a new global politics of Earth and its environments? The answer, like the digital itself, is necessarily multifaceted and multiscalar, and any attempt to come to terms with our present situation and future prospects will have to account for the various dimensions of the digital and its transformations of media, technology, the environment, and life itself. Both in our daily activities and in the organization of planetary-scale systems, the digital has become an unavoidable presence and infrastructure for **action**. A first step in articulating a politics for our planetary future therefore lies in identifying digitally mediated interlinkages among actions, agencies, and energies across multiple scales.

We might take a cue from Charles and Ray Eames's film *Powers of Ten* (1977), which famously zooms out from the human-centric **scale** of a 1 m^2 overhead view of a couple relaxing on a picnic blanket. The visual scope gradually widens, marking out each order of magnitude (10 m^2, 100 m^2, 1,000 m^2, etc.) as the camera rises to encompass the city, Earth, and ultimately the observable universe (at 10^{24} m^2). Then the camera reverses course and zooms back in at the level of the couple before plunging deeper, displaying microscopic views of skin cells and their structures, and then descending to the subatomic level of protons and quarks. *Powers of Ten*, an analog film produced with support from IBM, has inspired a variety of digital remakes and homages, including the computer-generated opening sequence of *Contact* (1997) and the iPhone app *Cosmic Eye* (2012), not to mention the interactive platform Google Earth. Apparently, digital media sees itself called upon to re-create the Eames's film in order to demonstrate the superior imaging power of computational visual effects. However, the power of the digital is not limited to re-creating an analog spectacle; indeed, when viewed today on a digital platform like YouTube rather than projected from celluloid, *Powers of Ten* evokes the transformative power of the digital as a radically multiscalar, ecological force.

Consider, therefore, this quotidian scene: I turn on my laptop, open an internet browser, navigate to YouTube, and play *Powers of Ten*. Beyond the visual spectacle itself, my viewing of the video sets a variety of things in motion, drawing a widening set of circles and pointing to a broad ecology of media, power, and politics. Simply using a computer requires energy; when my laptop's **battery** is depleted, I will have to reconnect it to a power source. Infrastructural dependencies start to multiply: the wiring in my house leads to a local transformer, which, by way of overhead lines or underground cables, is connected to a power plant that converts fossil fuels, nuclear reactions, or **renewable** resources into usable energy. The plant is in turn connected to a larger grid, regulated and coordinated digitally to ensure uniform regional coverage. Wide area synchronous grids provide power across multiple states, countries, or even continents, but they are susceptible to ecologically as well as politically induced failures, as evidenced in February 2021, when an ice storm brought the Texas Interconnection to its knees, in no small part because the grid is isolated from the larger Eastern and Western Interconnections in order to exempt the state from U.S. federal regulations on interstate power sharing.

As such networks are increasingly updated to smart grids, promises of more environmentally friendly systems of energy production and distribution are balanced by a greater reliance on IT infrastructures. This dependence opens the door to specifically digital vulnerabilities, including hacking and other cyberattacks like the Stuxnet virus that targeted the Iranian nuclear program or the ransomware attack that shut down the Colonial Pipeline on the U.S. East Coast in May 2021, when Russian hackers netted several million dollars in cryptocurrency—another vector in the increasing confluence of digital computation, ecological burden, and political-economic power. Cryptocurrencies are "mined" by computers competing against one another to solve mathematical problems, effectively turning **electricity** into digital money; in 2021, the carbon footprint of Bitcoin, the most prominent of these currencies, was more than twenty-two million metric tons of CO_2 per year.[1] Meanwhile, the annual emissions created by YouTube's servers and infrastructures account for another ten million metric tons of CO_2, to which I contribute when I play a video.[2] Silicon Valley, which promotes an image of technologies that are light and clean as a "cloud," is in fact very dirty: its energy demands contribute to ecological disasters like California's record wildfires of 2019 and 2020 while exacerbating the direct geological impacts of **mining**, fracking, and atomic waste.

Thus, zooming out from my individual interaction with *Powers of Ten* on YouTube reveals this action to be inextricable from the digital as an expansively planetary force. Indeed, the digital is an integral part of what environmental scientist Peter Haff has termed the "technosphere," an emerging quasi-autonomous system not unlike existing geological paradigms (including atmosphere, biosphere, or lithosphere), upon which it depends for resources.[3] The digital is part of a global "technological metabolism" that encompasses the entire planet—and beyond.[4]

But to fully understand the ramifications of this digital metabolism, we need to follow *Powers of Ten*'s lead and zoom back into our scenario. What else is happening, on a more minuscule level, when I watch a video on YouTube? Pixels flicker on my screen, while the video codec's algorithmic operations take place behind the scenes. Wi-Fi signals bounce around me before reaching my device, which communicates with the digital platform in a collaborative effort to optimize images via buffering and real-time adjustments to resolution and bitrate. Also invisibly relayed across the internet connection are valuable data about my viewing history, social media activity, recent purchases, and political affiliations. At the root of all this activity are streams of electrons, voltage differentials that course through my computer's mineral components, including rare earth materials mined in real-world war zones, thus anchoring the digital firmly in a material politics of *power* in its various senses.

These invisible operations bypass my subjectivity as viewer; they are both smaller and faster than I could ever hope to perceive, but they link back inevitably to the larger circles we have observed already, inscribing my actions into a new digital politics of the earth.

See also: **Documentary, Music, Online, Storytelling**

Notes

1 Reuters, "Factbox: How Big Is Bitcoin's Carbon Footprint?" (May 13, 2021), https://www.reuters.com/technology/how-big-is-bitcoins-carbon-footprint-2021-05-13/.

2 Chris Preist, Daniel Schien, and Paul Shabajee, "Evaluating Sustainable Interaction Design of Digital Services: The Case of YouTube" (paper, Conference on Human

Factors in Computing Systems, Glasgow, 2019), https://doi.org/10.1145/3290605.3300627.
3 P. K. Haff, "Technology as a Geological Phenomenon: Implications for Human Well-Being," *Geological Society, London, Special Publications* 395 (2014): 301–9.
4 Haff, 305.

DOCUMENTARY

Why is uncertainty useful?

THOMAS PRINGLE

Truth and falsity, fiction and nonfiction, are common terms in documentary studies. Revisiting theoretical questions about the medium's claim to observational veracity, documentary scholars point to debates about climate change as **evidence** of a post-truth era in the history of media.[1] Both documentary media and climate change are discourses that make evidentiary assertions about reality. They are also socially contested expressions of realism, for better and worse.[2]

Documentary studies and the energy humanities are both concerned with *uncertainty* as an epistemological concept that names incomplete public confidence in mediated expertise that people rely on to understand environmental risk. Uncertainty, as a problem for climate change criticism, creates demands for scholarship addressing *how* specific uses of fossil fuels contribute to global warming. Uncertainty is also at stake in questions about *how* documentary media navigates the camera's inscription of reality alongside the social construction of what counts as a true or false representation. In the convergence between these discourses, uncertainty involves theoretical questions about how energy use, and the naturalization of energy within culture, can be documented with veracity while also recognizing that the mediated realism of such evidence is social, situated, and motivated.

In "Documentary Uncertainty," Hito Steyerl asserts that "uncertainty" is the "principle" of modern documentary media, as "our belief in the truth claims articulated by anyone, let alone the media and their documentary output, is shaken."[3] Modern documentary recognizes the gradient veracity of living in uncertain anthropogenic environments, in which carbon emissions in one site can contribute to irruptive natural disturbances in places far away. Steyerl registers how the proliferation of **digital** devices capable of producing visible evidence democratizes—but also complicates—the ability to record, distribute, and view documentary records. For instance, when smartphone videos of wildfires and floods become interpreted as proxies for a changing climate, the evidence provided by the actual image is at best uncertain in linking an isolated biophysical event to changes in global atmospheric chemistry, much less the burning of specific fossil fuels.[4] The physical processes of global warming are difficult to represent visually due to their **scale**, temporality, and complexity.[5]

Uncertainty is thus a central term for the epistemology of both documentary and climate change. Uncertainty helps scholars describe how models of the climatic future are authoritative yet unverifiable.[6] Discourses of uncertainty within climate **science**, while a crucial part of the discipline's method, can also foment the political language of denial and create unrealistic cultural expectations of climate predictions as forensic.[7] Further, uncertainty complicates efforts to specify the effects of particular modes of energy use in relation to climate change risks at a global scale. Ursula Biemann's experimental documentary *Deep Weather* (2013) engages this problem formally. Biemann contrasts two locations linked by the unrepresented totality of a changing climatic system. First, she details extraction processes and experiences of workers at a tar sands site in **Alberta**. Interspersed with aerial footage of the damaged environment, a narrated, imagined future depicts Canada's intent to expand tar sands **mining**. The film then cuts to Bangladesh, with a fictionalized narrative of a future climate change disaster brought about by a massive storm surge, flooding the coast at night and drowning communities in their sleep. Attaching observational documentary images to fictive and uncertain futures, *Deep Weather* links the intended expansion of tar sands exploitation to the displacement of thirty million Bangladeshi people anticipated within current projections of three-foot sea level rise. The material relationship between tar sands extraction in Alberta and the climate vulnerability of people in Bangladesh is correlative and uncertain. However, Biemann's formal documentary conjecture responds to this epistemological challenge by narrating a history of the future that analogizes the plans of energy-intensive economies in the Global North with projected increases of ecological risk in the Global South.

Another approach, exemplified by Belinda Smaill and Molly Geidel, historicizes the relationship between oil and documentary by demonstrating that documentary practices during the postwar era helped naturalize societal reliance on petroleum.[8] This work explicates the media genealogy of "petrocultures." The recent dissemination of petroleum company documentary archives traces how the uncertain epistemology inherent to climate futures became socially and economically valuable to Big Oil. For instance, in Royal Dutch Shell's *Climate of*

Concern (1991), an unreleased educational film that explores the then-future global warming side effects of oil reliance, the filmmakers predict the emergence of climate refugees and the intensifying effects of uneven global industrialization on the inequities of carbon markets. While predating the liberal ecomodernist agenda of *An Inconvenient Truth* (2006), Shell's *Climate of Concern* remarkably promises tech-driven solutions for a catastrophe spurred by the business of the film's sponsor. In 1991, Shell's Film and Video Unit intended to assure audiences that future technological change would mitigate the coming, but still uncertain, problem. While never circulated, the documentary vividly illustrates how nonfiction corporate propaganda tactically presents an uncertain future as true in a manner that provides documentary pedagogy without presenting meaningful alternatives to fossil fuel dependency.

Speaking to such tactics, Imre Szeman's critical perspective on the documentary treatment of energy transition offers a theoretical typology for how documentary films seek—and fail—to describe the end of petroleum reliance: "strategic realism" utilizes media and government discourse to downplay environmental destruction and scarcity in favor of narratives emphasizing national energy security; "techno-utopianism" promises future scientific and technological fixes as alternatives to petrochemical dependency; and "eco-apocalypse" films depict debilitating catastrophic imagery as an alternative to confronting the immensity of change required to avert climate disasters.[9] These documentary modes each employ descriptive certainty to link the future of oil to the future of the climate. Accordingly, each mode can fail to map the fundamental contradictions of living in a petroleum-dependent culture. Utopian, verisimilar images of future fossil-fuel-dependent economies, replete with shining fields of solar panels, too easily displace less certain depictions of the inequality defining present political climates.

By contrast, in *I Have Always Been a Dreamer* (2012), Sabine Gruffat describes how the industries and global supply chains processing carbon-based fuels reconfigure urban landscapes. The film links the growth of Fordist automotive manufacturing in twentieth-century Detroit to racist labor conditions and housing policies. In a comparative gesture, the film shows how oil extraction in twenty-first-century Dubai entails the civic segregation of migrant laborers, partitioning urban ecology in ways similar to Detroit's racial divides. Through a transnational and transhistorical comparison of cities—a highly uncertain analogy—the documentary posits a methodology for representing the political impact of globalized oil-based industries that engineer unequal urban environments. Without asserting causation, Gruffat speculates about how the international links between oil extraction and automotive petroculture shape disparate civic pathways in comparably violent ways. While the racial segregation of labor in Dubai and Detroit may not be the first image of global warming that comes to mind, Gruffat's documentary embraces a set of uncertain relations to depict inequalities shared across the global social life of oil. In this way, uncertainty is less an impediment to documentary than the narrative fabric for showing the human toll of energy and the disparate environments it partitions and destroys.

See also: **Africa, Air Conditioning, Corporation, Fire/Bushfire, Lifestyle, Storytelling**

Notes

1 Erika Balsom, "To Narrate or Describe? Experimental Documentary beyond Docufiction," in *Deep Mediations*, ed. Karen Redrobe and Jeff Scheible (Minneapolis: University of Minnesota Press, 2020); Pooja Rangan, "Bad Habits, or, Can Reflexivity Be Good Again?," *Cambridge Journal of Postcolonial Literary Inquiry* 7, no. 2 (2020): 225–29.
2 Thomas Patrick Pringle, "Documentary Ascertainment: Climate, Risk, and Realism," in *The Documentary Moment*, ed. Joshua Malitsky and Patrik Sjöberg (Indianapolis: Indiana University Press, 2023).
3 Hito Steyerl, "Documentary Uncertainty," *Re-Visiones*, no. 1 (2011): n.p.
4 Thomas Patrick Pringle, "The Climate Proxy: Digital Cultures of Global Warming" (PhD diss., Brown University, 2020).
5 Julie Doyle, "Seeing the Climate? The Problematic Status of Visual Evidence in Climate Change Campaigning," in *Ecosee: Image, Rhetoric, and Nature*, ed. Sidney Dobrin and Sean Morey (Albany: State University of New York Press, 2009).
6 Paul N. Edwards, *A Vast Machine: Computer Models, Climate Data, and the Politics of Global Warming* (Cambridge, MA: MIT Press, 2013).
7 Wendy Hui Kyong Chun, "On Hypo-Real Models or Global Climate Change: A Challenge for the Humanities," *Critical Inquiry* 41, no. 3 (March 2015): 675–703.
8 Belinda Smaill, "Petromodernity, the Environment and Historical Film Culture," *Screen* 62, no. 1 (2021): 59–77; Molly Geidel, "Petrodocumentary and the Remaking of New Deal Culture," *American Quarterly* 72, no. 3 (2020): 797–821.
9 Imre Szeman, *On Petrocultures: Globalization, Culture, and Energy* (Morgantown: West Virginia University Press, 2019).

ECONOMY

Is economic expertise important?

KYLIE BENTON-CONNELL

When I lived in Naarm (known as Melbourne, on land claimed by Australia) in the first decade of the new millennium, a slogan towered over the central train station. Painted onto a smokestack of a retired fossil-fuel-fired power station, neat lettering spelled out a concise message to the city's daily commuter crush: "NO JOBS ON A DEAD PLANET."[1] Written during a flash point in the country's ongoing struggle between logging companies and forest defenders, its resonance has only grown. It serves as a pithy rhetorical short circuit, in an era when the world molded by fossil fuel industries and their supporters in the name of jobs and the **economy** has become more visibly and catastrophically unlivable.

The invocation of environmental protection as a threat to the economy or economic livelihoods is familiar to anyone who has worked for land defense. Honed for many years by extractive industries, this rhetorical move is one of the strongest tools deployed by fossil fuel companies to maintain and expand their political power. Proposed curbs on fossil fuel expansion become that greatest of threats, the job killer; this happens (albeit in a variety of complex and differentiated ways) in places ranging from the Ecuadorian Amazon to the tar sands of Turtle Island and the coal-rich regions of land claimed by Australia.[2]

Economic expertise is often marshaled in public discourse around fossil fuel expansion; economic impact studies are commissioned and new jobs and GDP growth percentage points resulting from the project in question are enumerated. These data are dutifully reported in broadsheet media and redistributed through social media clusters formed around both organic interpersonal connections and heavily manipulative astroturfing. Political commentators, spokespeople, and elected representatives magnify the effect. Pollsters enter the feedback loop, where political aides endlessly workshop and focus-group messaging to maximize perceived electoral appeal. The usual outcome is a set of policy decisions that favor fossil fuel expansion against the land, air, and **water**, insofar as extractive expansion is framed as an economic necessity.

Land defenders and those who seek to support them have various approaches to this conundrum. One approach is straightforward dismissal: in this view, concerns about jobs and/or economic impact are insignificant when faced with the planetary ecological emergency.[3] Another is to contradict economic claims on their own terms by producing economic counterexpertise: for instance, by producing economic modeling that shows fewer jobs and smaller GDP growth impacts of any given pipeline or coal mine. A third is to approach any pro-fossil-fuel economic rhetoric as simply capitalist mystification, and therefore refusing to engage on terms set by the ideology of capitalist accumulation.

More holistic and ambitious approaches try to grapple with the question of threats to economic livelihoods in other ways. The cluster of organizing and thinking that has formed around the concept of the **Green New Deal** aims to inoculate against the economy-fossil connection through organizing for social-democratic industrial policy. In this frame, transition away from economic life centered on fossil fuels does not have to mean austerity-driven inequality or sacrificing economic growth. We can indeed have it all and use the institutions in at least some of the contemporary world (elected governments, trade unions) to get to a post-fossil-fueled world there.[4] A more explicitly anticapitalist approach regards Green New Deal approaches as insufficient, advocating a break with economic growth per se, arguing for (and often modeling) decommodified ways of feeding ourselves, taking care of each other, and relating to the land. Versions of this vision are often largely decentralized and emphasize prefiguration in place of electoral strategies: rather than championing green jobs, they turn a critical eye on wage labor as a way of organizing social or economic life.[5]

Taking yet another approach to economic expertise—that is, considering it a site of close inquiry—yields political insights that strengthen organizing work to dismantle fossil fuel dominance. One might understand economic modeling not only as something to debunk (by mirroring its techniques, while attempting to displace its fossil-fueled content) or to ignore (by dismissing it as capitalist misinformation). We might instead, following Timothy Mitchell, conceive of economic expertise as what makes it "possible to conceive of a network, or market, or national economy, or whatever is being designed, and assist in the practical work of bringing it into being."[6] This approach to economic analysis as performative—that is, as bringing into being the world it purports to describe[7]—has gained significant traction in recent years within the academy. But if this approach has increased detailed, studious attention to how economic expertise and practice *make* economic realities, it has not produced a concomitant increase in attention to how these realities might be *unmade*. This is surely the most urgent question in the case of fossil fuels. What might a detailed engagement with economic analyses (within the dominant frames of economic knowledge, rather than counterhegemonic ones) tell us about the self-perceived strengths, vulnerabilities, and decision-making processes within fossil fuel industries and their funders?

Such an orientation is becoming more visible in social movement spaces relevant to the climate struggle. The linked campaigns to stop tar sands pipelines have taken the dominant economic analysis within the **Alberta** oil industry—that limited pipeline capacity is an economic constraint on extraction because shipping by rail is prohibitively expensive—and used it as the bedrock of a push to keep some of the most polluting oil in the world underground.[8] The analysis of West Coast ports by the anonymous Degenerate Communism collective in the context of ongoing labor struggles looked in detail at the throughput of various transport nodes, nudging readers toward identifying how various tactical interventions might affect profitability.[9] A website that enjoins readers to view it only using the Tor browser and a VPN provides a tabulated volumetric and financial account of the impact of approaches to disrupting **pipeline** infrastructure (from mass direct action to anonymous sabotage), urging people to draw their own tactical conclusions.[10] Movements for land defense—not just liberal technocrats—are grappling in some detail with how "the economy" is constructed, in order to intervene with maximal leverage. We are past "no jobs on a dead planet"; the task now is to shut down the fossil-fuel-powered apocalypse machine for good.

See also: **Blockade, Class, Corporation, Democracy, Expert, Industrial Revolution, Organize**

Notes

1 "Protester Gets High on His Message," *The Age*, September 5, 2003, https://www.theage.com.au/national/protester-gets-high-on-his-message-20030905-gdwagl.html.

2 Kimberley Brown, "How Will Ecuador's Elections Affect the Future of the Amazon?," *Al Jazeera*, February 6, 2021, https://www.aljazeera.com/news/2021/2/6/how-will-ecuador-elections-affect-the-future-of-the-amazon; Madeline Smith, "Keep Canada Working vs. Yes to TMX: How the NDP and UCP Pipeline Ad Campaigns Compare," *Toronto Star*, June 1, 2019, https://www.thestar.com/calgary/2019/06/01/keep-canada-working-vs-yes-to-tmx-how-the-ndp-and-ucp-pipeline-ad-campaigns-compare.html; Tom Rabe, "Fitzgibbon: Labor Must Set aside All Opposition to Coal Mining to Win Byelection," *Sydney Morning Herald*, April 2, 2021, https://www.smh.com.au/national/nsw/fitzgibbon-labor-must-set-aside-all-opposition-to-coal-mining-to-win-byelection-20210402-p57g7n.html.

3 Kate Aronoff, "Macron's Climate Tax Is a Disaster," *Jacobin*, December 11, 2018, https://jacobinmag.com/2018/12/yellow-vests-movement-climate-macron-cop24.

4 Thea Riofrancos, "Plan, Mood, Battlefield—Reflections on the Green New Deal," *Viewpoint Magazine*, May 16, 2019, https://viewpointmag.com/2019/05/16/plan-mood-battlefield-reflections-on-the-green-new-deal/.
5 Nicholas Beuret, "What Green Jobs Are They Talking About?," *The Ecologist*, June 15, 2021, https://theecologist.org/2021/jun/15/what-green-jobs-are-they-talking-about; "Between the Devil and the Green New Deal," *Commune* (blog), April 25, 2019, https://communemag.com/between-the-devil-and-the-green-new-deal/; Bue Rübner Hansen, "'Batshit Jobs'—No-One Should Have to Destroy the Planet to Make a Living," *openDemocracy*, June 11, 2019, https://www.opendemocracy.net/en/opendemocracyuk/batshit-jobs-no-one-should-have-to-destroy-the-planet-to-make-a-living/.
6 Timothy Mitchell, "Rethinking Economy," *Geoforum* 39, no. 3 (2008): 1118.
7 Michel Callon, *The Laws of the Markets* (Oxford: Blackwell, 1998); Donald A. MacKenzie, *An Engine, Not a Camera: How Financial Models Shape Markets* (Cambridge, MA: MIT Press, 2008); Koray Çalışkan and Michel Callon, "Economization, Part 2: A Research Programme for the Study of Markets," *Economy and Society* 39, no. 1 (February 2010): 1–32.
8 "Big Banks Admit No Keystone XL, Limited Expansion of Tar Sands Development," *350.Org* (blog), December 19, 2012, https://350.org/big-banks-admit-no-keystone-xl-no-expanded-tar-sands-development/.
9 Degenerate Communism, "Choke Points: Mapping an Anticapitalist Counter-Logistics in California," *libcom.org*, July 21, 2014, http://libcom.org/library/choke-points-mapping-anticapitalist-counter-logistics-california.
10 Stop Fossil Fuels, "Pipeline Activism and Principles of Strategy" (n.d.), accessed July 9, 2021, https://stopfossilfuels.org/strategy-principles/pipeline-activism/.

EDUCATION

How should energy educators respond to youth climate activism?

CARRIE KARSGAARD

Even before the **coronavirus** sent students home from school in March 2020, the planet saw a year of empty desks.

Schools emptied and streets filled in 2019 when students participated in a series of global youth climate strikes, following **Greta** Thunberg's *Skolstrejk för klimatet*. The climate strikes used empty desks to signal to political leaders and educators that preparation for the future was irrelevant in the face of inaction on the **Paris Agreement** and lack of attention to climate justice.[1] Nine-year-old Ridhima Pandey left an empty desk at BMDAV Centenary Public School, Haridwar, Uttarakhand, **India**, to join an international youth collective at the United Nations in New York in September 2019. Recognizing the uneven impacts on climate change on children's lives, she filed a lawsuit under the terms of the United Nations Convention on the Rights of the Child against five countries that failed to substantially reduce their carbon emissions.

And in the days before the COVID-19 lockdown, desks were **abandoned** in high school, college, and university classrooms across British Columbia, Canada, when Indigenous youth and their allies protested on the steps of the provincial legislature in support of the Wet'suwet'en Hereditary Chief's decision to block the $6.6 billion Coastal GasLink **pipeline**. This youth **action** drew attention to the intersection of fossil fuel extraction with issues of colonial injustice, including the denial of Indigenous peoples' sovereign authority to their land.

In each of these cases, youth left official institutions of education to mobilize campaigns for energy **justice,** using blockades, strikes, artwork, social media, and tool kits to educate one another about how to enact change. Their bold actions targeted not only climate change but also children's rights, decolonization, and social justice. And they were intended, too, to draw attention to the limits and problems of existing educational systems in teaching climate change, and their incapacity to empower students to do something about it.

Mainstream education is implicated in climate-harming processes and worldviews. Most blatantly, petro-pedagogies promote the fossil fuel industry's view of what constitutes appropriate climate action, whether overtly or through subtly individualized neoliberal education.[2] More insidiously, mainstream forms of education rely upon an extractivist worldview that has led to this moment of planetary crisis.[3] Transforming education in response to youth climate justice activism requires a confrontation with its extractivist foundations. **Youth** climate action has drawn attention to the need for educational systems to foster what Sheena Wilson calls "deep energy literacy"—a genuine engagement with alternative knowledge systems, including **Indigenous**, African, and alternative land-based cosmologies.[4]

What might deep energy literacy look like? A high school student from South Korea told me that schools need to replace an "individualist mindset," focused on personal educational success in isolated, core subjects, with a "truthful mindset" that would allow students to fully grapple with the challenges of energy transition; such a mindset could come about only by linking technical and scientific knowledge of energy to culture, politics, and economics.[5] For her, energy learning was necessarily relational. She understood herself as connected to young people around the globe by pipes, power lines, and transport vehicles; the movement of fossil fuels and electrical impulses; and a global **community** of humans and more-than-humans bound together by energy systems. In her description of energy entanglement, she moved beyond neoliberal conceptions of energy citizenship in schools, which frame climate action as primarily dependent on changes to individual habits (e.g., turning off lights and taking the bus to school) and thus fail to recognize the need to realize just futures by rejecting and re-creating social and cultural systems.[6] She pointed to the need to ask different questions, engage with a plurality of ways of knowing, and collaborate across contexts, disciplines, and social positions.

If education is to foster deep energy literacy, educators need to interrupt and destabilize energy education, which currently resides in a modern-colonial imaginary and is rooted in STEM subjects.[7] Imagine, for example, if youth were to engage with non-Western knowledge systems—to explore other "ways of relating to and with land/nature/one another on terms that are other than, or more-than modern."[8] A more relational education, engaged with alternative knowledges and using interdisciplinary forms of inquiry, would support the capacity of youth to conceptualize alternative futures and make concrete changes to our energy use. Such efforts would not pathologize youth frustration, anger, and despair at inaction and injustice—nor recommend pacifying practices and coping mechanisms—but instead listen to youth voices, address root causes, and confront the toxicity, displacement, and exploitation of our current energy realities.

Desks, school infrastructures, and commonly used educational practices can entrench hierarchical, segregational, and human-centric approaches to knowledge.[9] Following the lead of their students, and with the aim of creating just energy futures, educators might instead abandon desks altogether, locating learning in sites of/that matter, in order to reconnect youth to the land and to the global community of learners with whom they share this planet.

See also: **Animals, Civil Disobedience, Economy, Family**

Notes

1 Fridays for Future, "Our Demands" (last modified 2022), https://fridaysforfuture.org/what-we-do/our-demands/.

2 Lynda Dunlop, Lucy Atkinson, Joshua Edward Stubbs, and Maria Turkenburg-van Diepen, "The Role of Schools and Teachers in Nurturing and Responding to Climate Crisis Activism," *Children's Geographies* 19, no. 3 (May 4, 2021): 291–99, https://doi.org/10.1080/14733285.2020.1828827.

3 Amber Murrey and Sharlene Mollett, "Extraction Is Not a Metaphor: Decolonial and Black Geographies Against the Gendered and Embodied Violence of Extractive Logics," *Transactions of the Institute of British Geographers* 48, no. 4 (2023): 761–80, https://doi.org/10.1111/tran.12610.

4 Sheena Wilson, "Hacking the Techno-Transition: The Possibilities of Deep Energy Literacy," *Sociální Studia / Social Studies* 19, no. 1 (December 5, 2022): 29–53, https://doi.org/10.5817/SOC2022-29550.
5 Lynette Shultz and Carrie Karsgaard, "Energy Humanities News | Researchers Work with High School Students from 18 Countries to Address Energy Futures," *Energy Humanities*, December 10, 2020, https://www.energyhumanities.ca/news/researchers-work-with-high-school-students-from-18-countries-to-address-energy-futures.
6 Breffní Lennon, Niall Dunphy, Christine Gaffney, Alexandra Revez, Gerard Mullally, and Paul O'Connor, "Citizen or Consumer? Reconsidering Energy Citizenship," *Journal of Environmental Policy & Planning* 22, no. 2 (March 3, 2020): 184–97. https://doi.org/10.1080/1523908X.2019.1680277.
7 Sharon Stein, Vanessa Andreotti, Rene Suša, Cash Ahenakew, and Tereza Čajková, "From 'Education for Sustainable Development' to 'Education for the End of the World as We Know It,'" *Educational Philosophy and Theory* 54, no. 50 (October 18, 2020): 1–14. https://doi.org/10.1080/00131857.2020.1835646.
8 Lisa Tilley and Ajay Parasram, "Global Environmental Harm, Internal Frontiers and Indigenous Protective Ontologies," in *Routledge Handbook of Postcolonial Politics*, ed. Olivia U. Rutazibwa and Robbie Shilliam (London: Routledge, 2018), 306.
9 Hikaru Komatsu, Jeremy Rappleye, and Iveta Silova, "Will Education Post-2015 Move Us toward Environmental Sustainability?," in *Grading Goal Four*, ed. Antonia Wulff (Leiden: Brill, 2020), 297–321; Iveta Silova, Jeremy Rappleye, and Yun You, "Beyond the Western Horizon in Educational Research: Toward a Deeper Dialogue about Our Interdependent Futures," *ECNU Review of Education* 3, no. 1 (March 1, 2020): 3–19, https://doi.org/10.1177/2096531120905195.

ELECTRICITY

How does electricity organize society?

DANIELA RUSS

Electricity is the only form of energy crafted in the steam economy's machinery that will survive the end of fossil fuels. That the future is electric is an outlook shared by climate activists and by those who designed coal-fired power plants over a century ago. Electrification is a key element of the **Green New Deal** and other transitional projects that use energy from renewables—solar, wind, or water—to replace coal and oil in transport and industry. Electricity is so useful for transitions because of its modular nature. Electricity integrates the generation, distribution, and consumption of energy into one technical system, which must be balanced at every moment. Along the grid, any use of energy relates to every other. This technical challenge is also an advantage because it makes power systems into an infrastructure that can be used to coordinate various productions and uses of energy. In a fossil-free energy economy, where surplus energy is stored in electric cars and industry is fueled by green hydrogen, this infrastructure becomes even more all-encompassing. *Can we say, then, that electricity organizes a society's energy use?*

The emphasis on electricity as a collective organizer was nowhere stronger than in the early Soviet Union, when Lenin coined the slogan "communism is Soviet power plus electrification of the entire country."[1] The Soviets favored electricity (generated from peat, coal, and **water**) for many reasons, but particularly because it was produced most efficiently when it was produced collectively. Unlike petroleum lamps and the automobile, the production of electricity for individual use was extremely expensive and inefficient. Electricity consumption was to be either conspicuous or collective. By *collective*, the Soviets meant something like the **design** of early twentieth-century power plants, which were large, centralized systems that spanned a few hundred kilometers and served a variety of industrial and private consumers with diverse load patterns: varied patterns of use, at different times. The Soviets hoped that these systems, apart from providing cheap power, would allow for the coordination and control of energy flows in society as in a "single factory," where power is distributed so that various processes take place at the same time without offsetting each other.

The Soviet Union's appropriation of large-scale power production was a reaction to electric monopolists emerging in the West in the early twentieth century. General Electric and the German Allgemeine Elektrizitätsgesellschaft (AEG) became powerful corporations by operating utilities and selling electrical equipment. Their business model fed on the diverse energy use patterns of the social collective. Because profitability hinged on twenty-four-hour use of generation capacity, the bustling economic life of a city, region, or national **economy** enabled the profitable sale of electricity. These large, monopolistic power systems catalyzed debates about the collective ownership and coordination of electromagnetic waves—and not only among Bolsheviks. The idea that electricity organizes the collective or allows for its self-organization in a way different from other forms of energy rested on the argument that electricity was a natural monopoly with high infrastructure costs and low marginal costs, so that every additional kilowatt-hour reduces the cost of the energy produced. The cheap and abundant production of electricity tends toward a centralized supply of a diversity of loads, which is why most countries organized their industry around a powerful regulatory commission, large private corporations, or public utilities.

Today's grids work quite differently. Monopolies tend to be conditioned by markets rather than regulated by commissions. Enthusiasm about the socializing aspect of electrification waned in the seventies and eighties, particularly in the West. Regulated monopolies or national power companies were seen as locking consumers into a system that hampered the innovation and efficiency necessary to adapt to higher oil prices and interest rates. Many countries separated infrastructure from service and encouraged the competition of utilities on a still mainly monopolistic infrastructure. By breaking up large energy companies involved in all dimensions of power generation and distribution, governments encouraged the formation of energy service companies (ESCOs) that specialized in making certain processes more efficient. With renewables, however, a new age of monopolistic organization could be looming.

Among the services that ESCOs provide is load management, a method for documenting and shaping customers' consumption to reduce a system's overall cost. Data collection and load profile analysis have long allowed utilities to coordinate and profit from diverse forms of energy consumption. **Renewable** power grids will be more complex: more sectors will be electrified, renewable power follows a short-term seasonal or climatic rhythm, and the clear distinctions between producers and consumers are likely to vanish. Coordinating this complexity through big data collected from the grid will be an enormous challenge, which may lead to utilities and ESCOs being swallowed by companies more experienced in data processing.[2] Google and other tech companies have invested a staggering amount of money in renewable energy to run their servers on carbon-free power.[3] Google has also developed an algorithm based on the neural networks of DeepMind—the artificial intelligence company Google acquired in 2014—that has, according to the company, reduced energy consumption in one of its server farms by 40 percent.[4] In 2010, Google entered the power market directly by founding a subsidiary, Google Energy, which can act as a utility and sell energy, capacity, and services, including the use of AI to improve efficiency. Little is known about Google's energy conservation algorithm, and given AI's own **resource** footprint, doubts remain whether it can be called sustainable.[5] Additionally, the social consequences of AI-optimized grid management remain unstudied.

Electricity does not organize society's energy use by itself, although it was often employed to that end. In the early days, electrical grids facilitated the integration of water and thermal power, public transport, industry, and small businesses into a single technical system. Where utilities coordinated these uses according to their business interest, states made use of the system to allot power during wartime or to develop the economy in a planned way. By treating electricity like any other service, regulators have opened the business of coordinating energy use to corporations instead of states. In a renewable energy economy, these new monopolies would no longer need to own the material infrastructure in order to participate in the electrical grid. They would simply own the software and algorithms without which grids would no longer function. By continuing to strive for an "electrified world,"[6] we salvage one project of the fossil economy—the coordination of energy use through electricity—for the fossil-free afterworld. Let's make sure it's worth it.

See also: **Corporation, Digital, Music, Transitions**

Notes

1 Vladimir Illich Lenin, "Doklad Vserossiiskogo Centralnogo Ispolnitelnogo Komiteta i Soveta Narodnykh Kommissarov o Vneshnei i Vnutrennei Politike, 22 Dekabria," in *V. I. Lenin: Polnoe Sobranie Sochinenii* (Moscow: Izdatelstvo Politicheskoi Literaturoi, 1970), 159.
2 Sidney Sawaya, "Is Google the Future of Clean Energy?," *SDxCentral*, January 2, 2021, https://www.sdxcentral.com/articles/news/is-google-the-future-of-clean-energy/2021/01/.
3 "Google Enters the Power Market with Online Tools for Utilities," *Energy Digital*, May 17, 2020, https://energydigital.com/utilities/google-enters-power-market-online-tools-utilities.
4 Richard Evans and Jim Gao, "DeepMind AI Reduces Google Data Centre Cooling Bill by 40%," *Deepmind*, June 20, 2016, https://deepmind.com/blog/article/deepmind-ai-reduces-google-data-centre-cooling-bill-40.
5 Kate Crawford, *Atlas of AI: Power, Politics, and the Planetary Costs of Artificial Intelligence* (New Haven, CT: Yale University Press, 2021).
6 Carl Schmitt, *Roman Catholicism and Political Form* (Westport, CT: Greenwood, 1996), 13.

EVIDENCE

How is climate evidence used?

JAMES WILT

Since its election victory in 2015, Canada's Liberal government—led by Justin Trudeau, son of legendary prime minister Pierre Trudeau—has claimed that its policy agenda is motivated by evidence. The party's 2015 platform emphasized its "evidence-based policy" and its commitment to making decisions "using the best data available."[1] One election analyst concluded that "perhaps as much as anything, the Liberals' adoption of many evidence-based policies may have proven decisive."[2]

Evidence has since been leveraged by the government to justify and shore up legitimacy for policies on everything from the efficacy of carbon pricing to **coronavirus** responses; in late 2016, Trudeau even defended the approval of the heavily opposed Trans Mountain oil sands **pipeline** expansion as "based on debate, science & evidence."[3] Left unspoken, of course, is what actually constitutes evidence, including who produces it and how—questions that have formed the basis of disciplines such as science and technology studies (STS), political ecology, and histories of race and colonialism.

The unresolved conflict between claims and reality with regard to evidence emerged in full view during the 2021 election—called by Trudeau several years before it was required, amid the Delta-driven fourth wave of COVID-19—when renowned energy economist Mark Jaccard penned a supposedly neutral analysis about the "climate sincerity" of the four major parties for *Policy Options*.[4] Using a modeling tool called gTech, Jaccard compared each party's stated climate target with their policy commitments; he concluded that the pipeline-purchasing Liberals scored an 8/10 in climate sincerity, compared to 5/10 for the climate-change-denying Conservatives and a gutting 2/10 for the social democratic New Democratic Party (NDP). Regarding the NDP's plans—which pledged to force industry to pay for its emissions, eliminate fossil fuel subsidies, and implement carbon budgets—Jaccard projected that an "ambitious target combined with economically inefficient policies is devastating to the **economy**."[5]

Predictably, this dire conclusion was quickly and gleefully seized upon by the Liberal Party, generating many braggadocious social media graphics that included hyperbolic claims like "the NDP plan for the climate is to offshore jobs and pollution."[6] During televised leaders' debates, Trudeau frequently invoked Jaccard's analysis, telling NDP leader Jagmeet Singh at one point that "we need to talk about **science** and we need to talk about experts. So how is it that the experts rated our climate plan to be an A and rated your plan to be an F?"[7] Almost overnight, a single economist's assessment using a proprietary modeling tool was rendered the unanimous position of faceless experts and then weaponized in political debate.

To be sure, there was pushback to this process. In a lengthy critique for *National Observer*, Seth Klein from the David Suzuki Institute's Climate Emergency Unit advised that people take Jaccard's assessment with "a hunk of salt" as the ratings were developed based on a metric that rewarded lower ambitions: "It's like an Olympic diver getting the top score because she or he successfully nails the least complicated dive," Klein wrote, adding that "being 'sincere' about a climate plan that is clearly inadequate hardly deserves praise from anyone."[8]

This analysis, and the ensuing use of it by the Liberals as a rhetorical cudgel, spoke to a broader conflict concerning evidence in hegemonic energy politics. Profoundly ideological but supposedly neutral analyses by many economists and policymakers disparage and disregard direct government intervention a priori, with capitalist states relegated to creating the legal frameworks and financial incentives for green private investment to occur: carbon taxes, electric vehicle subsidies, and public-private partnerships. Within this episteme, direct state involvement in a rapid low-carbon transition is dismissed as inefficient or costly (except for solving so-called market failures in the fossil fuel sector by—for instance—nationalizing a pipeline).

But such a position represents a flagrant denial of the material realities of climate change, particularly given already-catastrophic and escalating impacts in the Global South, among **Indigenous** communities, and in poor racialized neighborhoods subjected to "organized abandonment and organized violence" by capital and state.[9] Every new **IPCC** report confirms that capitalism cannot solve this crisis within the urgent time frames required.

All real-world evidence points to the desperate need for every country to implement a radical **Green New Deal**–type response led by massive public funding and ownership of renewable energy, low-carbon public transit and housing, and electrification of industrial processes, as well as halting new fossil fuel infrastructures and phasing-down existing extraction with a genuinely just transition for workers. However, as Jaccard's analysis demonstrated in a roundabout way, capitalist growth is incompatible with a habitable planet.

This scenario means identifying and supporting real-world struggles: blockades, strikes, protests, sabotage. As prison abolitionist Mariame Kaba puts it, "My answer is always the same: collective organizing."[10] The only evidence-based policy remaining is revolution.

See also: **Class, Degrowth, Democracy, Documentary, Expert, Gaslighting**

Notes

1 Liberal Party of Canada, "Real Change: A New Plan for a Strong Middle Class" (2015), https://liberal.ca/wp-content/uploads/sites/292/2020/09/New-plan-for-a-strong-middle-class.pdf.

2 Charlie Smith, "Justin Trudeau's Emphasis on Evidence-Based Policies Paved the Way to Liberal Victory," *Georgia Straight*, October 19, 2015, https://www.straight

.com/news/559181/justin-trudeaus-emphasis-evidence-based-policies-paved-way -liberal-victory.

3 Liberal Party (@liberal_party), "The evidence is in: a price on carbon pollution works," Twitter, April 30, 2018, https://twitter.com/liberal_party/status /991008676424159232; Justin Trudeau (@JustinTrudeau), "These measures are necessary to protect our country from COVID-19, are based on the latest available evidence from our experts," Twitter, March 16, 2020, https://twitter.com /JustinTrudeau/status/1239679320601346052; Justin Trudeau (@JustinTrudeau), "If I thought this project was unsafe for the BC coast—I would reject it. Period. This decision was based on debate, science & evidence," Twitter, November 29, 2016, https://twitter.com/JustinTrudeau/status/803719850124255232.

4 Mark Jaccard, "Assessing Climate Sincerity in the Canadian 2021 Election," *Policy Options*, September 3, 2021, https://policyoptions.irpp.org/magazines/septembe -2021/assessing-climate-sincerity-in-the-canadian-2021-election/.

5 Jaccard.

6 Taleeb Noormohamed (@Taleeb), "The NDP's Swiss cheese plan isn't enough to even begin the fight against climate change," Twitter, September 12, 2021, https://twitter .com/Taleeb/status/1437233893694836745.

7 Justin Trudeau (@JustinTrudeau), "Mr. Singh, we need to talk about science and we need to talk about experts. So how is it that the experts rated our climate plan to be an A and rated your plan to be an F?," Twitter, September 9, 2021, https://twitter .com/JustinTrudeau/status/1436151781311094784.

8 Seth Klein, "If You Want Your Vote to Help the Climate, Here Are the Questions You Need to Ask," *National Observer*, September 9, 2021, https://www.nationalobserver .com/2021/09/09/opinion/election-2021-want-your-vote-help-climate-questions.

9 Ruth Wilson Gilmore, "Prisons and Class Warfare: An Interview with Ruth Wilson Gilmore," interview by Clément Petitjean (Verso Books, August 2, 2018), https:// www.versobooks.com/blogs/3954-prisons-and-class-warfare-an-interview-with -ruth-wilson-gilmore.

10 Mariame Kaba, "Free Us All," *New Inquiry*, May 8, 2017, https://thenewinquiry.com /free-us-all/.

EXPERT

What counts as effective climate expertise?

SHANE GUNSTER

Constrained by a neoliberal political imaginary, academic expertise about how to tread more lightly upon the earth has mostly been channeled into a "chimera" of behavioral change technologies that fetishize individualized lifestyle and consumption choices.[1] The inevitable failure of such interventions has generated skepticism and often outright dismissal of the relevance of behavioral change to the inescapably collective, political dimension of reining in capitalism's war on nature.

And yet political **action** *is* no less behavioral than driving, flying, or eating meat: building a broad-based climate movement will require countless numbers of individuals to change their **behavior**. Understanding the *how* (rather than the why) of such change is the most precious form of expertise we'll need in the months and years ahead. If we can justifiably indict most research into behavioral change for emphasizing the wrong kinds of behavior, then we should likewise demand that nominally radical calls for political engagement devote greater attention to fleshing out how people can actually move—and be moved—from personal concern to collective action. Expert knowledge in behavioral change has generally failed to engage with the broader structural, cultural, and material determinants of social practice; conversely, critical interventions that spotlight those determinants tend to neglect the granular, practical expertise that empowers people to actually change their political behavior.

The laudable shift from consumption to politics does not dissolve the infamously stubborn value-action gap with which behavioral change research has wrestled for decades: our proenvironmental attitudes and intentions often fail to translate into action. Indeed, the gap between intent and action emerges in stark relief when refracted through a political lens: a December 2020 survey, for example, noted that of those identified as "alarmed" by the climate crisis—over one-quarter of the U.S. population—only 4 percent reported participating in a campaign to take action to reduce global warming. On the other hand, an extraordinary 58 percent indicated they would "definitely" (23 percent) or "probably" (35 percent) be willing to join such a campaign.[2] Worried about the climate crisis; aware of the urgent need for political action; supportive of more aggressive policies; angered by a status quo dragging us ever closer to catastrophe; eager for bold, transformative change: increasing numbers of people stand poised on the edge of climate activism. Brushed against the grain, the vast body of research into behavioral change can yield valuable insights and practical expertise for accelerating the transformation of latent political desire into manifest political action.

Ripe for such tactical appropriation is the concept of eco-teams. Pioneered in the United States in the 1980s and later applied in the United Kingdom and the Netherlands, eco-teams consist of six to eight individuals who meet for monthly, well-resourced discussions about how to reduce their environmental impacts. Designed to foster competence in household and institutional behavioral change, eco-teams are built upon three core practices.[3] First, groups are given step-by-step, peer-delivered information about behavioral change that is carefully tailored to local conditions and uses social norm messaging to frame such change as easy, desirable, and popular. Second, individuals receive detailed feedback about the impacts of their behavior, thereby cultivating self-efficacy and the perception that their actions are meaningful and making a difference. Third (and most importantly), eco-teams offer a space to discuss the challenges of behavioral change and its relevance to ecological crises, to engage in collective problem solving, to become accountable to one another, and to celebrate achievements and build social identities that emphasize the intrinsic virtue, meaning, and pleasure of more sustainable forms of life and community.

While eco-teams have a strong track record in generating lasting behavioral change,[4] they inevitably bump up against the structural constraints that lock individuals and organizations into practices of high resource consumption and that make change difficult, expensive, and seemingly irrelevant. Interviews with eco-team members identified concerns "about the distributions and abuses of equity, **justice** and power that they see in their everyday lives, [including] . . . loss of community, loss of respect for each other and the environment, social forces that promote over-consumption, the lack of fairness implicit in economic systems, loss of positive social spaces for interaction, the continuation of colonial attitudes toward developing countries and fear of 'where is it all going.'"[5] Eco-teams necessarily catalyze *political* conversations about the entanglement of ecological crises and overconsumption with capitalism, consumerism, inequality, bureaucracy, and colonialism. Unsurprisingly, close to one-third of eco-team participants subsequently join "pro-environmental groups, campaigns or collective action."[6]

Hence, a modest proposal: the *political* eco-team. Imagine if the incidental exercise of the sociological imagination haunting the margins of eco-team experience was centered in groups devoted to such goals as nurturing pragmatic forms of civic literacy scaled to the needs, interests, and dispositions of particular groups; providing supportive spaces for the discussion, planning, and celebration of accessible forms of political engagement; and building the competence, efficacy, and **solidarity** for those with little political experience to experiment with novel political behaviors.

Beyond the salutary impact such teams would have upon the citizenship, advocacy, and activism muscles of their participants, consider how the task of resourcing such an enterprise might also bend the arc of academic knowledge toward the vernacular, thereby mandating the production of practical, accessible expert accounts of how to enhance political agency. Imagine the creative experimentation that might generate new—or rehabilitate old—forms of radical, practical expertise as thinking about political agency shifts from the (abstract) *why* to the (concrete) *how*.

Customarily, eco-team programs are divided into discrete segments such as waste reduction, energy use, water use, transportation, shopping, and so on. What areas of emphasis might political eco-teams explore? What bursts of procedural knowledge could spark the competence, efficacy, and **solidarity**—in short, the democratic expertise—for each of us to domesticate and activate politics as space, tool, and action for mastering our relationship with nature? My list of topics? It looks something like this:

- Carbon (footprint) mapping 101: Where do emissions come from?
- Sharing is caring: why talking about the climate crisis (and what to do about it) matters
- You are what you see: media dieting for a better world
- Public goods: how governments have been (forced to) make a better world
- A false dichotomy: how individuals make collective action
- Taking political stock: opportunities, pressure points, dead ends

What about yours?

See also: **Airplane**, **Communication**, **Democracy**, **Education**, **Organize**

Notes

1 Janette Webb, "Climate Change and Society: The Chimera of Behavior Change Technologies," *Sociology* 46, no. 1 (2012): 109–25.
2 Anthony Leiserowitz, Edward Maibach, Seth Rosenthal, John Kotcher, Xinran Wang, Jennifer Carman, Matthew Goldberg, Karine Lacroix, and Jennifer Marlon, *Climate Activism: A Six-Americas Analysis, December 2020* (New Haven, CT: Yale Program on Climate Change Communication, 2021), 8.
3 Scott Davidson, "Up-Scaling Social Behavior Change Programmes: The Case of Eco-Teams," in *Engaging the Public with Climate Change: Behavior Change and Communication*, ed. Lorraine Whitmarsh, Saffron O'Neill, and Irene Lorenzoni (Washington, DC: Earthscan, 2011), 181–83.
4 Henk Staats, Paul Harland, and Henk A. M. Wilke, "Effecting Durable Change: A Team Approach to Improve Environmental Behavior in the Household," *Environment and Behavior* 36, no. 3 (May 2004): 341–67.
5 Kersty Hobson, "Competing Discourses of Sustainable Consumption: Does the 'Rationalisation of Lifestyles' Make Sense?," *Environmental Politics* 11, no. 2 (2002): 109.
6 Davidson, "Up-Scaling Social Behavior Change Programmes," 189.

EXTINCTION

Is there a relation between energy and extinction?

ASHLEY DAWSON

Is there a relation between energy and extinction? On one level, this question may seem easy to answer. Although ecosystems have been changing with great rapidity since the advent of European colonialism roughly five hundred years ago, the extinction crisis in its current severe form is largely a product of the Great Acceleration of capitalist extraction, production, and destruction of natural resources that unfolded from the mid-twentieth century to the present. A report by the World Wildlife Fund found that overall population sizes of mammals, birds, amphibians, reptiles, and fish have dropped by 68 percent since 1970.[1]

It is no coincidence that exploitation of global natural resources also accelerated massively and uniformly in this period. This era was also, of course, the age of the now-famous hockey stick increase in greenhouse gas emission rates. The uptick of capitalist exploitation and what David Harvey called "accumulation by dispossession" of human populations during this period has been accompanied by underacknowledged but intense processes of extraction and what might be termed "accumulation by extinction."[2] While specific populations of plants and **animals** have been going extinct at increasing rates during this period, the abundance of whole ecosystems has also been decimated. Crashing biodiversity and extinction are, in other words, intertwined and mutually magnifying crises.[3]

Yet despite the clear historic convergences of spiraling extraction and extinction, the two have usually been viewed by many scientists as independent challenges. Conservation biologists have traditionally identified the five greatest threats to biodiversity using the HIPPO acronym: (1) habitat destruction, (2) invasive species, (3) pollution, (4) human overpopulation, and (5) overharvesting by hunting and fishing.[4] Anthropogenic climate change is conspicuously absent from this list.

Challenging this siloed approach, recent reports conclude that climate change and the rapid decline of not only individual species but also entire ecosystems are intertwined crises that should be tackled together.[5] Such recent efforts are a product of collaboration between the Intergovernmental Science Policy Platform on Biodiversity and Ecosystem Services (IPBES) and the Intergovernmental Panel on Climate Change (**IPCC**), scientific groups that traditionally worked in isolation. Overcoming these institutional and intellectual divides is imperative for correctly understanding the intersectional character of the present environmental crisis, one that extends beyond the human/demographic questions of race, class, and gender to which intersectionality has traditionally referred. Breaking down traditional institutional and intellectual walls is also essential to developing an accurate understanding of the struggles necessary to address this crisis.

For instance, measures to address the climate crisis that emanate from and fortify corporate interests in the Global North can have an extremely detrimental impact on the world's most biodiverse ecosystems and the people who have long stewarded them. Carbon mitigation projects that destroy biodiversity and provoke the extinction of specific species include the planting of vast tracts of land with crops for bioenergy and the creation of massive plantations of monocultural, nonnative trees in Global South countries that are intended to absorb carbon. Such greenwashing projects have been called out by groups like the Indigenous Environmental Network since they sacrifice biodiversity in order to allow corporations and affluent consumers to continue using energy and generating emissions at an ever-expanding pace.[6] Such greenwashing measures are an effective means to demobilize environmental action in Global North countries, channeling activism away from oligarchical power and into relatively ineffectual **lifestyle** politics.

The intersections of the biodiversity and climate crises are apparent even in sectors that may not immediately appear to be directly related to energy. Take the model of industrial agriculture spread around the globe by the Green Revolution since the 1960s. Food sovereignty organization GRAIN estimates that between 54 and 57 percent of all greenhouse gas emissions come from this form of agriculture.[7] This surprising statistic includes emissions produced throughout the capitalist food system, including the habitat destruction caused by industrial agriculture, the use of fossil-fuel-based inputs like chemical fertilizers and pesticides, processing and packaging of foodstuffs, and, finally, the astonishing amount of **waste** that characterizes the modern food system (up to 50 percent of the food raised by industrial agriculture is thrown away). As this report shows, the capitalist food system would be impossible without massive inputs of fossil fuels at every point in the supply chain. If, as the old saw has it, you are what you eat, people who eat the food produced by this system are quite literally made of oil. If fossil fuels are a form of solar energy congealed by millions of years of subterranean pressure, once extracted and incorporated into fossil capitalist production systems they become a kind of devil's blood, generating short-term vivacities that inevitably also provoke long-term mass extinction.

The inescapable conclusion is that addressing the extinction crisis would require a rapid transition away from the profligate combustion of fossil fuels that sustains the current global economic and political order. Hans-Otto Pörtner, a climatologist and leader of a collaboration between scientists working on biodiversity and climate change, has argued that "maintaining biodiversity and its functions relies on phasing out emissions from the burning of fossil fuels."[8] It also involves not simply reversing the degradation of carbon-rich ecosystems like savannahs and wetlands, but also ramping up non-fossil-fuel-based agriculture and forestry while slashing state subsidies to destructive industries. Although it is seldom mentioned in scientific reports, this approach to biodiversity means supporting the struggles of organizations like La Via Campesina, whose fight for agroecological food production challenges centuries of capitalist and colonialist exploitation. Finally, and most radically, it means giving land back to **Indigenous** people who sustained multispecies communities on it for thousands of years. As Red Nation activists put it in their *Red Deal* manifesto, it's **decolonization** or extinction.

See also: **Geoengineering, Net Zero, Settler Colonialism**

Notes

1 Marco Lambertini, "Living Planet Report 2020: Bending the Curve of Biodiversity Loss" (desLibris, 2020), http://www.deslibris.ca/ID/10104983.
2 David Harvey, *The New Imperialism* (New York: Oxford University Press, 2003).
3 Ashley Dawson, *Extinction: A Radical History* (New York: O/R Books, 2016).
4 Edward O. Wilson, *Half-Earth: Our Planet's Fight for Life* (New York: Norton, 2016), 57–58.
5 Georgina Gustin, "New Report: Climate Change and Biodiversity Loss Must Be Tackled Together, Not Separately," *Inside Climate News* (blog), June 11, 2021, https://insideclimatenews.org/news/11062021/biodiversity-climate-change-ipcc-forests-ocean/.
6 Global Alliance Against REDD, "REDD + Indigenous Peoples = GENOCIDE" (n.d.), https://no-redd.com/.
7 GRAIN, "The Great Climate Robbery" (December 7, 2015), https://www.grain.org/en/article/5354-the-great-climate-robbery.
8 Gustin, "New Report."

FAMILY

What is the environmental future of the family?

EVA-LYNN JAGOE

I contradict myself whenever I teach. On the one hand, I critique the family form, pointing out its heteronormative structure, patriarchal domination, and power imbalances. Whether I'm teaching Ursula LeGuin's *The Dispossessed*, Lucrecia Martel's *The Holy Girl*, or Leo Tolstoy's *Anna Karenina*, I ask students to interrogate the ways in which family constrains and incapacitates other ways of being in relation to each other. On the other hand, I tell funny stories about my family, making myself recognizable to the students as a mom who has watched her own kids go through university.

It makes sense that I'm conflicted in this way. Much as my affective language and behavior have been shaped by the intimacies of family, the critical thinking and cultural analysis that I have done for years as an academic have led me to question its form. I have taken seriously the work of theorists such as Edward Said, Lee Edelman, and Donna Haraway, each of whom, in distinct ways, champions kinships of choice and affiliation, rather than family of birth. The future that seems most free of heteronormative, racist, and xenophobic structures emerges as one that is queer, collective, or made of loose alliances and affinities between human and nonhuman species. In these views, the traditional family is not the way of the future; it's the weight of the past.

My colleagues seem to adhere to such ideas. In casual conversation, Professor A said to me and Professor B that her adult son was so frightened of climate change that he was considering not having children. Professor B, an established critical and cultural theorist, said that not procreating was probably the best thing the young man could do. Even though I have an arsenal of arguments against this kind of unexamined Malthusianism, I remained silent. I could have pointed out that the problem is overconsumption, not overpopulation, and that such statements express little more than xenophobic white Western anxiety.[1] I didn't respond because I want, on some internalized level, to fit in with the seemingly pragmatic claim there's nothing that can be done but to stop reproducing. However, I agree completely with Andreas Malm, who says that to not believe in future generations is to submit to a climate despair that is "a deeply reactionary type of ecology."[2]

What would have been the hardest thing for me to say is also what is closest to my heart: I think it would be a good thing if my kids had kids. I know, I know: future generations will suffer the effects of climate disruption, living lives that compound the hardships so many people in the world are already enduring. Yet I have hope that the children of the mid-twenty-first century will—with an irreverence that echoes **Greta** Thunberg's—be catalysts for an energy transition that heralds a more just and equitable future. Maybe some of them will even be my own grandchildren. If so, then the most pressing task is to create the best possible conditions for these subsequent generations. These conditions could arrive as early as the next decade.

Of course, with one son a socialist and the other an anarchist, my family's deepest hope is for complete revolution. But in the day-to-day, we're taking some steps toward transition. We recently bought a homestead in the Kootenay Mountains of Canada. It has been a difficult time of heatwaves, wildfire smoke, and doubt—doubt that we don't know what we're doing, that we shouldn't be private property owners in the first place. Yet it has also been a time of flourishing, in which we all work long hours on a project that is very much about the future.

I expected, when we moved to this famously alternative area, to find different intentional communities and communal living experiments. When I look around this valley, however, I see small family farms and homesteads, some of them still inhabited by the original families that immigrated here at the beginning of the twentieth century, others newly taken over by young couples with babies or intergenerational households like ours. Reading the sociologist Chris Smaje's book *Small Farm Future* resonates with what I've found on the ground. Smaje argues that some version of the privately owned family farm is the most viable way to keep a group of people committed to the land, to the labor, and to the time commitment that farming requires. Yes, he says, collective farms can also be successful, but there's nothing like a family unit in which each individual is personally invested in the successful future of a **farm**.[3]

There are obvious arguments against his proposal, ones that resonate with the critiques of patriarchy and heteronormativity that I articulated above. But what is compelling about Smaje's idea is that it does not imagine a complete overthrow of present-day realities. Most of the world, after all, still lives in family or household situations, ranging from an average of 6.9 people per household in sub-Saharan Africa to Europe's 3.1. The numbers tend to be larger in rural areas, where living in a larger intergenerational group is useful.

With the introduction of intensive fossil fuel fertilizers after World War II, farms became less about family holdings and more about monocrop production. More people moved to cities, where family is not so central in everyday life, though it still retains its paradigmatic power as an organizing principle of interpersonal relations. Now, as some farms begin to deindustrialize and heal the land through rewilding programs, restoration agriculture, and permaculture practices, the family farm, with its proven viability, could once again emerge as a possible site of energy transition.

As my sons and I prune cherry trees, we discuss the "Climate Genres" seminar I will teach next semester. My focus will be on critical agrarian studies, food forests, and our relation to the nonhuman. In our discussions, the students and I will interrogate the family unit in ways that expand upon what I have taught before. Family will emerge as a historically embedded and potentially generative form of **community** that can model forms of land stewardship, climate **action**, and change. Perhaps, as the theory I teach bridges the practices of my everyday life, I will begin to feel that I contradict myself less.

See also: **Animals, Commons, Extinction, Gender, Organize, Trans-, Youth**

Notes

1 Fred Pearce, "The Overpopulation Myth," *Prospect Magazine*, March 8, 2010, https://www.prospectmagazine.co.uk/magazine/the-overpopulation-myth; Banu Subramaniam, "Overpopulation Is Not the Problem," *Public Books*, November 27, 2018, https://www.publicbooks.org/overpopulation-is-not-the-problem/#fnref-25017-3.

2 Andreas Malm, *How To Blow Up a Pipeline* (New York: Verso, 2021), 153.

3 Chris Smaje, "Households, Families, and Beyond," in *A Small Farm Future: Making the Case for a Society Built around Local Economies, Self-Provisioning, Agricultural Diversity and a Shared Earth* (White River Junction, VT: Chelsea Green, 2020), 165–72.

FARM

Whose muscles power our food system?

EMILY PAWLEY

For many in the United States, the word *farm* evokes a comforting image: a black-and-white cow, a red barn, a cornfield, and a white man in overalls. It is always a dairy farm—they have **animals** but feel disconnected from slaughter.

Reproduced in picture books, toys, and food packaging, this image carries a powerful vision of white settler citizenship and a promised antidote to industrialization. It is also distanced from energy expenditure. No one is doing any work. The only visible energy source, the sun, smiles down crayon yellow, often with an actual face.

Reflecting on the same scene, environmentalists see an artifact of fossil energy. The cornfield is a monster grown fat on wars for cheap oil. Its proteins are likely made of nitrogen snatched from the atmosphere by natural-gas-hungry fertilizer factories. Its massive overproduction feeds climate-killing milk and beef, human-killing high-fructose corn syrup, and ethanol sold from the same pumps as gasoline. It is conventional to marvel that a system of production once based on sunshine, food, and work by living bodies (human and nonhuman) now converts fossil fuel BTUs into calories. As Michael Pollan wrote, humans now "sip petroleum."[1]

Frankly, there's a lot to this second way of seeing. Mechanized agriculture and industrial fertilizers have helped produce a global food system warped by surpluses and dumping—a glut of calories. Translations of fossil power into the language of muscles reinforce our sense that fossil fuels drive all food production. Commercial tractors offer "100-horsepower" engines, and experts describe oil barrels in terms of "energy slaves."[2] Machines replace bodily work from the combine harvester (replacing people), to the tractor (replacing horses), to the incubator (replacing hens).

But, if we look past our twinned dairy farm images, what other energies can we see? After all, there are not that many U.S. dairy farms—only 31,657[3]—many fewer than, say, the 3.5 million small coconut farms in the Philippines.[4] The ubiquitous dairy farm image hides millions of corporate farms and perhaps a billion smallholders. Within the wider world of farms, energy looks different than in the United States. In 2006 the Food and Agriculture Organization (FAO) estimated that 65 percent of the power on farms in developing countries still came from human and animal muscle.[5]

From the FAO's perspective, this was not supposed to have happened. Working animals in particular were supposed to be made obsolete by tractor-centered modernization.[6] Likely, though, they won't disappear. Animals can work in places tractors can't; they're self-reproducing, don't require imported parts, and can be fueled, sometimes, without cash.[7] In 2000 there were 26 million tractors in the world and perhaps 470 million working oxen.[8] When we include these working animals, we can tell different energy stories. To the pumping of oil, we can add the spread of clover, which, capturing nitrogen, builds muscle for animals around the globe.[9] To our list of energy crises, we can add the spread of rinderpest starting in 1888, which devastated sub-Saharan herding, plowing, and transportation and undermined resistance to colonialism.[10] To our climate fears, we can add the difficulty of fueling plant eaters through droughts, shading them in heat, and managing their methane.

Human muscle is also supposed to be insignificant in modern agriculture—an army of graphs show its decline as a percentage of energy through the last fossil-fuel-driven century.[11] During that same century, however, the actual number of people working in agriculture more than doubled; they now outnumber the entire global population when the steam engine was invented.[12] Even in the machine-rich United States, fruit rotting in the orchards during the **coronavirus** crisis revealed the limits of calorie-for-calorie comparisons of machine and human work.[13]

Fossil fuels are of course ineradicably political, a driver of global geopolitics awash in partisan rhetoric. But acknowledging the energy extracted from human farmers and farmworkers brings different political and economic relations to the fore: the intertwined systems of border control and intentionally absent documentation; the drivers of migration and mechanisms of debt bondage; and systems of land tenure, compensation, rent, and modern slavery. And, pushing in other directions, national and global movements of peasant and farmworker organizing. Narratives of farms focused on fossil fuels, awaiting or lamenting the replacement of human labor by "energy slaves," can't address these vast energy systems. They can't even see the undocumented immigrants who almost certainly work on our iconic U.S. dairy farm, hidden behind the white farmer and the red barn.[14]

What does an energy system look like when its sources can get tired? (Not the permanent exhaustion of extraction but the kind that requires sleep.) What does it look like when they can be poisoned by pesticides or killed by heat? When they resist, write, and **organize** politically? If we looked at farms in terms of their myriad realities, we might find out.

See also: **Africa, Class, Development, Extinction, Family, Industrial Revolution**

Notes

1 Michael Pollan, *The Omnivore's Dilemma: A Natural History of Four Means* (New York: Penguin, 2006), 46.

2 Bob Johnson, "Energy Slaves: Carbon Technologies, Climate Change, and the Stratified History of the Fossil Economy," *American Quarterly* 68, no. 4 (2016): 955–79, https://doi.org/10.1353/aq.2016.0074.

3 National Agricultural Statistics Service, USDA, "Milk Production" (February 23, 2021), 18.
4 Nithin Coca, "Coconut Farmers in Southeast Asia Struggle as Palm Oil Muscles in on Them," *Mongabay*, March 30, 2020, https://news.mongabay.com/2020/03/coconut-farmers-in-southeast-asia-struggle-as-palm-oil-muscles-in-on-them/.
5 Paolo Spugnoli and Riccardo Dainelli, "Environmental Comparison of Draught Animal and Tractor Power," *Sustainability Science* 8, no. 1 (2013): 61, https://doi.org/10.1007/s11625-012-0171-7.
6 Timothy Mitchell, *Rule of Experts: Egypt, Techno-Politics, Modernity* (Berkeley: University of California Press, 2002), 223–25.
7 Melaku Tefera, "Oxenization versus Tractorization: Options and Constraints for Ethiopian Framing System," *International Journal of Sustainable Agriculture* 3, no. 1 (2011): 11–21.
8 Giovanni Federico, *Feeding the World: An Economic History of Agriculture, 1800–2000* (Princeton, NJ: Princeton University Press, 2008), 48; R. Trevor Wilson, "The Environmental Ecology of Oxen Used for Draught Power," *Agriculture, Ecosystems & Environment* 97, no. 1–3 (2003): 21–37.
9 Maura Capps, "Fleets of Fodder: The Ecological Orchestration of Agrarian Improvement in New South Wales and the Cape of Good Hope, 1780–1830," *Journal of British Studies* 56, no. 3 (2017): 532–56.
10 Pule Phoofolo, "Epidemics and Revolutions: The Rinderpest Epidemic in Late Nineteenth-Century Southern Africa," *Past and Present* 138, no. 1 (1993): 112–43, https://doi.org/10.1093/past/138.1.112, 115.
11 Max Roser, "Employment in Agriculture" (Our World in Data, 2013), https://ourworldindata.org/employment-in-agriculture.
12 Federico, *Feeding the World*, 59.
13 Robert Suits, "Migrant Climates: Hobos, Energy, and Climate Precarity in the Great Plains, 1870–1940" (PhD diss., University of Chicago, 2020); Victor Seow, "Ideas in Motion: The Metabolism of Modern Migration," *Transfers* 8, no. 3 (2018): 123–29.
14 Margaret Gray, *Labor and the Locavore: The Making of a Comprehensive Food Ethic* (Berkeley: University of California Press, 2013).

FINANCE

Can we create an economy based on care?

MAX HAIVEN

The basic technologies of the thing we call *finance*—that is, binding agreements to make an exchange at some future point—are ancient.[1] But the modern capitalist financial system, represented by the finance, insurance, and real estate (FIRE) sector, emerged in relation to the global fossil fuel and extractivist system. Since capitalism's beginning, vast fortunes have been wagered and won on the world's stock exchanges through investments in lumber, peat, coal, crude oil, and **natural gas**.[2] This financial apparatus was essential to facilitating and insuring European colonialism, the slave trade, and the rise of private property as we have known it, all of which contributed to a world-historical shift in human relations with the earth.[3] Because concentrated carbon fuels are often located deep beneath the ground or ocean, extracting them has required ever more infrastructure, research and development, and specialized labor power. These changes demand huge outlays of capital for which financing has to be secured, increasingly through larger, more powerful financial institutions. These institutions, in turn, have enjoyed massive financial gains that afford them economic preeminence and political power.[4] The links between the capitalist realms of financial speculation and the extraction of hydrocarbons are difficult to overstate.[5] Since the twentieth century, the increasingly globalized FIRE sector has also been pivotal to the construction of hydroelectric dams and, to a lesser extent, to **wind** and solar energy.[6]

This process has largely served the already rich. While defenders of the financial system often argue that investments in extractive and hydrocarbon companies help to fund pension plans, the reality is that the benefits of the financial sector almost exclusively serve those who have wealth to invest.[7] Such is the FIRE sector's ideological and political influence that finance has been widely accepted as the best, if not the only, legitimate system through which to address the climate crisis—a crisis that finance helped to produce.[8] Carbon pricing and trading schemes, such as those negotiated in the Kyoto Protocol and refined at subsequent UN climate change conferences, assume that the only way to value the climate is to translate it into the language of financial markets.[9] This predicament is part a broader sociocultural shift, in which nearly everything is reinterpreted in the language and logic of finance. It's not simply that the financial sector uses its massive wealth to influence politics, media, and other democratic institutions.[10] Financialization means a broader shift at the level of ideas, dispositions, and subjectivities.[11]

Even resistance to the climate emergency and extractive industries must increasingly articulate itself in financialized terms.[12] The impressive success of efforts to encourage pension funds and public institutions to divest themselves of financial interests in fossil fuel companies suggests that financial arguments are an important tactic to garner public support for addressing a crisis that otherwise feels so vast and amorphous that it can elicit pessimism and despair. Yet if campaigners are not careful, this strategy can slide into a neoliberal narrative in which the most (if not only) effective solution to any problem is to leverage financial clout. This concession can foreclose a broader, more political strategy that sees energy and climate as matters of shared concern related to how humans interact with the "web of life."[13]

Notions of climate debt might offer a more radical and transformative discourse. Sometimes *climate debt* names a kind of ecological overshoot, in which **resource** consumption or carbon emissions will put us in debt to future generations doomed to inherit the legacy of our foolishness.[14] More conventionally, *climate debt* has become a politicized term related to the reparations owed by overdeveloped nations to developing nations.[15] While notions of climate debt go some way to addressing the role of colonialism and imperialism, they nonetheless frame the issue by invoking a financial discourse whose origins are rooted in extraction, competition, and accumulation. Instead, the climate crisis demands new forms of planetary cooperation that are antithetical to the fundamental logics of finance.

In an age of climate catastrophe desired by no one (even the very rich), it can be tempting to imagine that the world has been run off the rails by greed. But this conclusion would miss something important. Michael Hudson asks us to recognize that the financial sector is, essentially, the decentralized and internally competitive mechanism by which capitalism plans the global **economy**.[16] Tragically, it does so without any humanitarian or ecological goal, simply to perpetuate and expand the possibility of profit for the few. But its example demonstrates nonetheless that something like a global form of decentralized economic planning and coordination is possible. What would it mean to plan and coordinate differently, in the name of human and ecological care?

See also: **Bankrupt, Behavior, Corporation, Globalization**

Notes

1 Robert E. Whaley, *Derivatives: Markets, Valuation, and Risk Management* (Hoboken, NJ: Wiley, 2006).
2 Andreas Malm, *Fossil Capital: The Rise of Steam Power and the Roots of Global Warming* (London: Verso, 2016).
3 David Harvey, *Limits to Capital*, Essential David Harvey Series (London: Verso, 2018); Ian Baucom, *Specters of the Atlantic: Finance Capital, Slavery, and Philosophy of History* (Durham, NC: Duke University Press, 2005); Brenna Bhandar, *Colonial Lives of Property: Law, Land, and Racial Regimes of Ownership* (Durham, NC: Duke University Press, 2018).
4 Sarah Knuth, "Green Devaluation: Disruption, Divestment, and Decommodification for a Green Economy," *Capitalism Nature Socialism* 28, no. 1 (2016): 1–20.
5 Hannah Appel, *The Licit Life of Capitalism: US Oil in Equatorial Guinea* (Durham, NC: Duke University Press, 2019).
6 Mariana Mazzucato and Gregor Semieniuk, "Financing Renewable Energy: Who Is Financing What and Why It Matters," *Technological Forecasting and Social Change* 127

(2018): 8–22; Andrea Schapper, Christian Scheper, and Christine Unrau, "The Material Politics of Damming Water: An Introduction," *Sustainable Development* 28, no. 2 (2020): 393–95.

7 Grace Blakeley, *Stolen: How to Save the World from Finance* (London: Repeater, 2019).

8 Robert Guttman, *Eco-Capitalism: Carbon Money, Climate Finance, and Sustainable Development* (London: Palgrave Macmillan, 2018).

9 Sarah Bracking, "Financialization and the Environmental Frontier," in *The Routledge International Handbook of Financialization*, ed. Phillip Mader, Daniel Mertens, and Patricia van der Zwan (London: Routledge, 2020), 213–23; Emanuele Leonardi, "Carbon Trading Dogma: Theoretical Assumptions and Practical Implications of Global Carbon Markets," *Ephemera* 17, no. 1 (2017): 61–87.

10 Andreas Nölke, "Financialization and the Crisis of Democracy," in Mader, Mertens, and van der Zwan, *Routledge International Handbook of Financialization*, 425–36.

11 Randy Martin, *Financialization of Daily Life* (Philadelphia: Temple University Press, 2002); Max Haiven, *Cultures of Financialization: Fictitious Capital in Popular Culture and Everyday Life* (London: Palgrave Macmillan, 2014).

12 Naomi Klein, *This Changes Everything: Capitalism vs. the Planet* (New York: Simon & Schuster, 2014).

13 Jason W. Moore, *Capitalism and the Web of Life: Ecology and the Accumulation of Capital* (London: Verso, 2015).

14 Todd Dufresne, *The Democracy of Suffering: Life on the Edge of Catastrophe, Philosophy in the Anthropocene* (Montreal: McGill-Queen's University Press, 2019).

15 Rikard Warlenius, "Decolonizing the Atmosphere: The Climate Justice Movement on Climate Debt," *Journal of Environment & Development* 27, no. 2 (2018): 131–55; Jonathan Pickering and Christian Barry, "On the Concept of Climate Debt: Its Moral and Political Value," *Critical Review of International Social and Political Philosophy* 15, no. 5 (2012): 667–85.

16 Michael Hudson, *The Bubble and Beyond*, 2nd ed. (Dresden: ISLET, 2014).

FIRE/BUSHFIRE

What can we learn from fire?

MEG SAMUELSON

Australia's Black Summer presented a terrifying harbinger of a climate-changed world, as unprecedented megafires ravaged over twenty-four million hectares and killed or displaced approximately three billion critters.[1] So prodigious was this conflagration that it generated its own storm systems and metastasized through dry lightning, while NASA observed smoke plumes ascending into the stratosphere and circumnavigating the globe.[2] Yet, though fire now manifests as unequivocal and unchecked devastation on a planetary scale, it has been a profoundly ambivalent, situational, and even obliging companion through much of human history—as signaled in the doubled keyword *Fire/Bushfire*.

Bushfire, the term typically used for *wildfire* in the antipodes, is a descriptor springing from settler-colonial discourses that disposed of southern **Africa** and Australasia as bush, or uncultivated lands. Both regions had in fact long been fostered with anthropogenic fire: practices of what is sometimes called *firestick farming* extend some fifty thousand years in Australia and even longer in southern Africa.[3] Rebutting the myth of a precolonial wilderness, historian Bill Gammage identifies fire as the "ally" with which "Aborigines made Australia" into "the biggest estate on earth."[4] Victor Steffensen, in rekindling **Indigenous** practices of "caring for country through fire," describes how "cool" burning coaxed and maintained grasslands and open woodlands, encouraged the distribution and reproduction of various species, and mitigated fire's destructive potential by reducing fuel loads that would otherwise allow it to burn hot and run rampant.[5]

As John Durham Peters notes, "Fire teaches the deep historicity of what we think of as nature."[6] The geographic distribution of fire knowledges is equally instructive. Certainly, fire evokes the species talk that has come to dominate conceptions of planetary crisis as the Anthropocene: from nutritional impacts on the structure of the genome to the organization of social life around the brazier, humanity has been cast in the forge of fire. Yet, while fire is often imagined as the tool with which *Homo sapiens* hammered out its species exceptionalism, it is an interactive process rather than an object wielded at will: fire's behavior and effects are determined by contextual factors such as climate and the type and quantity of available fuel. Fire is, in other words, a situational phenomenon with particular distribution patterns: Mediterranean climate zones, which include the eponymous basin along with California and the southwestern regions of Australia, Chile, and South Africa, are eminently fire prone due to their hot and dry summers. Some biomes not only are more flammable than others but also have evolved to become fire-dependent, including fynbos in the Cape Floral Kingdom of South Africa, the eucalypt forests of southeastern Australia, and the savanna grasslands of central and southern **Africa**, Australasia, the Americas, and South/Southeast Asia. The geography of fire thus centers the southern latitudes along with the Indigenous lands of northwestern America that fire helped to make: it is here that deep familiarity with fire's regenerative and destructive potentialities forged cultures adept at living and working with its ambivalent powers.[7]

To better understand fire, then, we need again to "provincialize Europe."[8] Pyrohistorian Stephen Pyne is instructive in this regard.[9] He emphasizes how environments produced and nurtured with anthropogenic fire over thousands of years fell under the governance of people trained in the anomalously cool and damp climes of northwestern Europe. Commanding a different kind of firestick—its power entirely contained, tractable, and deadly—these intruders imposed fire practices developed in remarkably pyrophobic conditions onto a largely pyrophilic world. In a telling conjunction, the First Fleet arrived to establish the colony of New South Wales in Australia in the very same year that Boulton and Watt's rotative steam engine was built in Birmingham. The year 1788 is thus indicative of how a new fire regime bifurcated an ambivalent force into two opposed aspects, while also seeking to render fire itself placeless. On the one hand, the new arrivals responded with fear to the free-burning fires they encountered in southern lands and mandated the fire-suppression practices that remain hegemonic; on the other hand, they replaced the manifest agency of fire with the putatively controlled closed combustion of fossil biomass.

These two fire practices are now combining into megafires fed with fuel loads that have been left to accumulate in fire-deprived environments and turbocharged by the increasingly hot, dry, and windy conditions that industrial burning is producing on a planetary **scale**. The growing frequency, duration, and intensity of these fires is such that pyrophilic environments are now often ravaged by flames that had previously stroked them into life, while fire extends its reach into environments that have not evolved to embrace it. In this "pyric transition," the shapeshifting nature of localized fire is solidifying into the face of global catastrophe.[10]

The billowing smoke plumes that encircled the earth as southeastern Australia burned in January 2020 might implicate a new planetary subject—the *Anthropos*—but the geohistory of fire and its interactions with humans offer more grounded ways of thinking and living through scorching times. Just as it teaches the historicity of what appears as nature, fire illuminates the collaborative, responsive, and situated practices through which habitable places are composed—as well as those through which they are unraveled. The universalization of imperial and industrial fire practices stoked Australia's Black Summer. In contrast, practices from the pyrophilic South, such as those described in Steffensen's *Fire Country*, model vital modes of inhabiting a burning **planet**, while offering a training in how to live respectfully in relation and in place.

See also: **Animals, Community, Family, Industrial Revolution, Permafrost, Settler Colonialism**

Notes

1 Commonwealth of Australia, "Royal Commission into National Natural Disaster Arrangements—Report" (October 28, 2020), https://naturaldisaster.royalcommission.gov.au.
2 NASA Earth Observatory, "Australian Smoke Plume Sets Records" (February 3, 2020), https://earthobservatory.nasa.gov.
3 Andrew C. Scott, *Fire: A Very Short Introduction* (Oxford: Oxford University Press, 2020), chap. 3.
4 Bill Gammage, *The Biggest Estate on Earth: How Aborigines Made Australia* (Sydney: Allen & Unwin, 2011), n.p.
5 Victor Steffensen, *Fire Country: How Indigenous Fire Management Could Help Save Australia* (Melbourne: Hardie Grant, 2020).
6 John Durham Peters, *The Marvelous Clouds: Toward a Philosophy of Elemental Media* (Chicago: University of Chicago Press, 2015), 121.
7 Andrew C. Scott et al., *Fire on Earth: An Introduction* (Oxford: Wiley-Blackwell, 2014), 10, 11.
8 Dipesh Chakrabarty, *Provincializing Europe: Postcolonial Thought and Historical Difference* (Princeton, NJ: Princeton University Press, 2000).
9 Stephen J. Pyne, "Frontiers of Fire," in *Ecology and Empire: Environmental History of Settler Societies*, ed. Tom Griffins and Libby Robbin (Edinburgh: Keel University Press, 1997), 19–34; Scott et al., *Fire on Earth*, chap. 12.
10 Scott et al., chap. 12.

GASLIGHTING

How has gaslighting circumscribed our environmental imagination?

GRACE FRANKLIN

In 2022, *Merriam-Webster* named *gaslighting* its Word of the Year. The term originated in Patrick Hamilton's play *Gas Light* (1938), where a con man manipulates his wife's perceptions of reality in order to deflect attention from his heist and undermine her credibility as a witness. After the word reemerged during the 2016 U.S. presidential election, its currency has only proliferated. Gaslighting "now applies in both personal and political contexts," *Merriam-Webster* explained, citing an example of its use in a congressional investigation: "Big Oil is 'gaslighting' the public."[1]

This investigation into Big Oil's disinformation tactics was revelatory, but the tie between psychosocial gaslighting and fossil fuels was nothing new. Hamilton's marital melodrama was written during the 1930s Energy Battle, when Britain's coal gas and electric industries were competing over the domestic energy market. In ad campaigns, electric interests presented gas as old-fashioned and dangerous, while gas pitched itself as a stolid national treasure.[2] By setting his play in the Victorian period, when coal gas power was at its height, Hamilton seemingly favored **electricity**'s gothic portrayal of gas. Beyond recalling period details—blood-red curtains, a gaslight curfew—the play's eccentric circumstances evoked gaslight's history.

Britain's coal gas utility was a pathbreaking fossil fuel distribution system. Beginning in 1805, it initiated a transition from visible, self-tended candles and oil lamps to networked power supplied by faceless, privatized companies. In its formative years, urban-scale gaslight encountered resistance from citizens wary of explosions, leaks, and fires—which regularly occurred—and legal pushback from fishermen who reported oily deposits and dead **animals** in the Thames. Coal gas representatives responded with pamphlets that maligned detractors as delusional, unpatriotic, and at fault for spreading lies.[3] This cast-blame, concede-nothing approach came to define the industry, which evaded regulation until the 1860s.

In *Gas Light*, the villain searches his home's uppermost story for rubies owned by a former tenant, whom he murdered. Whenever he turns the attic gas lamps on or off, his wife notices a pressure change in the drawing-room mantles. Though Jack tries to isolate Bella by rattling her faith in her ability to interpret the world and by telling others that she's going mad, interhousehold servant gossip brings her back into community. When a neighboring detective visits, she confides her suspicions about something amiss upstairs, and he shares his memory of a cold-case murder that happened in the house. The play's climactic scene chillingly collapses the distinction between past murder and present heist.

For audiences in the Richmond Theatre in December 1938, this dramatization had additional implications. Theaters had been early adopters of electricity in the late 1800s, but Hamilton's production debuted at a rare remaining gaslit theater.[4] The stage's functioning gaslights, like garish set dressings, might initially have had a transporting effect, but the jets' significance grew as they transformed from background to focal plot point—the tell. When Jack turns out to be not only the unseen source of Bella's dimming lamps but also, in an on-the-nose reveal, an elusive murderer named Sydney Power, resonances between Mr. Power and the gas industry also become apparent. Like the gaslight utility, he menaces Bella with explosive outbursts and, in the manner of a leak, threatens to weaken her perceptive faculties. Moreover, Mr. Power's attempts to highlight Bella's purported unreliability in order to obscure his own crimes seemed lifted from the playbook of coal gas producers. In other words, Bella's coming to consciousness, predicated on her understanding of coal gas infrastructure's networked properties, also initiated an epiphany for the audience. If the stage lights connected not only to the villain's haunt but also, quite literally, to homes throughout contemporary London, then the industry's insidious maneuvers must also remain at play. One reviewer wrote that he "rush[ed] to settle my gas bill immediately."[5]

Gas Light's energy context has faded from memory, but Big Oil continues to deploy coal gas's originary strategies. Harvard sociologists recently analyzed ExxonMobil's rhetorical shift from climate denialism to acknowledgment of ill-defined risk. During the mid-2000s, advertisements began to transfer blame for potential risk outcomes onto consumers by framing energy expenditure in terms of demand rather than production, thereby presenting ExxonMobil as "a kind of neutral innocent."[6] Gaslighting by institutions or political entities attributes "harm to . . . individual character flaws and poor choices in an effort to conceal how the mechanisms of power function to asymmetrically distribute harms."[7] BP epitomized this deflection of culpability when it popularized the concept of a carbon footprint, an image that scales down the problem of climate catastrophe, while implying personal mess or traces of a crime.

Much as Hamilton's villain sought to isolate Bella—within herself, within their house—fossil fuel rhetoric has individualized responsibility in ways that inhibit activism. A recent study found that climate messages emphasizing personal **behavior** decrease willingness to take **action**.[8] Failure to act derives in part from feeling hypocritical about not doing enough to qualify as an activist. Gaslighting may make targets feel not psychically deficient but morally defective.

In 2021, a commercial opened on a father rolling out of bed to comfort his child. He carries his son back to the child's bedroom and switches on a nightlight. "It's only human to care for those we love," a voice-over says. The ad closes with a view of a single-family home. A light in one window glows as a tagline appears, "Chevron: The Human Energy Company." Consider what extends beyond that frame—the networks of extraction, the scope of the problem, the need for collective, structural solutions. Even with coal gas illumination now a distant memory, how has gaslighting circumscribed our environmental imagination?

See also: **Communication, Gender, Normal, Storytelling, Sustainability**

Notes

1 *Merriam-Webster*, "Word of the Year 2022," November 26, 2023, https://www.merriam-webster.com/wordplay/word-of-the-year-2022.
2 Ruth Hibbard, "Advertising Power and Modernity," *V&A Blog*, August 8, 2018.
3 Leslie Tomory, "The Environmental History of the Early British Gas Industry," *Environmental History* 17 (2012): 29–54.
4 Richmond Local Studies Archive, LCF/10247, item 27428, Richmond, UK.
5 "In the London Theatres," *Aberdeen Evening Express*, February 2, 1939.
6 Geoffrey Supran and Naomi Oreskes, "Rhetoric and Frame Analysis of ExxonMobil's Climate Change Communications," *One Earth* 4 (2021): 710.
7 Alison Bailey, "On Gaslighting and Epistemic Injustice," *Hypatia* 35, no. 4 (2020): 667.
8 Risa Palm et al., "Don't Tell Me What to Do," *Weather, Climate, and Society* 12, no. 4 (2020): 827–35.

GENDER

Why is gender important to energy transition?

PETRA TSCHAKERT

Just energy transitions can no longer afford gender-blind spots. And energy **justice** has no place for spiteful stereotypes, prejudices, and patriarchal power structures that fuel interlocking forms of subjugation. Gender matters in the daily politics of what energy transitions could look like. In striving for energy justice, we must ask, who has the power and agency to negotiate equitable and inclusive strategies, from the intimate scale of our bodies and households to municipalities, nation-states, and the influential fora of global energy policy?

What roles and responsibilities women and men play and shoulder in energy **transitions** and whether these transitions are fair and accessible are not matters solely of biological sex.[1] Decision-making power is largely determined by how male and female energy agents are constructed and perceived, often in problematic ways.

First, there is a lingering gender blindness in policy and praxis that disregards how relationships with energy are deeply gendered and intersect with other dimensions of inequality. Despite progressive **community** energy programs, solutions continue to be regarded primarily as technical, decoupled from gender roles at the household, community, or institutional level.[2] Gender-neutral claims that just energy futures are achievable by all members of society are both simplistic and dangerous: they mask biases and reinforce existing inequalities while also fueling ageism, ableism, and racism. Female board members in community energy projects in Scotland, for instance, feel well positioned to address such blinders, advance gender-specific perspectives, and foster diversity and inclusion.[3]

Second, gendered energy visions can perpetuate stereotypes about who can lead transition efforts. The myth of women in the Global South as intrinsically virtuous, green, caring, and responsible citizens is equally as damaging as notions of women in the Global South as vulnerable, helpless victims of deprivation. In solar energy communities in Sweden, women increasingly challenge prejudiced expectations that they embody righteous role models because of their alleged closeness to nature and virtuousness; they also challenge unpaid, do-it-yourself workloads and boards still dominated by men largely oblivious to inequities in access to energy sources and opportunities.[4] In the industrial coal culture of petromasculinity, such as in Poland's Upper Silesia, a different type of flattening occurs when women are seen as predominantly silent actors and moral guardians, relegated to out-of-sight **family** roles, or vilified as trespassers challenging the image of the heroic male breadwinner.[5] Yet another bias is at work when single mothers who face fuel poverty, as observed in Spain, are expected to give up their own thermal comfort and health to accommodate *more deserving* household members.[6]

What gives hope is the contestation of gendered stereotyping in energy transitions, particularly in contexts of entrenched disadvantage. For instance, women among South Africa's urban poor have repeatedly demonstrated their resourcefulness, with many excelling in fuel stacking (i.e., combining various energy options depending on accessible substitutes) to secure electricity for washing, cooking, lighting, and heating, even in female-headed households without property ownership.[7] I feel inspired by the countless, courageous women, **Indigenous** peoples, and minoritized groups who are charting empowering energy futures, again and again from positions of struggle and resistance against misrecognition, disrespect, and exploitation. As a middle-aged, well-educated, white she-earthling writing from the comfort and security of my temperature-regulated home in Perth, Australia, I bow in admiration before the Kenyan women solar engineers trained at the Barefoot College in **India**; they defy traditional gender expectations, installing solar panels while also assuming sales roles and transforming power relations in their homes and communities.[8] And I bow before the women of GRID Alternatives and Mothers Out Front in the United States who merge social justice activism with energy **democracy**, by recruiting people of color as trainees to generate solar power for disadvantaged groups.[9]

Myriad lessons learned from feminist movements and theories, as well as everyday battles over agency, capabilities, dignity, and rights, leave little doubt that oppression reinforces unequal participation and decision-making power in implementing low-carbon **development** pathways. A just and gender-attentive energy politics needs to confront unequal gender and power relations that perpetuate entrenched hierarchies. Such hierarchies make those in positions of authority even more domineering while replicating practices of exclusion, erasure, and disenfranchisement among the less fortunate, thereby sustaining energy injustices. A just and gender-attentive energy politics recalibrates transition trajectories, by making visible underlying gender and **class** struggles over visibility, skills, and power sharing to recalibrate just transition trajectories.

The list of women-led energy initiatives, intrepid female entrepreneurs, grannies' transition crusades, and global networks such as the Global Women's Network for the Energy Transition (GWNET) is growing and coordinated opposition to polluting extractive industries mounting. There is a deep sense of connectedness between denizens committed to doing the right thing and the more-than-human beings with whom we share an existence in this volatile world—a care ethics necessary to untangle knotty power relations on the path to just energy futures.

See also: **Animals, Black, Expert, Solidarity**

Notes

1 Judith Fathallah and Parakram Pyakurel, "Addressing Gender in Energy Studies," *Energy Research & Social Science* 65 (2020): 101461.
2 Carelle Mang-Banza, "Many Shades of Pink in the Energy Transition: Seeing Women in Energy Extraction, Production, Distribution, and Consumption," *Energy Research & Social Science* 73 (2021): 101901.
3 Bregje Van Veelen, "Negotiating Energy Democracy in Practice: Governance Processes in Community Energy Projects," *Environmental Politics* 27, no. 4 (2018): 644–65.
4 Daniela Lazoroska, Jenny Palm, and Anna Bergek, "Perceptions of Participation and the Role of Gender for the Engagement in Solar Energy Communities in Sweden," *Energy, Sustainability and Society* 11, no. 1 (2021): 1–12.
5 Iwińska Katarzynya and Xymena Bukowska, "Women's Agency in a World of Flux: On Silesian Energy Transition," in *Gender and Energy Transition*, ed. Iwińska Katarzynya and Xymena Bukowska (Cham: Springer, 2022), 1–16.
6 Marta Gayoso Heredia, Carmen Sánchez-Guevara Sánchez, Miguel Núñez Peiró, Ana Sanz Fernández, José Antonio López-Bueno, and Gloria Gómez Muñoz, "Mainstreaming a Gender Perspective into the Study of Energy Poverty in the City of Madrid," *Energy for Sustainable Development* 70 (2022): 290–300.

7 Josephine Kaviti Musango, "Assessing Gender and Energy in Urban Household Energy Transitions in South Africa: A Quantitative Storytelling from Groenheuwel Informal Settlement," *Energy Research & Social Science* 88 (2022): 102525; and Saul Ngarava, Leocadia Zhou, Thulani Ningi, Martin M. Chari, and Lwandiso Mdiya, "Gender and Ethnic Disparities in Energy Poverty: The Case of South Africa," *Energy Policy* 161 (2022): 112755.
8 Mang-Banza, "Many Shades of Pink," 101901.
9 Elizabeth Allen, Hannah Lyons, and Jennie C. Stephens, "Women's Leadership in Renewable Transformation, Energy Justice and Energy Democracy: Redistributing Power," *Energy Research & Social Science* 57 (2019): 101233.

GEOENGINEERING

Is geoengineering inevitable?

DREW PENDERGRASS

Geoengineering encompasses many proposed technologies for modifying the climate at continental or global scales. Although geoengineering remains controversial, people across the political spectrum are attracted to its varied possibilities. The political fight of the century, then, may be less about whether to open Pandora's box than about which form of geoengineering to deploy, at what **scale**. Geoengineering—specifically solar geoengineering—may be inevitable, for reasons discussed below.

Geoengineering proposals can be distinguished by the part of Earth's energy balance they target. The climate is determined by a balance between shortwave radiation into the Earth from the Sun and longwave radiation emitted back into space. Proposals for carbon dioxide removal would take greenhouse gases out of the atmosphere, allowing more longwave radiation to escape and cool the Earth. A more controversial category of proposals aims to reduce incoming shortwave radiation. The most prominent of these is solar geoengineering, where high-flying jets would spray particles in the upper atmosphere to partially blot out the sun.

The risks of solar geoengineering are many. Spraying sulfur aerosols into the stratosphere could bleach the blue from the sky, degrade the ozone layer, and disrupt global patterns of precipitation.[1] Any nation-state (or other actor) that launched a solar geoengineering program might provoke conflict among nuclear-armed states who disagreed on where to set the planetary thermostat.

And yet, despite these risks, and despite the feasibility of less dangerous decarbonization paths, solar geoengineering is the most likely intervention to be deployed soon at scale. Why might this technology, whose proponents liken it to nuclear weapons and opioids,[2] appeal to policymakers? In a word, *scale*.

Carbon capture quickly runs into problems with either energy or land efficiency when deployed at scale. Existing technologies require at least six gigajoules of energy per megagram of CO_2 removed from the air; capturing enough to make a dent in global concentrations would require a significant percentage of global electricity generation.[3] Such energy costs could be avoided by sacrificing land instead. Bioenergy with carbon capture and storage (BECCS) is commonly prescribed in models used by the Intergovernmental Panel on Climate Change (IPCC). The concept is to grow a plantation of trees, which are then burned for electricity; the emitted carbon is captured in the flue and buried underground, leading to net-negative emissions. Although the technology does not yet exist, **IPCC** scenarios propose dedicating land three times as large as India to BECCS in the latter half of this century to capture carbon at scale.[4] "Natural climate solutions" avoid many of the biodiversity and air quality costs of BECCS and are the only mature method we have for carbon capture at scale, but also require extensive land to be dedicated to rewilded ecosystems. Because 35 percent of habitable land on Earth is used for animal agriculture, scaling either land-based option quickly will need to confront the meat industry.[5]

A safe path toward a stable climate could come through deep and fast emissions reductions coupled with rapid deployment of carbon dioxide removal. Because the biosphere has a limited capacity to remove carbon, the longer it takes to reduce emissions, the more that unproven technologies like BECCS or direct air capture will need to be scaled. Any such path is expensive and rife with potential political and economic conflict: fossil fuel companies, agricultural interests, and many industrial concerns will be euthanized, while new powers will rise through their control of various parts of new supply chains. While a muscular climate movement, forged through alliances among the working class, feminists, and environmentalists, could build enough power to overcome these interests, the specter of a red-green alliance provides an incentive for the powerful to line up behind a dangerous alternative.

Solar geoengineering could cool global mean temperatures back to preindustrial levels for only a few billion dollars a year—cheap enough for most countries, or even a single wealthy "greenfinger"—thereby neatly solving the scale problems of other forms of geoengineering. Solar geoengineering also cuts through the intra-capitalist **class** conflict that a green transition presents, buying time for market-driven emissions reductions and carbon capture while protecting the interests of fossil capital.[6] Put bluntly, solar geoengineering smooths relationships among capitalists in service of a planetary class war. Rather than an emergency intervention, solar geoengineering proponents present the technology as a means to temporarily prevent temperature overshoot as emissions are cut and carbon capture deployed. As the framing of solar geoengineering shifts toward promising a prudent, cost-effective policy with global humanitarian benefits (a win-win), it fits neatly within the politics of, for example, the existing U.S. Democratic Party.

Outside the technocratic center, intellectual traditions of both right and left may foster support for solar geoengineering. On the neoliberal right, the Cato, Hoover, and American Enterprise Institutes have worked on geoengineering projects; they deny climate change with one side of their mouth while proclaiming an ultimate, cheap, and technical solution with the other. Cap-and-trade schemes can thus be understood as neoliberal tactics of delay until cheap aerosols become the last climate solution remaining.[7] On the left, a long-standing sense of humanity's capacity to master nature sometimes prevails. Contemporary writers like Holly Jean Buck have made a left case for solar geoengineering; the technology also had Soviet proponents, including a 1974 textbook by Mikhail Budyko that drew on the Soviet Union's long history of weather modification projects.[8]

Solar geoengineering, then, may be inevitable as the final climate fix. Some philosophers muse that the rise of solar geoengineering marks the end of the Scientific Revolution and the beginning of a new technical age of engineering. The Scientific Revolution's habits of incremental experimentation seem quaint compared to the ambition of engineering impossibly complex systems like the atmosphere—an intervention that cannot be simulated beforehand on a computer or in a laboratory.[9] The chaos produced may be unpredictable, but perhaps could be mitigated by another planetary intervention that begets its own symptoms. With many different forces converging on solar geoengineering, we may be witnessing the dawn of a world where **action** precedes—and replaces—knowledge.

See also: **Carbon Management**, **Science**, **Treaty**, **Vegan**

Notes

1 Alan Robock, Kirsten Jerch, and Martin Bunzl, "20 Reasons Why Geoengineering May Be a Bad Idea," *Bulletin of the Atomic Scientists* 64, no. 2 (May 1, 2008): 14–59, https://doi.org/10.1080/00963402.2008.11461140.

2 Frank Keutch, "Stratospheric Aerosol Injection Could Be a Painkiller, but Not a Cure," *C2G*, March 21, 2021, https://www.c2g2.net/stratospheric-aerosol-injection-could-be-a-painkiller-but-not-a-cure-and-more-research-is-needed/; Brad Plumer,

"Should We Use Geoengineering to Cool the Earth? An Interview with David Keith," *Washington Post*, October 30, 2013, https://www.washingtonpost.com/news/wonk/wp/2013/10/30/david-keith-explains-why-geoengineering-isnt-as-crazy-as-it-sounds/.
3 International Energy Agency, "Direct Air Capture" (September 2022), https://www.iea.org/reports/direct-air-capture.
4 Joeri Rogelj et al., "Mitigation Pathways Compatible with 1.5°C," in *Global Warming of 1.5°C: An IPCC Special Report*, ed. V. Masson-Delmotte et al. (Cambridge: Cambridge University Press, 2018), 93–174.
5 Hannah Ritchie and Max Roser, "Land Use" (Our World in Data, September 2019), https://ourworldindata.org/land-use.
6 Kevin Surprise and J. P. Sapinski, "Whose Climate Intervention? Solar Geoengineering, Fractions of Capital, and Hegemonic Strategy," *Capital & Class* 47, no. 4 (2023): 539–64, https://doi.org/10.1177/03098168221114386.
7 Phillip Mirowski, *Never Let a Serious Crisis Go to Waste* (New York: Verso, 2013), 337.
8 Holly Jean Buck, *After Geoengineering* (New York: Verso, 2019); Mikhail Budyko, *Climatic Changes* (Washington, DC: AGU, 1977), 236–47.
9 Simon Factor, "The Experimental Economy of Geoengineering," *Journal of Cultural Economy* 8, no. 3 (May 4, 2015): 309–24, https://doi.org/10.1080/17530350.2015.1039459.

GLOBALIZATION

How does energy matter to globalization?

TANNER MIRRLEES

Here, there, and everywhere, energy is global. It is integral to the worldwide market and the international state system; it is the power behind globalization and the fuel of global warming's threat to all life on Earth.

Energy is part of a worldwide market presided over by a number of border-crossing corporations. Though headquartered within specific countries, firms such as Sinopec, Saudi Aramco, ExxonMobil, Chevron, BP, and Royal Dutch Shell conduct their operations across many countries in pursuit of resources to own, extract, and refine; workers' labor to hire and exploit; and consumers to whom to sell energy commodities. The worldwide energy market encompasses all countries, but some of these countries produce and consume much more energy than others. In 2020, China and the United States were the globe's biggest energy producers; they were also the year's largest energy consumers (145.46 exajoules and 87.79 exajoules, respectively). The total consumption (233.25 exajoules) of these two energy empires exceeded the combined usage of the next eighteen countries (217.52 exajoules). The worldwide energy market is also characterized by asymmetrical trade relations. In 2020, Saudi Arabia, Russia, and the United States exported the highest dollar value of crude oil, and China, Russia, and the United States exported the most nuclear. Most countries are not energy independent but are instead on the receiving end of energy flows imported from elsewhere.

Even though energy corporations permeate borders and energy trade interlinks far-flung countries, no global sovereign exists to enact or enforce a single set of rules for the worldwide energy market. The United Nations, the World Trade Organization, the World Bank, and the International Monetary Fund have a stake in energy governance, but they cannot compel states to pursue national energy interests against their will. Trade agreements (for example, the U.S.-Mexico-Canada free trade agreement) and regional strategies (for example, the European Union's Energy Union Strategy) tie some countries together, but it is states that opt in or opt out of these energy arrangements. A post-Westphalian world is not imminent: energy fuels statism, **nationalism**, and even war.

Around the world, many states own public energy monopolies, while others superintend private energy corporations. In any case, states connect territorial energy ownership and governance to national sovereignty, and they possess the power to make, enforce, or ignore energy laws, policies, and regulations. Some states are intertwined with and influenced by blocs of national energy capital; for example, the American Petroleum Institute is a trade association for nearly six hundred oil and gas firms hoping to shape U.S. federal energy policy. Also, because national energy capital's growth may be a source of tax revenue and exports that increase GDP, states often see this industry's interest as identical to the national interest and subsidize it with bountiful tax breaks and cash transfers. From 1950 to 2016, U.S. federal subsidies to the energy industry were about $145 billion each year, and the G20 has allocated roughly $548 billion per year to the oil, coal, and gas sectors.

The state's geopolitical power has long been integrated with energy capital's conquest of the globe's power sources, so when states internationalize, they do so largely to protect and promote their own national energy interests. When pursuing those interests, states engage in energy diplomacy, broker energy trade deals, and form energy partnerships. Iran and the Congo are separated by over five thousand kilometers of land and sea, and 90 percent of Iranians are Shia Muslims whereas 95 percent of Congolese are Christians; nevertheless, this gulf does not diminish these states' national interest in being members of the Organization of the Petroleum Exporting Countries. Between a quarter and a half of wars since 1973 were related to oil, and the military that waged most of these wars used the most oil to power itself. The U.S. military uses more fuels and emits more climate-changing gases than most midsized countries.

Integral to the worldwide market and interstate system, energy makes the wheels of global capitalist accumulation and nationalist geopolitics turn. Even so, it is oddly a minor topic in the foundational discourse on globalization. The word *energy* is mentioned just ten times across fifty essays authored by globalization's leading social theorists included in David Held and Anthony McGrew's *Globalization Transformations Reader* (2005). It appears twenty times in the seventy-seven canonical works republished by Frank J. Lechner and John Boli's *The Globalization Reader* (2020). Energy is not an entry in Annabelle Mooney and Betsy Evans's *Globalization: The Key Concepts* (2007); Wikipedia's twenty-thousand-word page on "globalization" cites *energy* just four times. There is still no scholarly consensus on what globalization *is*, but never in any of the major works on the topic is energy conceived as the *power* behind globalization. None of the key definitions of globalization address energy or reckon with the fact that energy makes globalization go. Now and for the future, the concept of globalization itself must be energized, or redesigned, to foreground energy's global importance and ecological consequences.

Globalization has mostly been powered by nonrenewable energy. From the neoliberal turn of the 1970s forward, CO_2-emitting nonrenewables have driven globalization and accelerated global climate change. The global **coronavirus** pandemic briefly disrupted globalization, slightly reducing the energy required to make it go, and bringing the world's primary energy consumption of 581.51 exajoules in 2019 down to 556.63 exajoules in 2020. For the first half of that year, global CO_2 emissions fell by 6.4 percent, but when the world **economy** began to recover, emissions once again started to rise. In 2021, emissions were forecast to exhibit the second biggest annual spike in world history.

What's to be done?

From the inaugural Earth Day in 1970 to its fifty-first anniversary in 2021, a global environmental consciousness has existed—for some at least. But a global multitude is very far away from being, understanding, *and* acting to change the reigning cross-border energy relations that exacerbate global climate change.

The climate justice activism of **Greta** Thunberg, the many small groups pushing for a Global Green New Deal, and the work of larger NGOs to release press bites and make green media spectacles are inspirational. But these kinds of interventions have yet to come close to changing everything. From the streets to the state, neoliberal, democratic socialist, and fascist movements battle to create and win consent for newly imagined energy communities and populist resource (inter)nationalisms, while most corporations, states, and citizens continue to use up nonrenewables when doing what they do. The planetary energy interdependence regime keeps heating everything up.

Nearly two hundred states signed on to the 2015 Paris Climate Agreement, but most countries (even France!) remain guilty of failing to honor their commitments. Global investments in, subsidies for, and production and consumption of renewables are starting to increase, but the world still turns by burning fossil fuel. Between 2015 and 2021, banks dumped more than $3.6 trillion into fossil fuels—nearly three times more than investment in renewables. Just one hundred fossil corporations are responsible for 70 percent of global emissions, and during the COVID-19 pandemic, the G20 *increased* fossil subventions, allocating roughly $277 billion in public monies to these unsustainable industries.

For the foreseeable future, though hopefully not forever, the globe's major corporate and state power brokers will power globalization with dirty energy that intensifies the pace of global warming. As a consequence, wildfires, heat waves, droughts, hurricanes, and rising sea levels will threaten the security of all life on Earth, just as they already have begun to do.

See also: **Corporation, Decolonization, Democracy, Law, Populism, Treaty**

GREEN NEW DEAL

Will the Green New Deal save us from extinction?

TODD DUFRESNE

White

I recently overheard a colorful analogy. An older woman—well-kept, white—complained that "we need a Green New Deal like we need a hole in the head." This got me thinking. Because sometimes a hole in the head is exactly what we need.

Red

Capitalism is a cancer on the body politic. But the raging incoherence of a system that inflicts suffering upon working people everywhere, and that destroys the

planet upon which we all depend, isn't only a cancer. It's a brain bleed, a painful, abiding pressure in search of a solution.

One solution is the Green New Deal: a trepanation of capitalism, an attempt to relieve the traumatic pressure it has placed upon people and planet. But it's also an attempt—and here's the tricky part—to save this convulsive and final form of capitalism from itself. An attempt, in short, to treat the brain bleed of rampant inequality, racism, and injustice even as the underlying cancer rages on. What is this cancer? It's economic growth.

Green Green

In the United States, green is the color of both chlorophyll and money. The first makes the natural world go round. The second makes the U.S. dollar-driven socioeconomic world go round. The Green New Deal mashes these worlds together—or tries to—essentially putting the environmental back into the **economy**.

Putting green into green is an attempt to put matter—literally *mater*, the mother in Mother Nature—back into the ever-more specular and fantastical system of capitalism. Greenbacks were once a proxy for gold bullion, which is a product of the earth. But the gold reserve system was already an abstract raising of the stakes on Mother Nature: the purification and refinement, through mining and smelting, of its motley imperfections. The authority of the central bank, and of the government it represents in turn, is just another reflection of this process. So, too, is Protestant religiosity. For the invocation of God himself, stamped upon the complex materiality of cotton paper, has always been the *ne plus ultra* of a system meant to ensure the efficient flow of money into commodities and back into money again.

Capitalism is the way we have turned a green planet into greenbacks, and matter into abstraction. In the West, it's also the way many of us have made ourselves incredibly rich, powerful, and free—free to play, think, and work. Free to reimagine what it means to be human. And free, finally, to disregard altogether the sometimes harsh conditions of the natural world.

Our exit from the Holocene is the call of Nature, a tangible reminder that we're *from it* and *within it*. The birth of the Anthropocene is a call for us to be *for it*, too.

Orange Cyan

So is a Green New Deal a green-washing of capitalism? Despite the different versions proposed around the world, the basic answer is: Of course it is. Does that matter? The answer is: Yes and no.

Yes, a Green New Deal matters to working people. It matters to starving people. It matters to the most oppressed and marginalized people everywhere.

But no, it doesn't matter to the planet. The Earth, even Gaia, doesn't care if we suffer and die. Only human beings care. Only we can ventriloquize for life on the planet. Only we can reject the extraction of oil, gas, and coal. Only we can marvel at surface water that actually burns, as it did for five hours in the Gulf of Mexico on July 3, 2021.

Similarly, only we can stop the root cause of climate catastrophe: economic growth. A 4 percent annual growth rate will double existing production in eighteen years. That extra wealth, funneled to a vanishingly small percentage of people, will further bake the planet—effectively sealing the fate of billions of life forms evolved for very different environmental conditions.

Consequently, unless the proposed Green New Deals include a program of radical **degrowth**—and they don't—they won't save the planet from the damning effects of anthropogenic climate change. Simply put, decarbonization is impossible without cutting growth. The cancer cannot be stopped by trepanation alone.

A New Green Deal might delay the inevitable for a decade. Which isn't nothing. Palliative care isn't nothing. But the best of intentions won't save us or the planet from the end of civilization as we've known and enjoyed it; widespread, serial disasters that undermine our natural and social worlds, from our food supply chains to our essential utilities; and mass dislocation, death, and flirtation with human **extinction**.

Black

It's true that a Green New Deal is a step in the right direction. Unfortunately, we're long past once-radical—but actually perfectly sensible—steps forward. Today we require leaps. We require teleportation and worm holes. We require more from proposed Green New Deals than they can possibly deliver. They may therefore be ethical and necessary and yet, simultaneously, futile and inconsequential.

Faint hope: along with Mastini and colleagues we could push for a "Green New Deal without growth"—and without capitalism.[1] We could rewrite this movement and truly place people ahead of billionaires and corrupt politicians. We could stand with social **justice** movements, from Black Lives Matter to Friends of the Earth. And we could refashion the economy to enable a future, not only of bare existence but of *meaningful existence*. And we could do that now.

And there's no alternative. We need a truly radical New Deal.[2] Anything less than world-historical, revolutionary change to human society will put us all *in the black* for good—and leave the planet to a timelessness beyond our comprehension.

Climate revolution requires sociopolitical revolution. Anything less is no deal at all. It's just the husk of one.

See also: **Black, Class, Development, Gender, Normal, Trans-**

Notes

1 Ricardo Mastini, Giorgos Kallis, and Jason Hickel, "A Green New Deal without Growth?," *Ecological Economics* 179 (2021), https://doi.org/10.1016/j.ecolecon.2020.106832.

2 Kate Aronoff et al., *A Planet to Win: Why We Need a Green New Deal* (London: Verso, 2019).

GRETA

Why is intersectional agility Greta Thunberg's rhetorical superpower?

SHEENA WILSON

Greta Thunberg became a celebrity activist and person of historical importance in 2019 at the age of sixteen, as she led millions in global climate strikes to pressure the powers that be to act on climate change. The year ended with her being declared *Time*'s Person of the Year. Forecasting a dire future unless leaders created radical, rapid change, she accomplished more than moving the environmental agenda to the center of the conversation. She mobilized her identity as a young autistic woman to amplify her call for **justice**. Thunberg explicitly calls for action on climate science and implicitly demands a new politics of energy—one that threatens the dominance of pro-fossil futures. As a public figure, she represents the overlapping desires of various intersecting climate and justice movements as well as their converging challenges to the petrocultural status quo.

Societies are organized around the energy sources that power them.[1] New energy systems usher in social changes and struggles for power. With every energy transition, some demand more equitable and fair societies, while others clamor to accumulate greater wealth. In the late nineteenth century, when the age of oil was being rung in, many saw the promise for greater social justice—for instance, the potential in fossil power to replace slave labor.[2] However, the systems created around fossil fuels also gave rise to deepening imperialism, more deadly warfare, and new forms of capitalism that exacerbated wealth disparity and extended white supremacy, colonialism, and patriarchal power relations. Thunberg represents the varied and overlapping desires of the current moment, for decolonial, anti-ableist, anti-misogynist, antiracist, queer, and multispecies ways of thinking and being that the current transition away from global petroculture makes possible.

Thunberg's call to heed science and respond urgently to the climate crisis presents a multivalent threat to petrocultural power relations. When those profiting the most from the current system attack her, Thunberg responds with intersectional agility, hacking the tools that built the master's house. To be intersectionally agile is to apply intersectional theories to climate responses in ways that nimbly adapt to specific local contexts even while being internationally astute. For example, stepping aside in Canada in October and again at COP25 in Madrid in December 2019, Thunberg ceded the spotlight to young **Indigenous** activists and activists of color.[3]

Greta has refused to stay silent about the fact that protectors of the planet—often BIPOC and mostly women—are targeted by those whose interests are threatened by responsible climate action, which has prompted the ire of male leaders (such as Brazil's president Jair Bolsonaro) known for their misogyny and racism. When Greta tweeted "Indigenous people are literally being murdered for trying to protect the forrest [*sic*] from illegal deforestation. Over and over again. It is shameful that the world remains silent about this,"[4] Bolsonaro called her *pirralha*, translated widely as "brat."[5] But Thunberg refuses to allow attacks on her age, **gender**, or neurodiversity to succeed in deflecting attention away from her calls for climate action informed by science.

Neither diminished by public attacks nor bullied into submission or silence, Greta refuses to soften her climate messages through a smile, a joke, or a giggle—behaviors expected of girls and women. Repeatedly, when Donald Trump, Vladimir Putin, and others use sexist and ageist tropes to belittle her as nonnormative, she responds by incorporating their insults into her Twitter bio tagline. In Canada, leader of the far-right People's Party, Maxime Bernier, invoked Thunberg's Asperger's diagnosis and mental health struggles, proclaiming her "clearly mentally unstable. Not only autistic, but obsessive-compulsive, eating disorder, depression and lethargy, and she lives in a constant state of fear."[6] Amid public outcries in social and mainstream media against these attacks, Thunberg refused to be cast as a victim, either by the powerful men who castigated her identity or by the well-meaning who came to her defense. She tweeted, "When haters go after your looks and differences, it means they have nowhere left to go. And then you know you're winning! I have Asperger's and that means I'm sometimes different from the norm. And—given the right circumstance—being different is a superpower."[7] Likewise, she parries attacks by those who follow the rhetoric of leaders like Bernier and Trump, including the pro–oil and gas convoys of United We Roll truckers and X-Site Energy Services,[8] which produced a sticker featuring an illustration of Thunberg naked and being controlled from behind by someone grabbing her braids.[9] Responding to the sticker, Thunberg tweeted, "They are getting more and more desperate. . . . This shows that we're winning."[10] These attacks are themselves a warning—a wake-up call about a culture of sexual violence fueled by an extractivist, exploitative petrocultural politics that may outlast the era of fossil fuels.

Thunberg represents a threat in part because she "refuses to do what we have come to expect of girl activists who cross borders on our screens—making *us* feel good about a future, and in this case, one that relies on the global use of fossil fuels."[11] She refuses strategies of power that diminish and silence **youth**, women, neurodiverse, BIPOC, LGBTQ2+, and more-than-human relations to achieve capitalist ends. She redirects attention back to climate science and demands that equity and justice guide our climate responses. The climate crisis was caused not by fossil fuels on their own but by the extractivist systems that organize petrocultural life; the exploitations of the current system are laid bare through the varied discourses of oppression invoked to neutralize Thunberg. With her intersectional agility—a powerful tool necessary to trigger system change—Thunberg represents both a threat to the petrocultural status quo and the potential for a new politics of climate and energy.

See also: **Black**, **Civil Disobedience**, **Electricity**, **Normal**, **Trans-**

Notes

1 Sheena Wilson, Imre Szeman, and Adam Carlson, "On Petrocultures: Or, Why We Need to Understand Oil to Understand Everything Else," in *Petrocultures: Oil, Politics, Culture*, ed. Sheena Wilson, Adam Carlson, and Imre Szeman (Montreal: McGill-Queen's University Press, 2017), 3–19.

2 Andrew Nikiforuk, *The Energy of Slaves: Oil and the New Servitude* (Vancouver: Greystone Books, 2012).

3 Associated Press, "Greta Asks Media to Focus on Other Young Climate Activists," *Canadian Manufacturing*, December 9, 2019, https://www.canadianmanufacturing.com/environment-and-safety/greta-asks-media-to-focus-on-other-young-climate-activists-243683/.

4 Greta Thunberg (@GretaThunberg), "Indigenous people are literally being murdered for trying to protect the forrest [*sic*] from illegal deforestation. Over and over again.

It is shameful that the world remains silent about this," Twitter, December 8, 2019, https://twitter.com/GretaThunberg/status/1203732257401380869.

5 Bianca Britton, "Greta Thunberg Labelled a 'Brat' by Brazilian President Jair Bolsonaro," *CNN*, December 11, 2019, https://www.cnn.com/2019/12/11/americas/bolsonaro-thunberg-brat-intl-scli/index.html.

6 Peter Zimonjic, "Bernier Walks Back 'Mentally Unstable' Attack on Greta Thunberg—Then Calls Activist a 'Pawn,'" *CBC News*, September 4, 2019, https://www.cbc.ca/news/politics/bernier-climate-greta-thuberg-1.5270902.

7 Gabrielle Drolet, "Personal Attacks on Greta Thunberg Are Beneath All Politicians," *Toronto Star*, September 3, 2019, https://www.thestar.com/opinion/contributors/2019/09/03/personal-attacks-on-greta-thunberg-is-beneath-all-politicians.html.

8 Angela Wright, "United We Roll Wasn't Just about Oil and Gas," *CBC News*, February 26, 2019, https://www.cbc.ca/news/opinion/united-we-roll-1.5030419.

9 Sarah Smellie, "The Horrible Greta Thunberg Sticker Highlights Alberta's Toxic Oil Culture," *Vice*, March 10, 2020, https://www.vice.com/en/article/wxe7yw/the-horrible-greta-thunberg-sticker-highlights-albertas-toxic-oil-culture.

10 Premila D'Sa, "Greta Thunberg Responds to Graphic Sticker Linked to Alberta Oil Company," *HuffPost*, February 29, 2020, https://www.huffpost.com/archive/ca/entry/greta-thunberg-sticker-response_ca_5e5a9c3ac5b6010221117dae.

11 Jessalyn Keller, "'This Is Oil Country': Mediated Transnational Girlhood, Greta Thunberg, and Patriarchal Petrocultures," *Feminist Media Studies* 21, no. 4 (2021): 685.

HABIT

What kinds of friction can disrupt the status quo?

RHYS WILLIAMS

In debates over how to transition to a sustainable energy future, the energy-hungry habits of individuals in the Global North have emerged as a potential site of intervention. Can (and will) citizens in the Global North adopt less carbon-intensive patterns of **behavior**? The changes envisioned often include things like taking public transport or **cycling** rather than driving; eating locally grown food (or growing your own) to reduce food miles and dependence on global industrial agriculture; using **air conditioning** and heating and hot water more sparingly; flying less or not at all; and recycling, reusing, and repairing rather than buying new. Two assumptions underwrite these goals: first, that individual choice is a site where conscious change can occur relatively easily; and second, that consumer-led demand will trigger wider changes in the systems that still support energy-intensive habits. The end goal is often a green capitalism, where planet and profit will both be served by the same mechanisms.

Such hopes for habit-led change, however, rest on an outmoded understanding of human beings as rational decision-making animals. While the facts of climate change are now widely accepted, high-carbon habits have shifted only a little, if at all. An international survey conducted in 2021 noted this massive gap between values and actions, noting that respondents tended to overestimate the importance and efficacy of climate actions that they were already taking, or that "*required less individual effort, or for which they bore little direct responsibility.*"[1] In other words, respondents pinned their hopes on relatively minor things they were already doing or on things they could do nothing about. But given the scale and speed of the change necessary to meet the challenges we face, to rely solely on these individual efforts is clearly inadequate.

Habit-centered approaches sometimes also aim to influence the practices of corporations and industries. Even a quick glance at what the latter are up to is enough to bury the idea that such efforts will be adequate. Even companies whose agendas have a green gloss have no desire to reduce consumer demand: growth and profit remain at the core of corporate rationale. Renewable energy companies strive to grow the bottom line by replacing fossil-fuel energy production, car companies by making the switch to mass production of electric cars, and food technology companies by growing lab meat to replace animal-derived burgers and steaks. For these companies and their financial backers, there is no future reduction of demand; their efforts aim to maintain or expand current levels and kinds of consumption.

Perhaps a better approach would begin by recognizing that individuals and their habits are intimately entangled with the infrastructures that surround and facilitate them. It's hard to change from driving a car to using public transport when the former is so much more convenient; our cities, towns, suburbs, and countrysides have been designed around cars since the mid-twentieth century. It's hard to eat locally or grow your own food when you don't have the space, money, or time to do so, and especially when your eating habits have been formed since birth by a global food system that proffers ready meals, fast food, and a diversity of foodstuffs that cannot be matched by your local farmer. It's hard to change your reliance on domestic temperature control when your building lacks that capacity and has no shade in the summer or heat retention in the winter. Infrastructures and the built environment can obstruct the hard work of changing one's habits.

The word *habit* calls to mind an individual scale of action, but we are all habituated within a sociotechnical system whose infrastructures encourage some behaviors and obstruct others; they facilitate certain ways of life at the expense of others. This infrastructural embeddedness is why we continue acting habitually despite wanting to act otherwise: regardless of our consciously held beliefs or values, we are tethered to patterns of everyday behavior by sedimented path dependencies embedded into our built environment, technical systems, and expectations. To act against these patterns is to replace habit's quality of smoothness with the friction of acting against the grain of our sociotechnical system. This friction means living with the expensive, slow, multileg journeys to work or supermarket by bus and train; the expense, time, and limited choice of shopping local and organic; and the sweltering summer heat and shivering winters that are the outcome of reducing reliance on indoor climate control. In this way, we are all caught in the "chain of ease" and discouraged from acting otherwise;[2] this friction accounts for the stickiness of bad habits despite our attempts to act on our best eco-behavior.

To really change habit, one needs to change the world around habit. Superficial demands for behavioral change must give way to a reckoning with the ways that habits form within infrastructures and a recognition that our infrastructures have politics. In other words, infrastructures (and the habits they foster) are bound up with histories of arguments advanced, decisions made, and actions taken from within contested and unequal relations of power. For example, habits of sustainable mobility can emerge only from a commitment to extended, reliable, and affordable public transport, city planning that doesn't privilege the suburbs, new models of working remotely, and the provision of bicycle lanes and bikes to those who want them. To demand sustainable agriculture is to demand new models of landownership, a four-day week to give people time to care for their own plots, education programs, support for local businesses, and (potentially) the distribution of meat-producing technologies across local areas. To demand sustainable levels of domestic climate control is to demand the mass provision of retrofitted insulation, solar heaters, and cognate technologies and to demand new building codes that lessen the need in future. To change habits requires changing the dynamics and trajectories of friction-free smoothness, by creating the infrastructural, material, and political conditions in which sustainable actions can assume the unthought status of habit because they just *feel* easy and right, and previous energy-intensive habits are interrupted systemically, in ways that demand extra time, money, or effort to continue. To do that, however, requires producing friction: not in the individual sphere, but in the public sphere.

See also: **Degrowth, Design, Lifestyle, Transitions, Vegan**

Notes

1 Henley, Jon, "Few Willing to Change Lifestyle to Save the Planet, Climate Survey Finds," *Guardian*, November 7, 2021, https://www.theguardian.com/environment/2021/nov/07/few-willing-to-change-lifestyle-climate-survey, emphasis added.
2 Jennifer Wenzel, *The Disposition of Nature: Environmental Crisis and World Literature* (New York: Fordham University Press, 2019), 31.

INDIA

Can there be a just transition in India?

SWARALIPI NANDI

Where is India, with its massive population of 1.3 billion people, headed in its energy future? Discussions of energy in India are often driven by the rhetoric of growth. Fossil fuels are still among the cheapest ways to power economic growth, making them hard for developing countries to ignore. When it comes to the Global South, "Global warming typically takes a back seat to feeding, housing, and employing these countries' citizens."[1] The environmental impact of fossil fuels in India has been downplayed in light of aspirations for unhindered energy supply, especially in relation to the current government's aggressive growth policies, which are accompanied by technocratic agendas like universal **electricity** access for all citizens. Such grand narratives of growth are hinged on a nationalistic rhetoric—as proclaimed in Modi's famous slogans of *atmanirbhar Bharat* (self-reliant India) or "Make in India"—which prioritizes **local** interests over global environmental needs.

It's for this reason that **Greta** Thunberg's and Rihanna's tweets about farmers' unrest in India were vehemently opposed by government as "international interference" in India's internal matters and that India didn't fall in line behind calls to isolate Russia in the wake of its invasion of Ukraine. Amid a global effort to isolate Russia through economic sanctions, India continued to import cheap oil from Russia, its largest supplier in 2022. India also refused to support the price cap mechanism implemented by the other G7 countries, intended to halt the flow of revenue helping to sustain Russia's ongoing aggression. Defending its contentious stand, India insisted that it will continue to buy oil from wherever it gets a good deal to meet domestic demand. The government has acted out of "the interest of the Indian people," asserted external affairs minister S. Jaishankar, referring to the country's continued need for carbon-based fuel.

This putatively anticolonial gesture of prioritizing local growth over planetary concerns about carbon footprints (sometimes dismissed as "Western") is dubious, to say the least. India's high consumption of fossil fuels—the third highest in the world—involves a domestic energy gap, which is particularly acute for rural electrification. Moreover, with the state serving as the key facilitator of energy distribution in India, current exclusionary politics outlined in proposed amendments to the National Register of Citizens and the Citizenship Amendment Act raise serious questions about energy democratization.[2] It is also important to note the private players in the field of energy supply and distribution in India. Contrary to the government narrative of Russian oil being purchased in the interest of the people, nearly three-quarters of the imported oil went to private refineries such as those operated by Reliance Industries and Nayara Energy, while the public-sector refiners that supply more than 90 percent of the average Indian's fuel needs received only a small share.[3]

Given the politics of ethnic nationalism and rampant crony capitalism at work in energy decision making, how are we to read the government's public embrace of renewables? At the G20 summit in 2022, Prime Minister Narendra Modi declared that India will generate 50 percent of its electricity from **renewable** sources by 2030. And the Indian parliament passed the Energy Conservation Bill of 2022, which seeks to mandate the use of non-fossil-fuel sources not only in sectors like steel, refineries, fertilizer, and cement but also in large residential complexes. The seeming paradox of India's continued use of fossil fuels and expressed desire for renewables converges in the figure of Gautam Adani. The rise of this self-made billionaire was closely linked to the political rise of Narendra Modi. While Adani's coal conglomerate has reaped gigantic profits from imports to meet domestic demand, he has also aggressively ventured into the green energy sector and holds majority shares in the renewable sector, closely followed by another tycoon, Mukesh Ambani. With private players dominating the renewables sector, the government's public commitment to transition raises more questions than answers.

Trepidation about the direction of a green transition in India is also informed by the history of conflicts over the country's renewable energyscapes, which have often turned into neoextractivist projects resulting in displacement and oppression. Hydropower projects have long faced grassroots resistance, as exemplified by the Narmada Bachao Andolan (Narmada Protection Movement), which has demanded compensation and rehabilitation for communities impacted by dam construction. Similarly, new nuclear power plants at Jaitpur, Kovvada, Mithi Virdi, and Haripur, which require uranium mining and heavy water use and result in hazardous radioactive waste, have been met by local protests. India's status as the world's third largest producer of solar energy comes at the cost of encroachment upon the land of marginal communities. In Rajasthan alone, communities have filed claims in at least fifteen cases since 2011 against solar projects, while the Charanka solar plant in Gujarat has disrupted the traditional grazing routes of agro-pastoralist shepherd communities. In spite of their massive environmental impact on ground water, land, and local communities, solar power ventures are exempted from environmental impact assessments, thus facilitating unregulated operations. Like the grand narrative of progress, the grand narrative of green transition tends to mandate grand action without taking into account the small players—local communities, tribes, and grassroots voices.

The alternative to grand action is, however, not individual action. While inspiring, the efforts of small farmers like Kuldeep Singh Cheema, who have received the country's highest carbon credits, or the energy-efficient houses developed by individuals like Solar Suresh can make only minimal impacts on energy policy and practice in India. A just transition needs to include comprehensive, **community**-based policies to facilitate a true paradigm shift from fossil fuel dependence. The task is daunting and requires a collective effort supported by a robust democratic culture of bottom-up governance. But it needs to be done: the government's official renewable policy promises an energy future not much different from India's energy past or present.

See also: **Air Conditioning, Cuba, Development, Neoextractivism, Solar Farm**

Notes

1 Tucker Davey, "Developing Countries Can't Afford Climate Change," *Future of Life*, August 5, 2016, https://futureoflife.org/2016/08/05/developing-countries-cant-afford-climate-change/.

2 The Citizenship Amendment and National Register of Citizens acts were bills that proposed to restrict Indian citizenship to non-Muslim illegal immigrants while seeking citizenship documents from Muslim residents. The act was called out for its utterly discriminatory politics against the Muslims, a gesture that aligns with the Hindu right ideologies of the party in power.

3 S. Dinakar, "RIL, Nayara Account for a Combined 69% of Russian Oil Shipments to India," *Business Standard*, June 22, 2022, https://www.business-standard.com/article/companies/ril-nayara-account-for-a-combined-69-of-russian-oil-shipments-to-india-122062101322_1.html.

INDIGENOUS

How are Indigenous and extraction related?

DEENA RYMHS

The politics of energy in the present moment demands rethinking some key concepts and attending to modes of seeing and resisting modeled by communities fighting extractive projects around the world. Ecological crises can reveal the fault lines of conventional mappings of territory, exposing unsustainable fictions of borders and boundaries and challenging common understandings of global and local, near and far. The radical potential immanent in this moment of ecological, cultural, and political upheaval is a recognition of the ethical connections not only between human communities but also between human and other-than-human worlds. Indigenous political struggles have fostered such reimaginings in their decolonizing efforts to find sources of knowledge, meaning, and livability in the grip of extractive capitalism.

Indigenous people are organizing against resource extraction and related industries around the world. Standing Rock, Unist'ot'en, Elsipogtog, Aamjiwnaang, Sarayaku, San Miguel Ixtahuacán, Rio Tigre, Ogoniland, Baturraden, Kucheipadar: in these and many other places, Indigenous-led resistance to pipelines, fracking, chemical refineries, oil drilling, gas exploration, gold extraction, geothermal plants, and bauxite mining has challenged the authority of governments and companies to make unlivable the environments these communities call home. The persistent defiance of Indigenous people fighting the transformation of their homelands into sacrifice zones in the United States, Canada, Ecuador, Guatemala, Peru, Nigeria, Indonesia, India, Australia, and elsewhere has transformed environmental politics, even as such defiance is often met with state-sanctioned intimidation and violence. Their commitment to protecting Earth, along with an abiding conviction in the right to self-determination, challenges the authority of the nation-states, petrostates, and client states in which these communities are located. In Canada, for instance, Wet'suwet'en First Nation continues to challenge the legal reach of federal and provincial powers after *Delgamuukw v. British Columbia* (1997), a Supreme Court decision that established Indigenous title as inalienable and sui generis. Rejecting the doctrine of terra nullius, *Delgamuukw* was a landmark case for the Wet'suwet'en people and other Indigenous nations in Canada who were recognized as never having ceded title to their lands. The spur for the Wet'suwet'en people's legal actions was the flooding and deforestation of their traditional lands by a hydroelectric megaproject. Environmental battles—even those involving renewable energy projects—can catalyze shifts in power relations and renewed assertions of political sovereignty.

Such political victories are, however, relatively few in comparison to the countless struggles for bare survival undertaken and endured by Indigenous communities who disproportionately bear the toxic burden of extractive industries with their bodies, lives, and diminished prospects for future cultural survival. For many Indigenous communities living downstream from large-scale extraction projects—communities like Fort McKay in **Alberta**, located next to the third biggest crude oil reserve in the world and Canada's largest tar sands mining operations—the cost has been greater than the benefit as residents experience increasing rates of cancer within their community. Human and ecological sickness blur together as residents can no longer engage in traditional activities like hunting and berry gathering on land now poisoned by sulfur and nitrogen dioxide. In the Aamjiwnaang First Nation in Canada, recorded levels of sulfur dioxide in the air are ten times higher than in the neighboring cities of Ottawa and Toronto, and the amount of benzene (a carcinogen linked to leukemia and other cancers) is thirty times higher.[1] Surrounded by over fifty refineries and chemical plants, Aamjiwnaang residents live in one of the most polluted places in Canada. Nearly twelve thousand kilometers from Aamjiwnaang, Adivasi tribal people in Kucheipadar, **India**, contend with social and environmental devastation caused by bauxite (aluminum ore) mining. Deforestation, soil erosion, and biodiversity loss are the ecological legacies of opencast bauxite mining in the Kashipur region. Community members suffer health issues like brittle bones, tooth and gum infections, skin irregularities, and respiratory problems, along with noise pollution, increased violence, and goon terror.[2]

The remarkable frequency with which resource industries inflict dispossession and toxic burdens for tribal people may suggest that *Indigenous* and *extraction* are irreconcilable opposites. Can this assumption be troubled, however? One problem with this opposition is that it imagines Indigenous people ahistorically, standing outside of industrial capitalism and its revenue-generating economies. The frequency with which the term *resource curse* is heard in relation to Indigenous communities can be troubling for its ring of paternalism. Contrary to these logics, numerous Indigenous nations exercise an active role in the energy industry beyond collecting royalties from oil and gas output: the Osage of Oklahoma, Native corporations in Alaska, and the Mandan, Hidatsa, and Arikara Nation of **North Dakota** are but a few tribal groups in the United States who have leveraged ownership stakes in oil and gas projects for reinvestment in community infrastructure, health care, education, and trust funds.

Another reason why *Indigenous* and *extraction* might seem to be antithetical is Indigenous people's grounding in an ontological order that understands the environment quite differently from industry, government, and activist rhetorics. The #NoDAPL protests at Standing Rock witnessed this onto-epistemological shift as the Lakota and Dakota people affirmed their commitment to protecting their relative, *Mni Sose* (the Missouri River), in adherence to the principle of *Wotakuye* (kinship). The conception of *Mni Sose* as relation entails a profoundly different imaginary than government and industry discourse on natural resources. Affirming a kinship relation with *Mni Sose* also departs from mainstream environmentalism, which, in positing the environment as essential to human futurity, tends to instrumentalize it or advocate for democratic ownership of it ("the water belongs to everyone"). These assumptions differ from what *Mni Sose*'s relatives say about her: "She is alive. Nothing owns her."[3] Even if economically and politically marginal in the nation-states where they are situated, Indigenous groups are increasingly central to the politics of energy and environment. There are many obstacles to be overcome, but perhaps the greatest will be the capacity to imagine.

See also: **Blockade, Family, Neoextractivism, Organize, Settler Colonialism**

Notes

1. Kelly Anne Smith, "Regulatory Gaps Toxic to Aaamjiwnaang First Nation," *Anishinabek News*, March 10, 2021, https://anishinabeknews.ca/2021/03/10/regulatory-gaps-toxic-to-aamjiwnaang-first-nation/.
2. Abhijit Mohanty, "Photo Essay: Bauxite Mining a Curse for Adivasis in Odisha's Baphlimali," *The Wire*, August 23, 2017, https://thewire.in/rights/photo-essay-bauxite-mining-curse-adivasis-odishas-baphlimali.
3. Jaskiran Dhillon and Nick Estes, "Introduction: Standing Rock, #NoDAPL, and Mni Wiconi," Hot Spots, *Fieldsights*, December 22, 2016, https://culanth.org/fieldsights/introduction-standing-rock-no-dapl-and-mni-wiconi.

INDIGENOUS ACTIVISM

What is the power of radical care?

AMBER HICKEY

From Bougainville to Aotearoa to Turtle Island, **Indigenous** communities have long engaged in radical care for land, people, and waters. Hi'ilei Hobart and Tamara Kneese describe radical care as "a set of vital but underappreciated strategies for enduring precarious worlds."[1] Radical care mobilizes and sustains more recognizable forms of activism globally, including the spectacular convergence in public space of communities advocating for change, which has received the most media attention. It is nothing less than a form of energy produced to fuel the fight against what Justin McBrien calls the "Necrocene": the extractive worlds animated by ideologies of white supremacism, capitalism, and heteropatriarchy.[2]

The movement against the Dakota Access Pipeline (#NoDAPL) offers one example of the importance of radical care in Indigenous activism, particularly Indigenous activist media. Many who were not at Standing Rock in person followed the events by watching drone footage gathered by water protectors such as Shiyé Bidzííl and Myron Dewey. What viewers might not immediately notice is the care with which these water protectors collected their footage and engaged with their camera-equipped drones. Both Shiyé and Myron blessed their drones before they took flight, acknowledging the importance of these tools in the struggle against the Dakota Access Pipeline, as well as their need for protection from security officials.[3] Myron extended his care for the practice of drone piloting to the sphere of education. He practiced care for **community** members by passing on his knowledge of drone piloting, navigating the settler legal system, and resisting surveillance at the camps to support countless other water protectors.

Since his passing in September 2021, that care continues to manifest in the work of those he mentored. Shiyé approached his drone as if it were a bird and related to it as kin.[4] This extension of kinship networks to the technological nonhuman echoes work by artists and scholars such as Jason Edward Lewis,

Noelani Arista, Archer Pechawis, and Suzanne Kite, who ask what it might mean to "make kin" with machines, such as artificial intelligence technologies.[5] Shiyé began to imagine what this kind of engagement might look like with camera-equipped drones. Drone footage gathered during the movement against the Dakota Access Pipeline does not merely document the violence of TigerSwan, Morton County, and Energy Transfer Partners—all of which were central forces in the oppression of water protectors—but also instantiates these decolonial practices of radical care.

As with Myron's commitment to transmitting knowledge of drone piloting to his community, education and equipment were key to the work of Outta Your Backpack Media (OYBM), a youth media justice project launched in Occupied Kinłani (so-called Flagstaff, AZ) in 2004 and housed in the Taala Hooghan Infoshop. The initiative coalesced in response to demand from the Indigenous youth community in the area. OYBM's work was grounded in ceremony, collective agreements, and trust, all of which contributed to its longevity. The work of OYBM entailed much more than creating films. They also had a community garden that informed their process and the temporality of their work. **Youth** involved in the initiative mobilized the process of filmmaking to rise up against harmful stereotyping and resource colonialism.[6] According to organizer and mentor Klee Benally, the community garden "informed our approaches to education through understanding that **storytelling** should not just be mediated through colonial technology."[7] Participants resisted resource colonialism by using secondhand equipment and by fitting backpacks with solar panels to power their technology. Stereotypes were challenged by mentors like Klee at Taala Hooghan Infoshop, who spent countless hours devising intentionally "non-manipulative pedagogies."[8] These approaches to mentorship demonstrated trust in the youth involved. In this sense, trust, too, is a form of care. OYBM rejected notions of Indigenous youth as untrustworthy and instead saw them as fundamentally worthy of trust.

Together, OYBM's gardening practices, media choices, and mentorship processes contributed to the assertion of a temporality of radical care that defied the accelerated pace of corporate media and the temporality defining extraction, capitalism, and colonialism. Participants and mentors called this alternative temporality "slow media." For OYBM, slow media emerged in response to the constantly accelerating rate of consumption and rapid rise of social media platforms in the early 2000s.[9] Slow media is a key part of OYBM's understanding of media justice more broadly. This anticapitalist pacing allowed them to emphasize the importance of process. During workshops with elders, slow media functioned as a way to create intergenerational bridges between youth and elders.[10] OYBM's work is a model of how a temporality of radical care might be manifested more broadly.

These examples of Indigenous activism demonstrate how practices of radical care, with their emphasis on relationality, kinship, and transmission into a living future, can disrupt the nonfutures of the death-driven Necrocene. Radical care deserves to be recognized as more than something that happens in the background of organizing work. Rather, its sustaining energy catalyzes decolonial activist media that can counter extractivist economies.

See also: **Animals, Art, Blockade, Documentary, North Dakota, Organize**

Notes

I am grateful to Klee Benally, Shiyé Bidzííl, and Myron Dewey for speaking with me, reviewing drafts, and trusting me with this work. Thank you to Jordan Reznick, Chris Walker, Danila Cannamela, and members of Colby College's American Studies writing group (Laura Saltz, Laura Fugikawa, Ben Lisle, and Lisa Arellano) for invaluable comments.

1 Hi'ilei Hobart and Tamara Kneese, "Radical Care: Survival Strategies for Uncertain Times," *Social Text* 38, no. 1 (March 2020): 2; Melanie K. Yazzie and Nick Estes, "From the Red Nation to the Red Deal: A Conversation with Melanie K. Yazzie and Nick Estes," in *The Routledge Companion to Contemporary Art, Visual Culture, and Climate Change*, ed. T. J. Demos, Emily Eliza Scott, and Subhankar Banerjee (London: Routledge, 2021), 437–47.
2 Justin McBrien, "Accumulating Extinction: Planetary Catastrophism in the Necrocene," in *Anthropocene or Capitalocene? Nature, History, and the Crisis of Capitalism*, ed. Jason W. Moore (Oakland, CA: PM Press, 2016), 116.
3 Myron Dewey, interview with the author, June 13, 2017.
4 Shiyé Bidzííl, interview with the author, June 20, 2017.
5 Jason Edward Lewis, Noelani Arista, Archer Pechawis, and Suzanne Kite, "Making Kin with the Machines," *Journal of Design and Science* (2018), https://doi.org/10.21428/bfafd97b.
6 Resource colonialism occurs when it's not solely the land, but the extractable resources that the colonizer desires. Al Gedicks, "Resource Colonialism and International Native Resistance," in *The New Resource Wars: Native and Environmental Struggle Against Multinational Corporations* (Boston: South End, 1993), 13–38; Sean Parson and Emily Ray, "Sustainable Colonization: Tar Sands as Resource Colonialism," *Capitalism Nature Socialism* 29, no. 3 (2018), 68–86, https://doi.org/10.1080/10455752.2016.1268187.
7 Klee Benally, interview with the author, August 15, 2022.
8 Klee Benally, interview with the author, April 26, 2022.
9 Benedikt Köhler, Sabria David, and Jörg Blumtritt, "The Slow Media Manifesto," *Slow Media*, January 2, 2010, https://en.slow-media.net/manifesto.
10 Benally, interview with the author, August 15, 2022.

INDUSTRIAL REVOLUTION

What does a focus on jobs mean for energy transition?

CARA DAGGETT

While U.S. Republicans and Democrats generally disagree about energy policy, they do agree about one thing: more jobs are needed. Democrats and many progressives tend to accept Republicans' framing of energy issues in terms of potential job losses (their mainline defense of fossil fuels) by promising green jobs in their own climate and energy proposals. This bipartisan emphasis on jobs reflects a shared belief that work is a key public concern that must be considered in any energy transition. More fundamentally, these shared assumptions about the centrality of work reveal its status as a key moral tenet of Western culture: one must work to deserve the trappings of citizenship and life itself, and people and **planet** should both be put to work.[1]

In Western culture, the conjoined value of work and energy has become common sense, and more of both is understood to be a good thing. Energy—a foundational unity in physics, the hardest of hard sciences—infuses the long-standing Protestant work ethic with cosmic truth. However, the figure of energy is neither timeless nor universal. The energy-work nexus, as moderns know it, took shape relatively recently, in the middle of the nineteenth century, and lies at the heart of the Industrial Revolution. That is the name that Westerners give to an era spanning the late eighteenth century to the nineteenth, when mechanized labor transformed European imperial economies. Machines, coal and oil power, European empires, the work ethic, slavery, **mining**, capitalism, and private property laws—all these phenomena that arose before the Industrial Era were knit together into something new, with fossil fuels as a novel motive force.

Energy was one of those new things.[2] Prior to industrialization, energy had been mostly a poetic, philosophical word for vitality and change, stemming from Aristotle's *energeia*, a key term in his philosophy that classicist Joe Sachs translates as "being-at-work."[3] Aristotle coined *energeia* to talk about goodness as an ongoing, dynamic project, rather than a static achievement, a meaning far removed from modern work cultures. In the mid-nineteenth century, a relatively small group of white men enlisted *energy* to name the figure at the heart of the new science of thermodynamics, bringing together sometimes centuries-old experiments on heat, work, motion, and force. The ancient, metaphysical connection between dynamism and virtue, enshrined in the word *energy*, did not disappear when energy was translated into thermodynamic math.

In the laws of thermodynamics, energy names the unit that is conserved through all manner of transformations. That is the first law of thermodynamics, that energy cannot be created or destroyed. But the tidy math of energy conservation is complicated by the second law of thermodynamics, which states that although energy is conserved, it tends to become more disordered, or diffuse over time (i.e., the entropy of the system increases). Entropy's tendency to increase explains why the most interesting transformations, as in the work of steam engines as well as the activities of living bodies, are irreversible: energy tends to "run down," becoming less able to do work, as when it is "wasted" as friction.

Careful readers will notice that entropy creates an odd paradox, given that energy is still commonly defined as the capacity to do work. The second law reveals that the amount of energy is unrelated to its capacity to do work. But the descriptors of this energy that doesn't do work—*wasted*, *run down*, *degraded*, *lost*—are morally freighted. Physicist and Nobel laureate Percy Bridgman observed in the 1940s that "the laws of thermodynamics have a different feel from most of the other laws of physics. There is something more palpably verbal about them—they smell more of their human origin. . . . Why should we expect nature to be interested either positively or negatively in the purposes of human beings, particularly purposes of such unblushingly economic tinge?"[4] Leaving aside Bridgman's assumption that other laws of physics are somehow less attached to humans, with their slippery words and economic interests, the human touch—more accurately, the industrial man's touch—seems especially transparent in thermodynamics. This is no surprise, given that many of the self-proclaimed discoverers of energy were chiefly concerned with making steam engines work better and profiting from this knowledge. The economic tinge of thermodynamics is also evident when energy becomes an object of politics, as a shorthand for fuel, where energy means only that narrower category of things that are useful to human work.

That is not to say that modern energy and the thermodynamic laws that inaugurated it are false, but rather that they are partial, reflecting an engineer's understanding of fuel. In particular, they often still reflect Anglo-Protestant preoccupations with thrift and efficiency, set against a world of quarrelsome workers and wayward energies that seem to continually resist being set to work under white men's management. Engineering knowledge is useful to solve certain problems, as in the desire to increase a system's efficiency. But modern cultures too often adopt an industrial engineer's understanding of energy as a universal ethos by which to value change, such that to be in motion, to be doing things efficiently, to be active in a way that is useful (to capital) is imbued with the stamp of goodness and cosmic truth. Instead, political questions about whose efforts are valued, and for what purposes, should be at the heart of debates about energy-as-fuel systems.

That is why the call for more green jobs on the left reflects a limited imagination, plugging the horizon of public desire into the existing machine of exploitative wage labor, unpaid and underpaid reproductive labor, and cheap commodity rewards. Not to mention that the sunny goal of adding things—*more renewable energy! more jobs and unions!*—distracts from the tough political struggles that are needed to *remove* things: fossil fuels and the power of the fossil fuel industry, poorly paid work all along global supply chains, and the ability of corporations like Amazon to bust unions and take advantage of underpaid workers and to steal and misuse Indigenous and Global South land for cheap extraction and no-liability dumping.

Energy's multifaceted meaning, as a metaphor for change in the cosmos, lends the modern project of work a sense of universal rightness, while also obscuring the many ways that humans understand and value purposeful activity. If energy is parochial, if white man's energy was born out of the coal fires of the Industrial Revolution, it can be jettisoned in the next transition.[5]

See also: **Degrowth, Green New Deal, Justice, Neoextractivism, Sabotage, Transitions**

Notes

1 Kathi Weeks, *The Problem with Work: Feminism, Marxism, Antiwork Politics, and Postwork Imaginaries* (Durham, NC: Duke University Press, 2011).

2 Cara Daggett, *The Birth of Energy: Fossil Fuels, Thermodynamics and the Politics of Work* (Durham, NC: Duke University Press, 2019).

3 Aristotle, *Nicomachean Ethics*, trans. Joe Sachs (Newbury, MA: Focus, 2002), viii.

4 P. W. Bridgman, *The Nature of Thermodynamics* (Cambridge, MA: Harvard University Press, 1941), 3, http://archive.org/details/natureofthermody031258mbp.

5 Larry Lohmann, "White Climate, White Energy: A Time for Movement Reflection?," *Social Anthropology* 29, no. 1 (February 2021): 225–28.

IPCC

What can the IPCC tell us about the relationship between knowledge and action?

MARCELA DA SILVEIRA FEITAL AND JESSICA O'REILLY

The Intergovernmental Panel on Climate Change (IPCC) is the principal intergovernmental organization that assesses and reports upon the current state of knowledge about climate change. Established by the World Meteorological Organization and United Nations Environment Program, the volunteer IPCC authors provide policy-relevant **science** to aid in decision making. However, as social scientists studying the IPCC process, we, along with our interlocutors in the IPCC, know that the traditional scientific dispositions of neutrality and objectivity are increasingly untenable considering the climate crises that we face. Scientific expertise and climate science knowledge help us understand what climate actions we need to take. Translating climate science into climate **action**, however, is a contested practice that requires not only scientific acumen but also political, diplomatic, and **communication** skills.

The IPCC assesses already-published, peer-reviewed climate research. Its three Working Groups bring together scientists from multiple disciplines to broaden the potential acceptability of scientific knowledge in the political realm. From natural and physical science experts to anthropologists, psychologists, and economists, the IPCC has become increasingly diverse and interdisciplinary since it began in 1990. In its assessments, the IPCC defines climate problems and assesses a variety of possible solutions. However, it is political leaders who decide which solutions to implement.

Nowhere is this tension between science and policy more apparent than the IPCC's 2018 Special Report on Global Warming of 1.5°C (SR1.5).[1] In previous reports, the IPCC had held to the tacit global goal of limiting anthropogenic global warming to 2°C. Some policy actors, though, had been advocating for limiting warming to 1.5°C. This aspiration is reflected in the **Paris Agreement**. Policymakers realized that they did not have the scientific knowledge with which to reach this target, and so, for the first time, the United Nations

Framework Convention on Climate Change requested that the IPCC produce a focused report on the impacts of and options for limiting warming to this lower threshold.

This request immediately reorganized the research agendas of many climate scientists globally, giving them an opening to publish findings relevant for the assessment.[2] SR1.5, with its explicit alignment with the United Nations' Sustainable Development Goals, marked a transition in creating climate information that was politically useful and relevant to policy. SR1.5 brought together scientists with a specific research goal, promoting a special research agenda engaged with a directly policy relevant issue. The messages of SR1.5 were broadly circulated and interpreted by government actors as well as climate activists.[3] For example, the model projections assessed in SR1.5—showing a temperature threshold crossed in 2030, a mere twelve years into the future—were misconstrued by activists as the world having "twelve years left" to limit warming to 1.5°C. Not only is that deadline misleading, but it also obscures the fact that the report calls for action now, not in twelve years. Scientists, though, are not decision makers—nor are most of them **expert** public communicators. While this is not a problem per se, because scientists are trained as scientists and not as politicians, activists, or journalists, it can and does generate problems that must be addressed.

The coronavirus pandemic proved that some challenges cannot be solved through individual choices; we must act collectively to solve societal-scale problems. We may have the best available science, but if we do not have collective action, the problems remain unsolved or can worsen. Likewise, climate change is a wicked challenge that demands collective, integrative, multilevel, and multiactor efforts. The most recent assessment report from the IPCC has proven that the changes necessary to limit global warming to 1.5°C do not stem from a lack of climate knowledge.[4] Humankind already has abundant knowledge to undertake the transformation of energy infrastructure; what we lack is the political will or capacity to implement these changes. Humankind has the ability to add time to the climate clock, but only if we work collectively and measure our progress against defined targets.[5]

To produce policy-relevant knowledge capable of being translated into action, scientists need to reconsider their performance of science as objective and politically neutral; indeed, many already have. As SR1.5 shows, the bridges between science and policy are a matter of planetary survival. Even though the IPCC considers the separation of science from policy as an essential element to its work, the knowledge underpinning the IPCC assessments helps us understand the necessary paradigm shift we need to turn information into action: connecting research agendas to policy needs, developing **local** solutions to global responses, and linking climate issues to other ongoing social and political dynamics.

See also: **2040, Evidence, Nonlinear**

Notes

1 Intergovernmental Panel on Climate Change, "Summary for Policymakers," in "Global Warming of 1.5°C" (2018), https://www.ipcc.ch/sr15/chapter/spm/.
2 Jasmine E. Livingston and Markku Rummukainen, "Taking Science by Surprise: The Knowledge Politics of the IPCC Special Report on 1.5 Degrees," *Environmental Science and Policy* 112 (June 2020): 10–16.
3 Livingston and Rummukainen.
4 Intergovernmental Panel on Climate Change, "Climate Change 2021: The Physical Science Basis" (2021), https://www.ipcc.ch/report/ar6/wg1/.
5 Concordia University, "Climate Clock: Adding the Metric of Time to the Global Warming Conversation" (last modified July 25, 2021), https://www.concordia.ca/news/climateclock.html.

JAPAN

Why do fears of energy scarcity obstruct energy transition?

HIROKI SHIN

With its bullet trains, consumer electronics, hybrid cars, and robotics, Japan has long enjoyed a reputation as a high-tech nation. Japan's energy-efficient economy and its innovation of advanced energy conservation technology have been widely admired by both experts and consumers since the 1970s. Quite perplexing, then, is Japan's poor performance in climate change mitigation in the twenty-first century. Despite intensifying international criticism, Japan has maintained its dependence on fossil fuels while making a limited commitment to **renewable** energy. The 2011 Fukushima nuclear meltdown dealt a further blow to Japan's green credentials, as consumption of coal, oil, and natural gas increased in order to compensate for the loss of nuclear power.

Why has Japan failed to reinvent its energy **economy** over the past two decades of escalating climate emergency? What does the nation's dual attachment to nuclear power and fossil fuels tell us about its political culture? And how does Japan's eco-technological impasse illuminate broader issues pertaining to technology, politics, and nationhood in the present age of climate crisis? One way to address these questions is to historicize the sociotechnical imaginaries of energy in Japan because visions, expectations, and hopes about technology—rather than the material affordances of resources—have long defined energy politics in a nation that has depended on imported sources for most of its energy needs. In other words, technology has served as a surrogate for plentiful resources that the nation has inherently lacked.

The pursuit of technological advancement has been a dominant ideology in modern Japan, constituting what Richard Samuels called "techno-nationalism."[1] In the early twentieth century, techno-nationalism in this fledging industrial power hinged on a search for self-sufficiency through militarily applicable technology. After 1945, this emphasis shifted toward a vision of technology-driven economic **development** (*gijutsu rikkoku*).[2] Through economic policy, public education, and mass media, techno-nationalism has been hammered into the nation's collective psyche from the 1950s onward.

Championed by Japan's ruling triad—the Liberal Democratic Party, bureaucracy, and big business—techno-nationalism manifested itself in postwar Japan's energy politics as the nation navigated hydrocarbon-fueled industrialization (or the Great Acceleration). In a country poorly endowed with fossil fuel reserves, technology represented a far more tangible asset than energy resources. Japan's energy conservation technology and its acceptance of commercial nuclear power had their roots in a chronic fear of energy shortage, which was so powerful that it largely silenced the horror of wartime destruction by the atomic bomb. The Japan Atomic Energy Commission's long-term plan (1956) stressed the potential for nuclear power to become a domestic energy source through scientific research, envisaging a nuclear fuel cycle from domestic uranium mining to the disposal of nuclear waste. The Japanese perceived the promise of future energy affluence in nuclear power, but energy security remained elusive; after the hydrocarbon transition in the 1960s, it became an unattainable goal for a country with minuscule hydrocarbon reserves. Since then, barely more than 20 percent of Japan's primary energy has been supplied by domestic sources, despite the nation's pride in its advanced energy conservation technologies and increasingly domesticated nuclear power. Japan's techno-nationalism stood tenuously perched between the illusion of energy affluence and the reality of energy insecurity.

In the 1990s, when the international policy community increasingly turned its attention to limiting the world's carbon emissions, Japanese politicians, technocrats, and engineers were ill-prepared for the shifting global priorities; their vision of a post-scarcity future was at odds with the ecological imperative of decarbonization. As the pressure of carbon reduction mounted due to new international agreements such as the 1997 Kyoto Protocol, Japan's commitment to climate change mitigation visibly waned. Rather than restructuring the energy industry around renewables, the ruling political triad favored incumbent regional monopolies and centralized energy supply, arguing that decentralized renewables were unstable, unreliable, and unfeasible to support economic growth.

Japan's post-2011 move to a high-carbon economy showed the country's deep-rooted phobia of energy scarcity and its continuing infatuation with a post-scarcity future. The government has refused to remove nuclear energy from the energy mix, claiming that tougher regulations and new safety technologies would prevent any future accidents caused by nature or human error. A similar technological optimism was mobilized to legitimize Japan's renewed fossil fuel dependency; many advocates argued that clean coal technology would render fossil fuels innocuous—a controversial position that endorsed the continued extraction, consumption, and international trading of carbon fuel.

The near absence of a civil society critique of Japan's continuing fossil fuel binge demonstrates how deeply techno-nationalism is woven into its national identity. There are protests against nuclear power and other technological issues, but no one doubts that Japan's future lies in technological progress. This expansionary technology-cum-energy strategy is presented as the *only* way that the nation can survive, while no alternative is explored. Recently, a popular topic of public discussion has been the weakening—rather than the excess—of techno-nationalism. Exponents of *gijyutsu rikkoku* have ascribed the nation's loss of technological leadership to the decline of state investment in research and development, but it is debatable whether the developmental state model would save Japan's declining green credentials. Indeed, the experience of other techno-nationalist countries, such as China, seems to suggest that although state investment may encourage the growth of renewable energy, it does little to accelerate the retirement of fossil fuels. Where the state takes the upper hand, technology rarely challenges the policy status quo.

Technology is vital for mitigating the impact of global warming, but excessive techno-optimism has often led, as in Japan's case, to techno-utopian inertia: responses to an immediate problem are deferred to the future, in the form of technological innovations that have yet to occur. Breaking such an inertia would involve a rigorous critique of modern technology, which would embrace and encourage an epistemological shift regarding the place of technology in the world, making technology a matter "less of expansion than of repair, less of growth than of consolidation, less of disruption than of healing," as Bill McKibben writes.[3] Once expansion and growth are no longer the ultimate purposes of technological development, Japan may finally begin to free itself from the long-standing fear of **resource** scarcity, a fear that is a product of the carbon-intensive regime.

See also: **Degrowth, Local, Nationalism, Transitions**

Notes

1 Richard Samuels, *Rich Nation, Strong Army: National Security and the Technological Transformation of Japan* (Ithaca, NY: Cornell University Press, 1994).
2 Morris Low, "Displaying the Future: Techno-Nationalism and the Rise of the Consumer in Postwar Japan," *History and Technology* 19, no. 3 (2003): 197–209.
3 Bill McKibben, *Falter: Has the Human Game Begun to Play Itself Out?* (London: Wildfire, 2019).

JUSTICE

What kind of justice is energy justice?

PAULINE DESTRÉE AND SARAH O'BRIEN

Energy has emerged as a key ethical dilemma of our time. No longer a technical or geopolitical matter left to engineers and policymakers, energy is now widely recognized as constitutive of society's moral zeitgeist. Whether as a source of malaise or of hope, the fact that energy has acquired this ethical valence is worth mentioning and celebrating. It has not always been so.

In the ethical turn of energy studies, the concept of energy justice has gained traction, influenced by long-established movements for environmental and climate justice. The concept focuses on equity and access through four dimensions of justice: distributive, procedural, recognition, and restorative.[1] These dimensions are important, but what often disappears from view is the role and agency of energy itself. What kind of ethical response does *energy* demand? What kind of justice is *energy* justice?

The energyscapes of the present, fueled by hydrocarbons and nuclear power, challenge our notions of responsibility, risk, harm, and redress. Energy systems operate through what Jane Bennett calls "distributed agency": they are "assemblages" of material parts, operators, users, technicians, electrons, and rules, which makes it difficult to identify causality and responsibility.[2] How do we seek justice in the absence of liability? In the case of nuclear fallout and petrochemicals, contamination may spread across generations or indeed eons. The causes of harm can be centuries in the making and extend into the far future. How do we think about justice on timescales we can't even fathom?

Consider carbon emissions. A growing movement of NGOs and civil society groups are calling for "fair shares" in emissions reductions by taking historical responsibility into account in the nationally determined contributions.[3] By this accounting, Canada would have to pledge reductions of greenhouse gas emissions 140 percent below 2005 levels by 2030, and the United States, 195 percent. These figures include full domestic decarbonization and additional emissions reduction abroad via international initiatives providing financial and technological support. Due to cumulative emissions since industrialization, such pledges reflect moral responsibilities toward the international community, due to cumulative emissions since industrialization.[4]

Challenging historical geographies of energy privilege in this way offers a *reparative* transition pathway. But what happens when mechanisms of climate accountability stop short of questioning the fundamental shifts needed in energy production and consumption? What happens when high carbon emitters can pay their way out of financialized carbon debts through tradeable rights to pollute, or when geoengineering and techno-scientific fixes fail to address the expansion of the fossil fuel industry?

Energy justice needs to be foundational—not merely corrective or reactive—to our policies and infrastructures. This approach means rethinking energy planning altogether, rather than relying primarily on technological complexity and carbon accounting schemes. While counterintuitive, inaction (that is, no further extraction and a reduction in consumption) should be included within the justice framework. Ending fossil fuel lock-in—the default assumption that fossil fuels are inevitably here to stay—has long been a crucial pillar of environmental justice campaigners. Even the IEA recognizes that the pathway to **net zero** by 2050 cannot include "new oil and gas fields approved for development."[5] In Lancashire, England, the development of unconventional gas extraction via hydraulic fracturing was challenged by a long decade of strenuous grassroots campaigning. The moratorium imposed on fracking by the U.K. government in 2019 proved that our current "lock-in" with fossil fuel infrastructures around the world is not irreversible: what seems locked in can actually be unlocked. In this framework, keeping fossil fuels in the ground is the real innovation, unmooring our political and financial imaginations from hydrocarbons and unlocking the potential for just energy **transitions**.

But for countries that have contributed very little to global CO_2 emissions and want to prioritize industrialization, keeping fossil fuels in the ground may strike a disingenuous note. Senegalese president and African Union chairperson Macky Sall has criticized the "hypocrisy of the West" imposing moral moratoria on hydrocarbons in places that struggle with energy underdevelopment.[6] Senegal and other African nations where fossil fuel deposits have recently been discovered (Ghana, Uganda, Kenya, Mozambique) are claiming a moral right to oil in the name of development and carbon justice. Many of these countries have historically relied on low-carbon electricity such as hydropower and pursued alternative development pathways.

While the moral right to oil is a valid claim as a form of reparative justice, hydrocarbon's myth of inevitability can and should be dismantled. There is nothing inevitable about our energy infrastructures and the ways that we consume energy, nor about the seemingly sedimented drive for hydrocarbon-fueled growth. A foundational conception of energy justice could finance the transition to a green economy while simultaneously ending hydrocarbon extraction. For instance, this approach could entail compensating countries with low historical emissions for forgoing carbon-intensive extraction—as attempted (unsuccessfully) in 2007 by Ecuador in an effort to sustainably protect its oil-rich Amazon territories.[7] But the current failure of climate **finance**, such as the unfulfilled pledge of $100 billion in annual funding for climate adaptation for the Global South, sends a clear message: when it remains optional, justice is just too expensive to achieve.[8] This is not because of a lack of funds but because energy—its *global* consequences and slow violence—remains peripheral. The dispersed, unpredictable, and unequally distributed effects of climate change and energy production have made clear that acting in isolation—as individuals, nation-states, and species—will only jeopardize collective flourishing, let alone survival.

In the case of carbon emissions reductions targets or bans on new oil and gas extraction, a just energy framework should aim not only to mitigate the failures of our current energy system but also to transform it. Energy demands new ways of enacting justice across time, space, and species. In attuning to energy and its particularities, we contend, justice may become transformative.

See also: **Action, Africa, Autonomy, Degrowth, India, Keep It in the Ground**

Notes

1 Benjamin K. Sovacool and Michael H. Dworkin, "Energy Justice: Conceptual Insights and Practical Applications," *Applied Energy* 142 (2015): 435–44.

2 Jane Bennett, *Vibrant Matter: A Political Ecology of Things* (Durham, NC: Duke University Press, 2009).
3 See AFP, "Growing Movement for 'Fair Share' Climate Commitments," *France 24*, May 11, 2021, https://www.france24.com/en/live-news/20210511-growing-movement-for-fair-share-climate-commitments, and the report from Christian Holz, "Deriving a Canadian Greenhouse Gas Reduction Target in Line with the Paris Agreement's 1.5°C Goal and the Findings of the IPCC Special Report on 1.5°C" (Climate Action Network Canada, 2019), https://climateactionnetwork.ca/wp-content/uploads/CAN-Rac-Fair-Share-%E2%80%94-Methodology-Backgrounder.pdf. Also see Hannah Ritchie, "Who Has Contributed Most to Global CO2 Emissions?" (Our World in Data, October 1, 2019), https://ourworldindata.org/contributed-most-global-co2 for data on historical emissions.
4 For information on the underlying methodology, see Holz's report and "CAT Rating Methodology—Overview," *Climate Action Tracker*, September 2021, https://climateactiontracker.org/methodology/cat-rating-methodology/.
5 International Energy Agency, "Net Zero by 2050: A Roadmap for the Global Energy Sector" (2021), 21, https://www.iea.org/reports/net-zero-by-2050.
6 David Pilling, "Africa Resists Pressure to Put Emissions before Growth," *Financial Times*, August 26, 2022, https://www.ft.com/content/c4ff5997-2b04-4f07-9252-12da0d2f8b02.
7 Brad Plumer, "Ecuador Asked the World to Pay It Not to Drill for Oil. The World Said No," *Washington Post*, August 16, 2013, https://www.washingtonpost.com/news/wonk/wp/2013/08/16/ecuador-asked-the-world-to-pay-it-not-to-drill-for-oil-the-world-said-no/.
8 Jocelyn Timperley, "The Broken $100-Billion Promise of Climate Finance—and How to Fix It," *Nature*, October 20, 2021, https://www.nature.com/articles/d41586-021-02846-3.

KEEP IT IN THE GROUND

How can we build momentum to keep fossil fuels in the ground?

ANGELA CARTER

Oil, gas, and coal are the main sources of emissions causing the climate crisis. To avoid total climate catastrophe, fossil fuel production must decline by 3 percent each year until 2050.[1] However, international climate negotiations have until recently been silent on fossil fuels—even the 2015 **Paris Agreement** includes no mention of them. Until the 2021 UN Framework Convention on Climate Change Conference of the Parties (COP), fossil fuel companies and associations had unparalleled access to state leaders through which they deflected calls for curtailing fossil fuel production and instead urged negotiators to focus on carbon pricing and technological solutions. In this way, the global community has lost decades focusing on solutions that have not delivered the required emission reductions while sidestepping an obvious approach: phasing out the fossil fuel supply.

In contrast, communities around the world have long understood the need to rein in fossil fuel production to protect human health, community sovereignty, and the environment. "Keep it in the ground" movements appeared in the mid-1990s in Ecuador and Nigeria. Inspired by this community activism, Oilwatch advocated for fossil fuel extraction moratoria during the 1997 Kyoto Protocol talks. Today, these movements are blossoming around the world and are finally being reflected in international climate negotiations.

How can we build global momentum to keep fossil fuels in the ground? There are four pathways for supply-side climate action: stop the digging, compel governments to act, hit firms where it hurts, and build the future we need.

To stop the digging means contesting the extraction of fossil fuels at their source. Germany's Ende Gelände movement ("here and no further") uses **civil disobedience** to block coal mining; Kenya's "deCOALanize" campaign successfully fought the construction of a new coal plant in Lamu. This approach means joining or supporting **Indigenous** actions to obstruct the web of pipelines delivering oil to market, like Wet'suwet'en Peoples who opposed the Coastal GasLink **pipeline** in Canada and the Standing Rock Sioux Tribe who fought the Dakota Access Pipeline. It means participating in actions to prevent the export or import of fossil fuels—like groups in Québec who successfully fought GNL Québec's export terminal for fracked **natural gas** from western Canada and the hundreds of Pacific Climate Warriors and "kayak-tivists" who barred coal ships from leaving New South Wales, Australia, in an effort to "Make Coal History."

To compel governments to act means pressuring states to ban new fossil fuel extraction and wind down existing production, following the example of social movements in Belize, Costa Rica, Denmark, France, Germany, Ireland, and New Zealand. Here, focusing on fossil fuel supply means demanding that governments stop subsidizing fossil fuel projects; it also means challenging banks, pension funds, and insurance companies that prop up the fossil fuel sector—like the Stop the Money Pipeline campaign calling out the financial sector in North America for funding climate chaos. It means suing governments that neglect their climate commitments by continuing to extract fossil fuels.[2] Citizens in France and Germany have sued their governments for insufficient climate action and won; when she was seven years old, Raba Ali sued the federal government of Pakistan in 2016 for having infringed on her human rights by developing coal fields. To compel government action also involves calling for the end of fossil fuel advertising in cities. Amsterdam bowed to public pressure and banned ads for gas-powered cars and flights in the subway system; Dutch citizens have sought to extend that ban across the country and to have warning labels added to gas pumps, as reminders that fossil fuels cause climate chaos.

To hit firms where it hurts means holding corporations to account. Ecuadorians sued Chevron for enormous toxic waste spills in the Amazon; climate activists in the Netherlands brought Royal Dutch Shell to court, forcing the company to align with Paris Agreement targets. This pathway also involves shareholder activism: pressuring major oil companies from the inside to cut emissions, enhance lobbying transparency, and transition to clean energy. The Engine No. 1 hedge fund led an effort to elect directors to Exxon's board who have subsequently pushed the company to "reenergize" by transitioning to renewable energy. Pressure can be exerted on the industry by joining divestment campaigns at universities, workplaces, banks, and places of worship and by calling on public pension funds to get out of fossil fuels. Institutional investors including the British Medical Association, the World Council of Churches, the University of California, and the Norwegian Sovereign Wealth Fund Investors have divested assets valued at nearly $40 trillion.[3]

To build the future we need means participating in "green recovery" or just transition efforts that shift to a low carbon energy system while both protecting the workers and communities most impacted and inviting them to reap the rewards of building an equitable, decarbonized society. A first step is to endorse the global Fossil Fuel Non-Proliferation **Treaty** and then to encourage community groups, cities, and national governments to do the same. The Dalai Lama and a hundred other Nobel Prize laureates have already signed on, alongside hundreds of researchers and scientists and cities from Melbourne to Barcelona and Los Angeles.

Until recently, few people in major fossil-fuel-producing countries would have dared to advocate a fossil fuel production phaseout. Today, calls to curtail supply are growing across the globe; they are echoed by international climate negotiators who are finally acknowledging that "we simply have to stop digging and drilling."[4] COP26 in 2021 marked a breakthrough. Denmark and Costa Rica launched the Beyond Oil and Gas Alliance to set hard deadlines on ending fossil fuel extraction, and early drafts of the Glasgow Climate Pact called on member states to accelerate phasing out coal power and fossil fuel subsidies.[5] The global community is finally focusing on the root cause of the climate crisis. This change is propelled by countless multifaceted efforts around the world to keep fossil fuels in the ground; this supply-side climate activism invites us to get involved.

See also: **Alberta, Blockade, Green New Deal, Indigenous Activism, North Dakota**

Notes

I thank Nadine Fladd, Amy Janzwood, and Valerie Uher for their guidance.

1 Dan Welsby et al., "Unextractable Fossil Fuels in a 1.5 °C World," *Nature* 597 (2021): 230–40.
2 See Climate Change Laws of the World, "Search over 5000 Climate Laws and Policies Worldwide" (2024), https://climate-laws.org.
3 Global divestments are tracked at the Global Fossil Fuel Divestment Commitments Database, https://divestmentdatabase.org/.
4 United Nations, "UN Secretary-General's Remarks" (December 1, 2019), https://www.un.org/sg/en/content/sg/press-encounter/2019-12-01/un-secretary-generals-remarks-pre-cop25-press-conference-delivered.
5 United Nations, "Draft CMA Decision Proposed by the President. Draft Text on 1/CMA.3" (United Nations Framework Convention on Climate Change, November 10, 2021), https://unfccc.int/documents/309006.

LAW

How can we change the law to address the climate crisis?

MICHAEL B. GERRARD

Can the gap be closed between where the law is today and where it should be in order to solve the climate crisis? If so, by whom?

I can imagine a global law that requires all nations to slash their use of fossil fuels and massively build out **wind**, solar, and other clean energy sources, and that also tells rich nations they must compensate the poor for the damage they have done to the climate. I can also imagine a unicorn. Neither will ever be. A basic tenet of international law is that countries cannot be bound by treaties to which they do not consent. That is why the United Nations climate agreements, including the one forged in Paris in 2015, are so soft, full of words like *should* and *aim* and *encourage*, and short on *shall* and *enforce* and *penalty*. Anything harder would have been rejected by the countries that are major users or suppliers of fossil fuels.

The International Court of Justice sitting in The Hague can wield only the power that states have given it. In the realm of climate change, it could issue advisory opinions, but nothing binding on the major countries. The real legal action—legislative, executive, and judicial—is at the national level. And there, it is driven by domestic politics.

In the United States, all the great environmental laws were passed between 1970 and 1990, and all but one were signed by Republican presidents. But the era of bipartisan consensus on the environment ended a generation ago. The divide between the Democratic and Republican members of Congress has grown steadily ever since.

So the Environmental Protection Agency (EPA) and other agencies have been forced to use old laws to address new problems. Every time that party control of the White House flips, EPA is told to plow forward or to fall back.

Here, the courts play a central role, both positive and negative. In 2007 the Supreme Court issued a landmark decision, *Massachusetts v. EPA*, finding that the Clean Air Act of 1970 empowers the EPA to regulate greenhouse gas emissions. When President Obama's EPA used that power, the lower courts mostly went along, but the Supreme Court twice said it had gone too far, and it crushed the main effort to move away from coal. When President Trump's EPA aggressively retrenched, the courts repeatedly held it back (mostly because Trump's people were so sloppy in following the required procedures). The courts have also blocked some destructive projects like pipelines and highways.

Efforts to get the U.S. courts to forge new rules on climate change, and not just apply those provided by Congress and EPA, have not gone well, at least so far. Numerous lawsuits have been brought against fossil fuel companies and major greenhouse gas emitters seeking monetary damages or other relief, but they have been held up, first, by a unanimous 2011 Supreme Court decision finding that it's the responsibility of the EPA, not the federal courts, to set appropriate emissions levels, and more recently by still-unresolved procedural wrangling over whether such cases belong in federal or state courts.

The boldest quest for judicial intervention was a suit brought by twenty-one young people in 2015, *Juliana v. United States*, seeking a court order that the federal government "prepare and implement an enforceable national remedial plan to phase out fossil fuel emissions." That case found a sympathetic trial-level judge in Oregon and attracted a great deal of public attention, but ultimately (after some negative signals from the Supreme Court) the Ninth Circuit Court of Appeals threw the case out in January 2020. The court said it agreed with the plaintiffs that climate change is a grave threat and that humans are mostly responsible, but that it was the job of the executive and legislative branches of the federal government, and not the courts, to fix it.[1]

This tension among the branches and this fealty to the separation of powers is not felt by courts everywhere. The Supreme Court of the Netherlands, in a case brought by the Urgenda Foundation and others, found in 2019 that the Dutch government's pledge to the **Paris Agreement** was so weak that it violated human rights and ordered the government to do more. Similar decisions followed in Germany and France. And it is not only Europe; courts in Australia, Brazil, Colombia, Mexico, Nepal, and Pakistan are flexing their muscles in the fight against climate change, and momentum is building.[2] But the courts of the world's two largest emitters—China and the United States—are not on board.

In 2021 a court in the Netherlands ordered Royal Dutch Shell to slash its greenhouse gas emissions 45 percent by 2030. That is the first decision of any court in the world holding fossil fuel companies or major emitters accountable for climate change without a legislative mandate. Shell was ordered to reduce not only its own emissions (such as from its refineries and leaky pipelines) but also those of its customers. Shell may try to comply (I'd say cheat, but it's probably legal) in part by selling off some of its assets to others who will do the polluting and by buying "offsets" that pay someone else to reduce their emissions. Shell can be expected to make full use of the many legal tools that entrench the use of fossil fuels.

But to the extent that Shell (or anyone else) actually lowers the demand for oil, gas, and coal, that would be real progress. Reducing the supply of fossil fuels is important, but entities that are largely immune from climate laws—such as the national oil companies of Saudi Arabia, Russia, and China—will keep the fuel flowing. Demand must also be reduced. In this way, fossil fuels are like drugs, guns, and undocumented labor—we can try to stanch the supply, but so long as there is a demand, someone will provide them.

Meanwhile most economists agree that a price on carbon, while not a silver bullet, would have the broadest impact on both the demand for and the supply of fossil fuels.

A U.S. law could accomplish this. But not one from the courts; it needs to come from Congress. For that to happen we need to constrict the role that money—including fossil fuel money—plays in politics and in our endangered **democracy**.

See also: **Corporation, Design, Justice, Scale**

Notes

1 *Juliana v. United States*, 947 F.3d 1159 (9th Cir. 2020). The case remains active, with a decision by the Oregon federal court that the plaintiffs could continue to seek redress; see Climate Change Litigation Databases, "Juliana v. United States" (2024), https://climatecasechart.com/case/juliana-v-united-states/.

2 Michael Burger and Daniel Metzger, "Global Climate Litigation Report: 2020 Status Review" (United Nations Environment Program and Sabin Center for Climate Change Law, 2020), x.

LIFESTYLE

Are lifestyle changes a waste of time?

MARK SIMPSON

Ursula Biemann's *Deep Weather* (2013) is a minor masterpiece of cognitive mapping.[1] A dissymmetrical diptych, this video essay unfolds a vivid reckoning with contemporary fossil-fueled climate emergency. "Carbon Geologies," the film's first part, gives 120 seconds of a bird's-eye perspective on Albertan tar sands extraction in its vast, awful, alienating scale. "Hydrogeographies," the film's second part, dwells for seven minutes on one consequence of such extraction: the urgent, arduous efforts of Bangladeshis to build embankments against rising seawaters, rendered in imagery shot at the surface and amid the tumult. Threading together the diptych's two frames is Biemann's whispered voice-over, which imbues the description of these linked endeavors with the sound and sense of incantatory ritual. The film invites viewers to confront the world in sand:[2] those innumerable tar-soaked grains filling truckloads of bitumen gouged out of the Athabasca Basin; those innumerable water-logged grains filling sandbags piled against surging tides along the coasts of Bangladesh; and so forth.

Is it perverse to consider *Deep Weather* a lifestyle film? No and yes. No, because the video intimates that today's petroculture—those habits, desires, and beliefs seemingly unrelated to energy—supplies the diptych's hinge: this geological lifestyle force, irresistible because ubiquitous, is deep weather's absent cause. But yes, because the *style* in lifestyle connotes a sense of living that reductively misunderstands human agency as a matter of individual choice, eliding the pressures of the petrocultural system that shapes the contradictory conditions within which human agents must live. *Lifestyle* cannot grasp the import or consequence of the terraforming practices depicted in *Deep Weather*—much less their juxtaposition.

This contradiction reveals the difficulties posed by *lifestyle* as a concept associated with energy politics today. Everywhere we are hailed to confront questions of energy in terms of lifestyle. Such hailing takes two forms: either that fossil fuels enable the freedoms characterizing the modern good life, with any prospective change to this mode of living posing an existential, even ontological threat; or that the fossil curse is the curse of modern consumption, which a shift to more sustainable, frugal lifeways alone can dispel. Antithetical in their commitments, these competing lifestyle visions nonetheless share a protagonist: the entrepreneurial self as privileged political subject, whose style of life requires endless self-fashioning—or what Michel Feher would term a perpetually fungible *self-appreciation*.[3] Such difference without difference might suggest that, as a basis for politics now, lifestyle instantiates what Imre Szeman and I call *impasse*: "a continuation of the same wherein the overcoming of blockages cannot solve—and may in fact compound—the abiding stuckness."[4] As impasse, the world in sand becomes a mire, engulfing the multitude thanks to the obtuseness or indifference of the privileged.

If the entrepreneurial self is the political subject proper to the *lifestyle* concept—and therefore symptomatic of impasse—then what framework might reckon the energy problem otherwise? Critiquing "the concept of *lifestyle*" as a form of "individualization," "a moment of freedom of choice that abstracts from **class** structures, gendered and racialized relations, as well as from the organization of capitalist societies as nation-states," Ulrich Brand and Markus Wissen analyze the contemporary conjuncture in terms of "the imperial mode of living"—"a paradox located in the very centre of multiple crisis phenomena" that, propelling and intensifying climate catastrophe, ecological damage, social fracture, economic precarity, and geopolitical friction, nonetheless serves to steady social dynamics in those settings where its benefits tend to accrue.[5] For Brand and Wissen, the force of such paradox means that "the imperial mode of living is based on exclusivity; it can sustain itself only as long as an 'outside' on which to impose its costs is available. But this 'outside' is shrinking as more and more people access it and fewer people are willing or able to bear the costs of externalization processes."[6] Such arguments resonate with Biemann's report on deep weather in *Deep Weather*, especially its intimation of ever-intensifying turbulence in the forecast.

Brand and Wissen underscore the importance of a politics that attends to structural contradictions and dissymmetries, rather than focusing merely on individuating lifestyle choices. Materialist geographer Matthew Huber advances a provocative account of such a politics, which rejects truisms about **degrowth** in favor of a program of decommodification. For Huber, the two regnant environmental paradigms—"lifestyle" and "livelihood"—are not opposed but instead bound together by austerity's choking thread: an ethic of universalized reduction and restraint unable to reckon the global social dynamics in which a very few benefit from profligate abundance while the great many subsist on much too little.[7] An austerity politics, whether neoliberal or ecological, only obscures such fissures. Rather than taking human need "as a source of 'footprints' that must be reduced," Huber contends, "we should acknowledge [that] the majority of people in capitalist society need more and secure access to these basics of survival. To make this political we need to explain how human needs can be met through ecological principles."[8] At stake is nothing less than "a politics of building and enlarging the zone of social life where capital is not allowed": a daunting, urgent endeavor in which the energy question is always and everywhere a matter of power, determined not through choice as individuating style of life but through struggle as collective practice ventured across material, geographic, ecological, and experiential **scales**.[9]

See also: **Alberta, Commons, Documentary, Habit, Justice**

Notes

1 Ursula Biemann, *Deep Weather* (2013), https://vimeo.com/90098625; Fredric Jameson, "Cognitive Mapping," in *Marxism and the Interpretation of Culture*, ed. Cary Nelson and Laurence Grossberg (Urbana: University of Illinois Press, 1988), 347–57.

2 The allusion invokes William Blake's image of a "World in a Grain of Sand" in "Auguries of Innocence." See *The Complete Poetry and Prose of William Blake*, ed. David V. Erdman (New York: Anchor, 1988), 490–93.

3 Michel Feher, "Self-Appreciation," trans. Ivan Ascher, *Public Culture* 20, no. 1 (Winter 2009): 21–41.

4 Mark Simpson and Imre Szeman, "Impasse Time," *South Atlantic Quarterly* 120, no. 1 (January 2021): 77–89, 80.

5 Ulrich Brand and Markus Wissen, *The Imperial Mode of Living: Everyday Life and the Ecological Crisis of Capitalism*, trans. Zachary Murphy King, ed. Barbara Jungwirth (London: Verso, 2021), 44, 5.

6 Brand and Wissen, 7.

7 Matthew Huber, "Ecological Politics for the Working Class," *Catalyst* 3, no. 1 (Spring 2019): 7–45, 13.

8 Huber, 36.

9 Huber, 38.

LOCAL

How do local climate politics work best?

BRIAN COZEN AND DANIELLE ENDRES

The local is often articulated as an important site of energy politics. Examples of local energy actions abound, situated across a variety of intersecting vectors of the personal and political: these include individual energy efficiency and reduction actions one can do at home; microgrid and rooftop solar campaigns seeking **community**-based democratic control over **electricity**, including efforts in **India**, Germany, Australia, and the United States; state- or province-level regulations that counter national-level inaction (e.g., California's energy policy during the Trump presidency); and water protectors mobilizing at sites of proposed or existing oil pipelines (such as Dakota Access Pipeline and Line 3). Most often used as a noun or adjective to denote energy actions and politics undertaken in a particular place, *local* remains a polysemous term that can mean different things in different contexts.

Although particular places across multiple scales must be involved in just and equitable energy transitions, we argue that *local* should function in energy scholarship and activism not only as a neutral, geographic adjectival descriptor of the specific location of any given contest over energy politics, but also as a heuristic for understanding the dynamic characteristics of a given set of multiscalar relations at a given moment in time. In other words, *local* should also be understood as a process or action: as a verb. In the context of energy politics, we ask: why, and how, should scholars and activists localize the local? By shifting from *local* to its verb form *localize*, we can challenge the assumption that *local* (or locality) denotes simply a pin on a map, a static location, or an empty vessel in which to add or subtract fuels or energy technologies. Instead, to localize energy politics means to consider how intersections of time, rhythm, bodies, and place are at stake in energy's varied societal forms and functions. Localizing entails recognizing that energy politics are always in process, as moments in time that involve diverse ways of engaging productive tensions between the particular (local) and the universal (global). The global consequences of the Anthropocene—a precarity that manifests inequitably across positionalities—call for this sort of attention to local action while consistently linking that focus to broader, systemic foci.

To illustrate by way of analogy, consider how *local* is defined in medical discourse: "Of a remedy or treatment: acting upon or administered (esp. applied outwardly) to a limited area or particular part of the body; topical; not systemic."[1] Think of a local anesthetic and a broken arm bone. The anesthesiologist isolates the target area, localizes the anesthetic, and interrupts sensorial circulation in order to treat the arm bone. In this sense, localizing is a temporary action—the anesthetic wears off, the bone heals—that briefly isolates one part of a system to heal or repair that system. Localizing entails an approach to politics in which location is but one node in a complex interplay, such as when tracing the circulating impacts of and resistances to slow violence.[2]

Localizing energy politics is valuable because it can focus one's attention by provisionally narrowing in on a particular place (and moment in time) while not erasing or ignoring the larger system. Localizing focuses attention on specific enactments of energy politics in order to account for differences across locations, structures, and histories of power. These differences, in turn, point toward incongruities and complexities within a larger, ongoing, and ever-changing system. For instance, energy justice activists emphasize that there is no silver bullet: no single **renewable** energy technology that can be implemented equitably and effectively across all locations. This observation challenges the adequacy of universal solutions and insists instead on the need for a mix of localized solutions to ongoing energy transitions away from fossil fuels. Attending to the specific dynamics of a particular place can reveal issues that would otherwise be rendered imperceptible by top-down approaches, with local voices and interests often supported in theory and stymied in practice. To localize means to understand how place-based attachments can mobilize individual and collective actions that speak back to global approaches.

Localization, therefore, is a productive way to avoid the twinned pitfalls of universalizing top-down remedies and overemphasizing the boundedness and fixity of locality. *Local* is sometimes reduced to a mere buzzword for unilateral and top-down enactment of national or global policies. *Local* can obscure systemic relations of power and inequity by placing blame on particular places/populations for energy usage; it can exclude by means of endless NIMBY battles used to maintain unjust status quo energy systems, and it can be invoked to write off potential solutions as impractical or unscalable. (Consider, for example, claims that California is so unique that its energy politics cannot be a model for the rest of the United States, let alone other countries.) Too much focus on the local can fail to capture the snapshot aspect of localization, thereby obscuring the systemic nature of climate change and energy transitions. As Christina Demski and Sarah Becker argue about the challenges posed by climate emergency, the importance of local energy **democracy** campaigns does not obviate the need for a set of national and international policies and agreements that can work across a variety of locations.[3] Arturo Escobar seeks "strategies of localization" that can grasp the potential of the local in "all of its multiplicity and contradictions."[4] A localizing approach can foster understanding of part/whole relations within our energy systems by imagining energy politics as constantly in flux and reliant on local knowledges. Localizing attention to a variety of places, as opposed to focusing on one locale, might have a better chance of productively navigating these tensions among local, national, and global energy politics.

See also: **Autonomy, Indigenous, Japan, Scale, Solidarity, Transitions**

Notes

1 *OED Online*, "Local, Adj. and n.," accessed July 15, 2021, http://www.oed.com/view/Entry/109549.
2 Rob Nixon, *Slow Violence and the Environmentalism of the Poor* (Cambridge, MA: Harvard University Press, 2011).
3 Christina Demski and Sarah Becker, "Energy Security: From Security of Supply to Public Participation," in *Routledge Handbook of Energy Democracy*, ed. Andrea M. Feldpausch-Parker et al. (London: Routledge, 2022).
4 Arturo Escobar, "Culture Sits in Places: Reflections on Globalism and Subaltern Strategies of Localization," *Political Geography* 20, no. 2 (2001): 171, https://doi.org/10.1016/S0962-6298(00)00064-0.

MINING

Why are the ways that we imagine mining important?

GIANFRANCO SELGAS

Metals and minerals such as cobalt, coltan, copper, gold, graphite, iridium, iron, lithium, manganese, rhodium, and tantalum have been essential to the industrial, technological, and **digital** development of modern societies. But they also tell us something about the intrinsic relationships between technology and the geopolitics of mining. In the absence of such minerals, modern life would cease to function.[1] Lithium and coltan, for example, have increased their market value because of their indispensability for energy transition systems. While a radical change in energy consumption is imperative for a fair and livable future, an energy transition enabled by means of mineral extraction will be anything but clean and fair without significant changes in capitalist supply chains and the governance of the extractive industries, not to mention addressing the environmental impacts of dragging matter out of the ground.

Furthermore, mining in late capitalism involves more than the extractive industries with which we are familiar, involving coal, heavy metals, and minerals. It also names a dense network of material and symbolic operations dispersed across time and space. Mining brings to the fore deep entanglements between social and geological forms of exploitation. As Kathryn Yusoff puts it, "Geological extractions constitute a geopolitical field that organizes social fields—from the globalizing geopolitical power of oil and gas extraction to the specific ways in which the bodies of **Indigenous** women are trafficked to become the social-sexual frontier of that extraction."[2] While nineteenth-century geological sciences played an important role in molding our imaginaries about the earth, our expanded entanglement with the earth now supersedes actually existing extractive zones. Today, mining must be acknowledged as a contrapuntal ecology—not only as Earth plundered in relation to ethnic, racial, gendered, and colonial arrangements of socioecological life, but also as a symbolic operation within and through which socioecological life is produced and reproduced.

As a material operation, mining unfolds at the convergence of geological and sociopolitical formations. It involves global production, material exchange, and technologies connecting nations around the planet.[3] Entangled with almost every aspect of social life, mining and metallurgical technologies have shaped the world as we know it. The economic dimension of silver and gold during colonial times transformed the global economy. But nonprecious metals have also played a key role in the development of the industrialized world.[4] Mining's impact is far-reaching, and often devastating. The commodification of rocks and minerals has disrupted the functioning of Earth systems. Under capitalism, mines have produced (and reproduced) colonial modes of exploiting various life forms. The environmental damage caused by mineral extraction ranges far beyond the pit or the shaft. It entails widespread deforestation, soil erosion, and contamination of water and more-than-human bodies.

Mining not only transforms the earth by means of extractive intervention but also expands beyond the subsoil to become an operative metaphor structuring political and economic discourses. It is a symbolic operation that creates and destroys worlds through the logic of extraction. Mining consolidates a mode of appropriation and exploitation of the earth, which has become materially, socially, and ideologically intertwined with the most basic notions of freedom, mobility, and care. As a symbolic operation, mining animates ideas of growth and **development** alongside promises of sustainable and clean futures in a mode of life that has become generalized in both the Global North and the Global South. But mining's symbolic allure deploys these social imaginaries by jeopardizing the very scope of life: it is often imposed against the will of the population, thereby reinforcing cultural dispossession and environmental deterritorialization.[5] More insidiously, this symbolic allure can lead communities and citizenries to regard extraction as the only available path to the future, even while the material operations of mining foreclose possibilities and jeopardize livelihoods.

From a material and symbolic standpoint, minerals, mines, and mining bring together local and planetary histories. The history of mineral-rich nations in Global South regions like Latin America is telling. Latin America has been a profitable laboratory for the mining industry since colonial times. The exploitation of subsoil deposits has played a preponderant role in the region's economy for more than five centuries, ranging from sixteenth-century Mexican silver to twentieth-century Venezuelan oil and twenty-first-century Argentinian and Chilean lithium.[6]

Supported by governments, the mining industry has superimposed extractive infrastructures through global financial circuits enabling geological exploitation and the proletarianization of labor. Weak environmental controls and a lack of accountability have come together with the criminalization of social-environmental struggles and human rights violations.

The imperative of energy transition from fossil-based systems of energy production and consumption to renewable energy sources has only intensified this history. Replacing gasoline-powered combustion engines with batteries requires supplies of critical minerals such as lithium. But what we are witnessing in the Lithium Triangle deposits in Bolivia, Chile, and Argentina, or in the gold extraction and oil-backed cryptocurrency issuance in the Venezuelan Orinoco Mining Arc, tells a different story about climate futures. Mineral-rich countries are suffering ecosystem degradation, health crises, the flourishing of illicit businesses, and violations of human rights and sovereignty. The transition to a more sustainable future is coming at the cost of rampant ecocide and plundering Indigenous land.

As a new political geography of extraction unfolds in the shift from fossil fuels to renewables, a critical engagement with mining as a material and symbolic operation is vital. How can we develop an ethical relation to the materials we extract from the earth and the ways in which they are extracted? It is imperative to uncover mining's unrelenting symbolic operation by proposing alternative paradigms centered on decarbonization, decapitalization, and **decolonization**. We cannot keep turning a blind eye to spaces scarred—and benefited—by mineral and hydrocarbon extractivism. We must contest our high-energy modernity in order to better understand geology's constitutive relation to the socioecological world. And we must interrogate the latter if we want to decenter the conceptual and cultural categories that keep isolating humans on a biodiverse planet and that stand in the way of political and transformative action.

See also: **Battery**, **Black**, **Electricity**, **Gender**, **Neoextractivism**, **Pipeline**

Notes

1 Jussi Parikka, *A Geology of Media* (Minneapolis: University of Minnesota Press, 2015).
2 Kathryn Yusoff, "Geosocial Strata," *Theory, Culture & Society* 34, no. 2–3 (2017): 125, https://doi.org/10.1177/0263276416688543.
3 Martín Arboleda, *Planetary Mine: Territories of Extraction under Late Capitalism* (London: Verso, 2020).
4 Adam Bobbet and Amy Donovan, "Political Geology: An Introduction," in *Political Geology: Active Stratigraphies and the Making of Life*, ed. Adam Bobbet and Amy Donovan (Cham: Palgrave Macmillan, 2019), 2–3; Jason W. Moore, *Capitalism in the Web of Life: Ecology and the Accumulation of Capital* (London: Verso, 2015).
5 Maristella Svampa, "Commodities Consensus: Neoextractivism and Enclosure of the Commons in Latin America," *South Atlantic Quarterly* 114, no. 1 (2015): 65–82, https://doi.org/10.1215/00382876-2831290.
6 Kendall W. Brown, *A History of Mining in Latin America: From the Colonial Era to the Present* (Albuquerque: University of New Mexico Press, 2012), 128–29.

MUSIC

How are listening and energy related?

SHERRY LEE AND EMILY MACCALLUM

Music, as sound, is a form of energy. Sound waves are immaterial: we don't see the sound we hear, and it's difficult to pinpoint its precise moment or site of production. The energy of music moves invisibly among us. Air is the medium through which music reaches human ears; since temperature affects the speed of sound, music will move faster on a hotter future Earth. Sound also moves much more quickly in water, so whales' songs travel more swiftly—and over far greater distances—to reach nonhuman listeners underwater. In our daily perception, however, we tend to experience sound as near-instantaneous communication among bodies in the world.

While sound waves are immaterial, they always have material origins, travel through physical mediums, and have material effects. We set them in motion with our hands, feet, and voices; they exert pressure throughout our bodies, not just on eardrums and auditory nerves. Listening, therefore, is an attentive bodily engagement with sonic energy. But music making has almost always been entangled with technology, too. Most traditional musical instruments are biotic in substance, made from plant matter, particularly wood, and innumerable animal bodies—skins, hair, intestines, horns, tusks.[1] Making music entails energies of production and consumption *before* sounding.

Despite these material origins, musical sound remains ephemeral until captured once again by technology. Early recording cylinders combined substances produced by live and fossilized organisms, in the form of beeswax and paraffin; the disc revolution inaugurated by shellac was animal-mineral, mixing insect secretions imported from colonial South Asia with limestone and carbon. The subsequent shift back to petroleum set in motion a long trajectory of music's plastic dependency: vinyl discs, polyester strips for magnetic tape, cascades of plastics for cassettes, CDs, computer sound cards, and hard drives. Each format implies many energy forms: extracting and processing resources, producing sonic content in studio, and manufacturing and distributing products for worldwide consumption. It's clear that music culture is petroculture—a history of energies and materials that begins with hand-cranked wooden boxes of metal components and gains momentum from electric power and synthetic polymer innovations.

The twenty-first-century privilege of listening to nearly any music, anytime, anywhere is rooted in a history of intertwining sound with energy-intensive resources and modes of production on a global **scale**, throwing into relief the desires driving our fossil-energy economy. Yet how many listener-consumers understand the energy-rich history behind each advancement in storing, accessing, and listening to recorded sound? This surfeit of energies and matter is largely undetectable in musical end products, and industry actors rarely acknowledge the life cycles of music as material thing, from recording production through ultimate disposal. Once-loved vinyl records are burned or buried, surplus CDs crushed and landfilled. Since the early 2000s, the everyday listening experience has reverted to a seemingly intangible sphere of **digital** files, enabled by access to a perpetually expanding global music archive of disembodied 1s and 0s. But digitization is not dematerialization: data infrastructure is concrete and requires massive amounts of energy, and the hardware for streaming, too, eventually becomes trash.[2]

Scholars, musicians, and consumers alike are increasingly aware of the industry's predication on energies of global cultural mobility and format succession, and these trajectories of obsolescence have increasingly garnered environmental concern. In response, some independent record producers distribute packaging handmade from postconsumer waste materials. Meanwhile, an aesthetic fascination with old, discarded music resonates with trash artists who press reclaimed waste into reenvisioned recording formats.[3] Fantasies of a renewable grid and future fossil-free formats float alongside these imaginative economies of reuse. Yet another promise of such DIY sound-art collectives arises from their tendency to emerge as local communities. If the local reach of sound waves can define an acoustic **community**, acts of sound making and listening can also return music production from the global to the **local**.

There's no shortage of ecologically themed music and sound **art** that evokes acoustic landscapes with instruments, sonifies geological data, or amplifies field recordings captured deep within forests, oceans, or glacial crevices. Music can also amplify protests against structural inequity and climate injustice. The political relevance of this music is undeniable, given its capacity for awareness raising: listening in and to crisis is profoundly affective and powerfully motivating. But broader practices of individual or collective listening and sound making as lived engagement with environments, and with others—including those more-than-human—offer additional opportunities for making change. Mitchell Akiyama and Brady Peters's *Distribution of Sensation* (2019) installed surplus road-drainage pipes as tuned resonators under Toronto's Gardiner Expressway, harmonizing traffic frequencies that invited passersby to listen.[4] And apartment-bound Montreal residents sounded musical drones from their balconies during the pandemic-imposed lockdowns of 2020, enacting an acoustic community reliant not on language but on sonic-vibrational attunement.[5] In contrast with the global music industry, this phenomenon was a uniquely localized musicking in response to global crisis. For those within earshot, these experiments amplified possibilities for being in and knowing the world through a form of energy that is politically salient precisely because of its resonant capacity to instantiate a **commons**.

Is music, then, indispensable, or is it nonessential in comparison with more science-driven conversations about energy transition? Ethical music making and listening demand a new consciousness of the inextricable intertwinement of sound, energy, and material. Attending to music and sound with transformed awareness of the multiple energies they imply could inspire new modes of listening, sonic expressivity, voicing, and being within a perceptual commons.[6] Such future listening could reposition music as part of vital energy transitions.

See also: **Decolonization, Globalization, Online, Scales, Transitions**

Notes

1 For instance, Chris Gibson and Andrew Warren, *The Guitar: Tracing the Grain Back to the Tree* (Chicago: University of Chicago Press, 2021).
2 Kyle Devine, *Decomposed: The Political Ecology of Music* (Cambridge, MA: MIT Press, 2019), 28.
3 Elodie Roy, "'Total Trash': Recorded Music and the Logic of Waste," *Popular Music* 39, no. 1 (2020): 100–104.
4 The Bentway, "Events" (n.d.), https://www.thebentway.ca/event/the-distribution-of-sensation/.
5 Hubert Gendron-Blais, "A Community Attuned to the Outside: Reverberations of the Montreal Balcony Drone," *Journal of Sonic Studies* 22 (2021), https://www.researchcatalogue.net/view/1475487/1475488/0/0.
6 Anja Kanngieser, "Geopolitics and the Anthropocene: Five Propositions for Sound," *Geohumanities* 1, no. 1 (2015): 80–85.

NATIONALISM

What is the impact of transition on nationalism?

ZEYNEP OGUZ

The use of fossil fuels is rightly understood as one of the major causes of climate emergency, which has spurred an ongoing transition from coal- and oil-powered energy sources and systems to renewable and decarbonized ones. Less well understood are the ways that fossil fuels have been central to national imaginaries over the past century. Our moment of ecological crisis is also characterized by a resurgence of nationalism—whether antiglobalist, anticapitalist, right-wing, and/or authoritarian—which threatens the global cooperation that climate change mitigation and adaptation require. Given the affinities between fossil fuels and national imaginaries, what will become of nationalism during and after a transition to a different energy future? What are the political imaginaries on which **resource** nationalisms draw, and to what kinds of political arrangements are they linked? Answering these questions can help to shed light not only on the future of nationalism but also on whether the political dynamics and imaginaries of petromodernity might persist into a putatively green future.

Resource nationalism designates a set of political discourses and material practices that link the imaginaries of a nation to nature and natural resources. Natural resources are central to constituting and reproducing the "imagined community" of the nation.[1] Resource nationalism becomes entangled with questions of territory, sovereignty, and cultural identity. In many places, notions of territory are linked to the delimitation of resources as natural. Nation building translates geology and physical geography into natural resources, whose territorialization is seen by the state as essential to security and prosperity.[2] In Venezuela, for example, oil has been fetishized as a "magical" commodity and cast as an integral part of the nation's body while also legitimizing the state.[3] Similarly, in Kazakhstan, oil plays a key role in "confirming *the state* as a coherent actor and naturalizing its control over bounded, abstract space as well as the people and resources located therein."[4]

Like other nationalisms, resource nationalism is produced through its constitutive others: foreigners, threats, and enemies. Geographer Tom Perreault argues that "all resource nationalisms are fundamentally expressions of anxiety over control of economically, politically, or culturally important resources by a threatening other, either domestic or foreign."[5] Resource nationalists often claim to defend national sovereignty against foreign involvement, (neo)colonial domination, and the theft of national wealth, arguing that the people of a given country, rather than private corporations or foreign entities, should benefit from the resources of a territorially defined state. Ongoing histories of colonial occupation, imperialist intervention, and neoliberal policies that have systematically underdeveloped and impoverished the Global South play a significant role in nationalist mobilizations over resource extraction and ownership.

Resource nationalisms have often allied with anti-imperialist, anticolonial, and anticapitalist politics.[6] Across Latin America, populist-left governments have advocated for the collective ownership of fossil fuels and minerals, succeeding in reducing income inequality and poverty and improving health and education outcomes in the face of imperialist and neoliberal attacks. Anti-imperialist forms of resource nationalism have been a prominent feature of resource-poor countries, too. In Turkey, for instance, the National Petroleum Campaign of the 1960s united the Turkish left against the licensing and extraction privileges that foreign oil companies Shell and Mobil enjoyed at drill sites in southeastern Turkey.[7] The campaign's supporters argued that the government represented the interests of international oil companies rather than "the people" and called for the nationalization of oil.

The anxieties that animate resource nationalisms have not always fostered emancipatory and justice-driven goals. They have been entangled with colonial occupation and anti-Indigenous racism. In Bolivia, for example, although resource nationalism might appear to be a progressive effort to redistribute resource wealth and thereby counter neoliberalism, such extractive projects also threaten **Indigenous** territorial rights, compromising the purportedly anticolonial goals of pluralist states.[8] In Turkey, the petronationalism of the National Oil Campaign focused on questions of class rather than race; the campaign thus disavowed Kurdish demands for political freedom, cultural identity, and collective sovereignty and justified the ongoing exploitation of Turkey's Kurdistan.[9]

Resource nationalisms have often striven for national **autonomy** against global capital and imperial intervention over resource control, management, and distribution of gains. However, both their nationalist and extractive aspects can make them an inadequate and even counterproductive response to the entangled problems of colonialism and ecological emergency. One salient example is the discourse over recent offshore oil and natural gas discoveries in Cypriot waters. Mainstream political parties have largely ignored the ecological and political stakes of offshore gas extraction. Supporting state-led hydrocarbon exploration in the eastern Mediterranean Sea, they not only take the extractability of resources for granted but also legitimize the ongoing occupation of northern Cyprus by the Turkish state.[10] In Latin America, the anti-Indigenous overtones of state-sanctioned resource nationalism have led Indigenous and environmental groups to start targeting the extractive model itself. As political scientist Thea Riofrancos observes, a "post-extractive economy—not socialized extraction—is their utopian vision."[11]

The resource nationalisms associated with petromodernity posit the underground as a nationalized territory and geological matter as extractable; in the current transition beyond fossil fuels, resource nationalisms continue to fuel territorial imaginaries and political futures. Research on lithium, gas, sun, and wind reveals that so-called new energy regimes continue to translate geological matter and environmental forces into resources to be extracted for the sake of national progress and profit, while portraying these extractive regimes as ecologically friendly. Yet none of these substances and forces are resources in and of themselves. As the works of Zoe Todd and Jeremy Schmidt remind us, for example, oil and coal are in fact "weaponized fossil kin" in settler-colonial land.[12] But they can also be valued and reconfigured otherwise. In the face of intensifying resource nationalisms, anti- and decolonial projects might animate radically different political-ecological futures, where capitalism's hold over how nature is valued might be replaced by a planetary ethic and politics of care, justice, and collective flourishing.

See also: **Decolonization, Democracy, Japan, Populism, Transitions**

Notes

1 Benedict Anderson, *Imagined Communities: Reflections on the Origin and Spread of Nationalism* (London: Verso, 1983).
2 Nisha Shah, "The Territorial Trap of the Territorial Trap: Global Transformation and the Problem of the State's Two Territories," *International Political Sociology* 6, no. 1 (2012): 66.
3 Fernando Coronil, *The Magical State: Nature, Money, and Modernity in Venezuela* (Chicago: University of Chicago Press, 1997).
4 Natalie Koch, "Kazakhstan's Changing Geopolitics: The Resource Economy and Popular Attitudes about China's Growing Regional Influence," *Eurasian Geography and Economics* 54, no. 1 (2013): 127.
5 Tom Perreault, "Materializing Space, Constructing Belonging: Toward a Critical-Geographical Understanding of Resource Nationalism," in *The Routledge Handbook of Critical Resource Geography*, ed. Matthew Himley, Elizabeth Havice, and Gabriela Valdivia (New York: Routledge, 2021), 126.
6 Matthew Himley, "Mining History: Mobilizing the Past in Struggles over Mineral Extraction in Peru," *Geographical Review* 104, no. 2 (2014): 174–91; Thea Riofrancos, *Resource Radicals: From Petro-Nationalism to Post-Extractivism in Ecuador* (Durham, NC: Duke University Press, 2020).
7 Zeynep Oguz, "The Unintended Consequences of Turkey's Quest for Oil," *Middle East Report* 296 (October 13, 2020), https://merip.org/2020/10/the-unintended-consequences-of-turkeys-quest-for-oil.
8 Andrea Marston, "Strata of the State: Resource Nationalism and Vertical Territory in Bolivia," *Political Geography* 74 (2019): 102040.
9 Oguz, "Unintended Consequences."
10 Zeynep Oguz, "Harnessing Indeterminacy: The Technopolitics of Hydrocarbon Prospects," *Platypus* (blog), July 20, 2021, http://blog.castac.org/2021/07/harnessing-indeterminacy-the-technopolitics-of-hydrocarbon-prospects.
11 Thea Riofrancos, "Digging Free of Poverty," *Jacobin* 26 (Summer 2017): 36.
12 Zoe Todd, "Weaponized Fossil Kin and the Alberta Economy," *zoestodd* (blog), January 19, 2021, https://zoestodd.com/2021/01/19/weaponized-fossil-kin-and-the-alberta-economy; Jeremy Schmidt, "Settler Geology: Earth's Deep History and the Governance of in Situ Oil Spills in Alberta," *Political Geography* 78 (2020): 102132.

NATURAL GAS

Can natural gas be made visible, or viable?

JOHN SZABO

Natural gas was first noticed when it seeped between geological formations to the surface of the earth and, ignited by a spark, the invisible gas became a "sacred eternal flame." People congregated around such burning springs, understood to be manifestations of deities. The methane molecule dominant in natural gas remained hidden behind the veil of the flame and the religious experience it provoked. Thousands of years later, natural gas has become deeply intertwined with a different sort of deity, fossil capitalism, which many still worship but an increasing number of movements contest.[1] While environmental movements have targeted coal and oil, natural gas has largely eluded scrutiny. What explains this oversight? And how might natural gas be made a matter of public concern?

Natural gas has always been the smaller—and generally unwanted—sibling of oil. The modern history of natural gas was enabled by the invention of the **pipeline**. But even after pipelines facilitated the transport of natural gas, its uptake as a fuel was slow. When the frenzy for oil began during the second half of the nineteenth century, discoveries of oil deposits often also uncovered reserves of natural gas. Given the significant infrastructural investments necessary to capture and exploit natural gas, methane was frequently vented or flared at sites of extraction, reproducing the eternal flames of an ancient era. Oil became the prize; natural gas was the burden.[2]

In the twentieth century, investors began to harness the potential of natural gas for lighting, household appliances, industrial applications, and space heating. It became a convenient source of energy in towns that invested in the necessary infrastructure, although its utilization remained the exception rather than the rule, with 1.47 units of natural gas wasted for every unit consumed in 1930.[3] Only gradually would producers begin to harness it. And as with oil, "users . . . gained the benefit of cheap energy without assuming responsibility for its environmental damage."[4] Consumers tended not to perceive its environmental impact since the gas burners they encountered seemed to burn the fuel cleanly, leaving no visible trace behind.

The gaseousness of natural gas lends it an ethereal unworldliness that has allowed it to be culturally inscribed as the most exquisite alkane. Until it is refined, oil is a useless dark sludge of chains and hexagons, while coal is a pulverulent mess that burns dirtily. Compared to these fuels, natural gas has been described as the "modern" and "clean" alternative, given the relatively minimal particulate matter and other pollutants emitted during combustion.[5] For this reason, natural gas has been posited as a bridge from the fossil-fueled present to a future powered by green renewable energy.

The urgent necessity of climate **action** has slowly begun to challenge this (mis)understanding of natural gas as **clean** fuel. In fact, not only is natural gas a source of CO_2 emissions upon combustion, but methane (a molecular component of natural gas) is a greenhouse gas eighty-six times more potent than CO_2 over a twenty-year period and thirty-four times more potent over the span of a century.[6] The powerful and pernicious effects of methane on the climate have turned the airiness of natural gas into a liability since it leaks continuously into the atmosphere.

The recognition that what is invisible can nonetheless be harmful applies equally to natural gas infrastructure, which lies hidden in the deeply concealed arteries of urban and rural dwellings. This embedded infrastructure surfaces into visibility and public discourse for only two reasons: cooperation or conflict. The development of natural gas infrastructure typically requires cooperation and investment on a grand scale. Upon completion, projects are briefly celebrated but soon thrust into the background of fossil capitalism, of interest only to traders and a handful of engineers. The pipeline is relegated to its intended role as a means for time–space compression,[7] enabling the continuous delivery of the fuel to sites of demand, but hidden from plain sight.

Natural gas projects are also sources of conflict. Attempts to resist the construction of pipelines have mushroomed in recent years, as social movements, frequently led by **Indigenous** nations, have sought to deter the expansion of this network that precipitates global heating. Their success has varied, but they have mobilized broader social support and **solidarity** for environmental resistance in countries from Canada to Italy and Nigeria. Investors may have wanted to keep infrastructure mostly out of sight, but this invisibilization is being reversed by social movements that bring the hidden circulatory system of this elusive fuel to public attention, prompting awareness, criticism, and resistance to existing and proposed projects. In turn, developers have sought to shield their projects from public scrutiny (and thereby avoid contestation) by siting pipelines and other infrastructures for liquified natural gas (LNG) further from reach, in the oceans. The "invisibility of pipelines has long served an ideological function,"[8] whose epitome is offshore operations that benefit capital with even less vulnerability to disruption than on land, so that natural gas might remain the dominant fuel of fossil capitalism during the energy transition and, possibly, after.

This spatial offshoring has occurred simultaneously with discursive attempts to depict natural gas as the cleanest fossil fuel, which can therefore serve as an indispensable bridge to energy transition and a low carbon future; such arguments even promise the development of new technologies to decarbonize this hydrocarbon.[9] Even as opposition to pipelines has grown, natural gas has still not fallen out of favor with companies, governments, and consumers who see natural gas as a way to meet energy demand with a fuel cleaner than coal or oil. Nonetheless, the appeal of natural gas is on the decline. The struggle between climate action and fossil capitalist interests has begun to bring natural gas out of the hidden depths of circuits of production to the forefront of public attention. This contestation has created an increasing awareness that methane leaks pose the risk of accelerating global heating, as well as an emergent recognition that, rather than serving as a bridge to a renewable future, the lock-ins of existing and new infrastructure will perpetuate reliance on fossil fuels for years to come.

Political contestation may take different forms depending on the setting. Countries and regions reliant upon natural gas that can afford to execute an energy transition, such as the EU and the United States, should be pressured to end support for the fuel. It may be permissible for coal-dependent countries to use it as a transition fuel, but only briefly, diligently, and with the consent of the broader populace. Areas without natural gas supplies should be supported to explore renewable alternatives. Countries like Albania and Montenegro, which are considering whether to develop domestic industries, must be offered support to develop alternative sustainable energy systems. Each of these scenarios will demand open political engagement and strategies to make visible the costs of this **resource** that, along with its infrastructure, are too often invisible and overlooked. After all, the ability of the global community to meet warming targets hinges on this task.

See also: **Indigenous Activism, Neoextractivism, Protest**

Notes

1 Elmar Altvater, "The Social and Natural Environment of Fossil Capitalism," *Socialist Register* 43 (2007): 37–59, https://socialistregister.com/index.php/srv/article/view/5857.

2 Daniel Yergin, *The Prize: The Epic Quest for Oil, Money & Power* (New York: Simon & Schuster, 2011).
3 Vaclav Smil, *Natural Gas: Fuel for the 21st Century* (West Sussex, UK: Wiley, 2015).
4 Christopher F. Jones, *Routes of Power: Energy and Modern America*, repr. ed. (Cambridge, MA: Harvard University Press, 2016), 143.
5 John Szabo, "Natural Gas' Changing Discourse in European Decarbonisation," in *Energy Humanities: Current State and Future Directions*, ed. Matúš Mišík and Nada Kujundžić (Cham: Springer, 2020).
6 Robert W. Howarth, "Methane Emissions and Climatic Warming Risk from Hydraulic Fracturing and Shale Gas Development: Implications for Policy," *Energy and Emission Control Technologies* 3 (October 2015): 45–54, https://doi.org/10.2147/EECT.S61539.
7 David Harvey, *The Condition of Postmodernity: An Enquiry into the Origins of Cultural Change* (Cambridge, MA: Blackwell, 1990).
8 Imre Szeman, *On Petrocultures: Globalization, Culture, and Energy* (Morgantown: West Virginia University Press, 2019), 247.
9 John Szabo, "Fossil Capitalism's Lock-Ins: The Natural Gas-Hydrogen Nexus," *Capitalism Nature Socialism* 32, no. 4 (2021): 91–110, https://doi.org/10.1080/10455752.2020.1843186.

NEOEXTRACTIVISM

What comes after neoextractivism?

DONALD V. KINGSBURY

Neoextractivism in Latin America was less a coherent ideology than a moment in the ongoing reconfiguration of state-society-nature. An expansive concept also known as "progressive extractivism," *neoextractivism* refers to left-of-center governments that used revenues arising from commodity booms in the early twenty-first century for expanded social welfare systems, infrastructure investment, and attempts to build South-South solidarities and complementary economies. Neoextractivism was thus also a social-ecological contract. *Somos gente de petróleo*, to take one example from Venezuela: here neoextractivism promised expanded social rights contingent upon the extraction and export of nature.[1]

This contract was shot through with ambivalences. Some of the most extractives-reliant states made the most sweeping and popular denunciations of ecocidal capitalism, (neo)colonialism, and extractivism itself.[2] Common public goods to be expanded included the environmental rights of citizens—freedom from pollution, protection from environmental racism, and equal access to and enjoyment of the natural world—and, in some cases, even proposed legal rights for nature itself.[3] After decades of neoliberal privatization and austerity, neo-extractivism's resuscitation of the developmentalist state entailed a centralization of power, even as it promised direct and participatory **democracy**. Many spoke the language of socialism while deepening macroeconomic dependence on the carbon capitalist world system. Neoextractivism's social-ecological contract negotiated these ambivalences by ignoring them, a resolution that has not survived the commodity busts of the mid-2010s and the deepening climate crisis.

Social contracts are founding fictions of the liberal imagination. They propose a mutually beneficial exchange between individual and collective agents that delineates the spatial and social limits of a polity. Often invoked in moments of origin, crisis, or collapse, social contracts distribute risk and protection while establishing meanings and practices of membership. In Latin America, a region with a long history of collective life along liberalism's "edges,"[4] the neo-extractive era put the often-backgrounded notion of the social contract front and center as states promised a greater share of collective well-being to marginalized populations. It also highlighted, forcefully, that (late) modern social contracts rely upon, and presume, ecological ones.

Specific commodities, practices, and trading patterns varied from country to country and from project to project. In all cases, however, neoextractivist states intensified and revised export-oriented development models established by their predecessors.[5] Oil and **natural gas** exports defined neoextractivism in Venezuela, Ecuador, and Bolivia. Agribusiness—especially soy—expanded the fiscal capacities of administrations in Argentina, Brazil, and Paraguay. When established extractive sectors faltered, due to the inevitable bust of commodity booms or more idiosyncratic internal pressures, new extractive frontiers were pursued. After rescinding an offer to leave oil in the ground in exchange for development funds in 2013, Ecuador supplemented petroleum exports with megamines in the Andes. In 2016 Venezuela opened 12 percent of national territory in the Arco Minero del Orinoco to **mining** exploration for diamonds, gold, and coltan while openly lamenting its dysfunctional petrostate. Argentina, Chile, and Bolivia have discussed the formation of a lithium cartel like OPEC to improve their position as exporters of a critical mineral for decarbonizing energy **transitions**. Overreliance on commodity exports follows the demands of "comparative advantage" long imposed on the region. Neoextractivism's novelties were the increased social orientation of extraction and a democratization of consumption aimed explicitly at righting generations of wrongs.

The neoextractivist turn was couched in a language of national sovereignty and identity aimed against neoliberalism's lost decades. In human terms, neoextractivism's successes were uneven and incomplete but noteworthy. Rates of inequality, poverty, and extreme poverty fell; access to education, health care, housing, and food increased. Reforms meant to increase the scope and practice of citizenship were introduced, particularly in countries that rewrote their constitutions during the neoextractivist moment. Development was again deemed a matter of public policy rather than a by-product of the market's invisible hand.[6] The resulting "commodities consensus," writes Maristella Svampa, reaffirms extraction's inevitability while making a virtue of necessity by tying that extraction to social programs.[7] In neoextractivism, populations were recruited as eager supporters and beneficiaries of extraction while resistance was criminalized.[8] Full citizenship hinged on extraction. True citizens supported the neoextractivist endeavor, contradictions and all. Nature was legally codified in the imagined community, but as a second-class citizen.

Even before the early 2010s commodity collapse, neoextractivism's social-ecological contracts had begun to fray. Tracking the emergence of *extractivismo* as a critical discourse in Ecuador, Thea Riofrancos identifies an intra-left split between "radical resource nationalists" and critics of ecocidal developmentalist models pursued throughout the region.[9] Criticism of extractivism increasingly takes the form of resistance to the developmentalist worldviews it relies upon and reproduces. For Maristella Svampa, this regional "eco-territorial turn" resists the limits of extractivism and neoextractivism, building from the experience of the poor and marginalized—workers, Indigenous peoples, women, and nature itself—who continue to bear disproportionate burdens of resource-export-led development.[10]

The sacrifice of nature for unequal human progress is as old as modernity itself. It is the exclusive purview of neither reactionaries nor leftists. Neoextractivism's variation on this theme emerged in the context of what Arturo Escobar optimistically considered the "last gasp" of Eurocentric modes of development,[11] but what follows could be worse. New social-ecological contracts will be overshadowed by geopolitical realignments, deepening climate crises, decarbonization, and new struggles against inequality in Latin America and beyond in the middle years of the twenty-first century. The contestation that will inevitably accompany these new formational fictions will also, unpredictably, be shaped by neoextractivism's reordering of state, society, and nature.

See also: **Battery**, **Commons**, **Decolonization**, **Extinction**, **Globalization**, **Populism**

Notes

1 Penélope Plaza, *Culture as Renewable Oil: How Territory, Bureaucratic Power, and Culture Coalesce in the Venezuelan Petrostate* (New York: Routledge, 2019).
2 Emiliano Terán Mantovani, *El Fantasma de la Gran Venezuela: Un Estudio del Mito del Desarrollo y los Dilemas del Petro-Estado en la Revolución Bolivariana* (Caracas: Fundación Celarg, 2014).
3 Todd Eisenstadt and Karleen Jones West, *Who Speaks for Nature? Indigenous Movements, Public Opinion, and the Petro-State in Ecuador* (New York: Oxford University Press, 2019).
4 Benjamin Arditi, *Politics on the Edge of Liberalism: Difference, Populism, Agitation, Revolution* (Edinburgh: Edinburgh University Press, 2007).
5 Eduardo Gudynas, *Extractivismos: Ecología, Economía, y Política de un Modo de Entender el Desarrollo y la Naturaleza* (Cochabamba: Centro de Documentación e Información Bolivia, 2015).
6 Steve Ellner, ed., *Latin America's Pink Tide: Breakthroughs and Shortcomings* (Boulder, CO: Lynne Rienner, 2019).
7 Maristella Svampa, "Commodities Consensus: Neoextractivism and Enclosure of the Commons in Latin America," *South Atlantic Quarterly* 114, no. 1 (2015): 67.
8 Teresa Kramarz and Donald Kingsbury, *Populist Moments and Extractivist States in Venezuela and Ecuador: The People's Oil?* (New York: Palgrave, 2021); Thea Riofrancos, *Resource Radicals: From Petro-Nationalism to Post-Extractivism in Ecuador* (Durham, NC: Duke University Press, 2020).
9 Riofrancos, *Resource Radicals.*
10 Maristella Svampa, *Las Fronteras del Neoextractivismo en América Latina: Conflictos Socioambientales, Giro Ecoterritorial, y Nuevas Dependencias* (Bielefeld: Bielefeld University Press, 2019).
11 Arturo Escobar, "Latin America at the Crossroads: Alternative Modernizations, Post-Liberalism, or Post-Development?," *Cultural Studies* 24, no. 1 (2010): 1–65.

NET ZERO

Do we really want to live in a net-zero world?

MIJIN CHA

The idea of net-zero emissions is seductive. Put simply, *net zero* means that overall emissions of climate pollution, such as carbon dioxide and other greenhouse gases, would be zero. Some places would continue to produce emissions, but these would be offset by reductions elsewhere so that net total emissions would be zero. In theory, the idea works perfectly. For instance, while technological advances are needed to decarbonize carbon intensive sectors like the petrochemical industry, total emissions reductions could continue by offsetting these industry emissions elsewhere until the appropriate technology develops. Rich countries could pay developing countries to preserve and/or build forests, a natural carbon sink, which could bring financial resources to the developing world, disincentivize deforestation, and preserve a vital natural resource. Net zero is market-driven climate policy at its finest. Paying for emissions reductions in another part of the world while continuing fossil activity in the developed world allows for Western markets to continue business as usual.

The problem is that net-zero emissions programs not only are failing to preserve forests and reduce emissions but also continue the very practices that caused the climate crisis by valuing profits over people. For developed countries and private companies, net zero allows the fossil economy to continue unabated with the accompanying profit margins. Yet, the reality is that faulty offset programs and problematic emissions accounting mean emissions never reach zero, net or otherwise.

That the climate crisis is driven by greenhouse gas emissions is uncontroverted.[1] There is also widespread agreement that burning fossil fuels is the leading cause of the high concentration of greenhouse gases in the atmosphere. However, the most effective pathways to reduce emissions are hotly disputed. Some advocates focus on scientific and technical aspects: how much emissions must be reduced to avoid climate catastrophe, what types of replacement fuel are needed, and so on. The policy recommendations that follow this line of thinking are technocratic, focused on producing more renewable energy and capturing emissions. Within this world, net-zero emissions make sense. The sole concern is emissions reductions; if they can be bought through offsets, then the end goal is met.

Another approach argues that the climate crisis is the manifestation of exploitative economic practices. Without addressing these practices, little to no advancement can be made toward meaningful emissions reductions. In this case, the policy recommendations stipulate not only how much renewable, **clean** energy is produced, but also how it is produced and for whom. For example, solar energy produced with exploited labor and environmentally destructive practices is not desirable even though the fuel source has changed. In this world, net-zero emissions continue the likelihood of harmful practices because these injustices are not factored into emissions accounting.

And, as it turns out, existing net-zero policies do continue extractive and exploitative practices—without the emissions reductions promised. Forests are the main source of economic livelihood for many communities globally, and this reality runs in direct conflict with offset programs. An investigation by independent media outlet ProPublica analyzed offset and carbon credit projects over two decades and found that "the polluters got a guilt-free pass to keep emitting CO_2, but the forest preservation that was supposed to balance the ledger either never came or didn't last."[2] Selling forested land for offset programs neither stopped nor stemmed deforestation because its causes are complex and not addressed through offsets. Without offering other economic opportunities for countries and communities reliant on forests, offset programs pay for temporary carbon storage. As trees are cut, their role as carbon sinks ends, and the carbon they captured is released back into the atmosphere. Net zero works only if the carbon is permanently offset, not just stored temporarily.

In addition to not reducing emissions, offset programs have violated human rights around the world. ProPublica analyzed programs dating back to the first global offset effort under the Kyoto Protocol's Clean Development Mechanism, where the European Union eventually stopped accepting most credits due to the technical and human rights scandals that arose from the program. Globally, Indigenous and local communities are evicted from land and face violence, even death, in the name of offset programs. An analysis of carbon offset programs in Panama found the destruction or displacement of ancestral sites and a lack of protections for local and **Indigenous** communities.[3] An offset effort by Norwegian companies led to villagers in Uganda being denied access to land to grow food and graze livestock, both essential to their subsistence livelihood.[4] Villagers were evicted from their land and faced physical violence from police and private security forces. Twenty-three local farmers in Honduras were murdered when they tried to recover land illegally sold to big palm oil plantations used as part of the European Union's carbon credit scheme.[5]

If net zero is not the way, what other emissions reductions paths exist? The answer is deceptively simple: direct emission reductions. In an evaluation of the first Kyoto Protocol's Joint Implementation commitment period (2008–2012), research found that 600 million tons of global carbon dioxide emissions could have been avoided if emissions were reduced on-site, rather than relying on offset projects.[6] Reducing potential carbon dioxide emissions where they would have been produced, rather than trying to offset them elsewhere, results in *actual* emissions reductions. It also improves the health of communities living next to carbon polluting activities by curtailing the pollution. Direct emissions reductions allow a path from the extractive, exploitative present to a more just, regenerative future.

See also: **Carbon Management, Gaslighting, Indigenous Activism, Justice, Nonlinear, Solidarity**

Notes

1 Scientific consensus overwhelmingly agrees with this assertion, and I do not engage with the small fraction of climate deniers who may oppose this assertion.
2 Lisa Song, "An (Even More) Inconvenient Truth: Why Carbon Credits for Forest Preservation May Be Worse Than Nothing," *ProPublica*, May 22, 2019, https://features.propublica.org/brazil-carbon-offsets/inconvenient-truth-carbon-credits-dont-work-deforestation-redd-acre-cambodia/.
3 Mary Finley-Brook and Curtis Thomas, "Renewable Energy and Human Rights Violations: Illustrative Cases from Indigenous Territories in Panama," *Annals of the Association of American Geographers* 101, no. 4 (2011): 863–72, http://www.jstor.org/stable/27980233.
4 Oakland Institute, "The Darker Side of Green: Plantation Forestry and Carbon Violence in Uganda" (October 23, 2016), https://www.oaklandinstitute.org/darker-side-green.
5 *Guardian*, "EU Carbon Credits Scheme Tarnished by Alleged Murders in Honduras," October 3, 2011, http://www.theguardian.com/environment/2011/oct/03/eu-carbon-credits-murders-honduras.
6 Anja Kollmuss, Lambert Schneider, and Vladyslav Zhezherin, "Has Joint Implementation Reduced GHG Emissions? Lessons Learned for the Design of Carbon Market Mechanisms" (SEI Working Paper No. 2015-07, 2015), 128.

NONLINEAR

Can nonlinear imaginative practices help us survive nonlinear catastrophes?

KATY DIDDEN

To begin his 2018 Nobel Prize lecture, economist William Nordhaus projected an image of Francisco de Goya's *The Colossus*. In the foreground of that shadowy landscape, people and horses flee down the valley as a muscled giant erupts across the stratosphere. In his account of the Dynamic Integrated Model of Climate and the Economy for which he won the Nobel, Nordhaus turned to metaphor to convey an emotional truth about rising temperatures: "I think of climate change as a menace to our planet, to our future. . . . And so, like the people huddling there, some of us are very frightened by the prospect."[1] If, for Nordhaus, *The Colossus* embodies climate fear, others interpret the figure differently, "as a symbol of Napoleon, of the whole of Humanity, and of War; as Prometheus; as a Spanish Rain-God."[2] When Nordhaus's interpretation is juxtaposed with these others, climate change emerges as a metaphor of metaphors that reveals additional dimensions of our predicament. What we fear when we fear climate change is all of it: war, drought, dictators, the devil, panic, storms, revolution, and the impact of these threats on humankind's current condition and future prospects.

Nordhaus introduced the idea of 2°C as a climate tipping point in a 1975 working paper published for the International Institute for Applied Systems Analysis.[3] While it has become a rallying cry, 2°C originated as an economic theory with an economic solution. Nordhaus theorized that, if average global surface temperatures rose by 2°C or 3°C, the costs of mitigating climate disasters would devastate the global **economy** and therefore threaten the survival of the human species. For him, the fix was to price fossil fuel emissions to reflect their actual social cost, although he thought this measure would, for political reasons, be unlikely to be adopted. He lamented in 2018 that "we're in for changes in the Earth's system that we can't begin to understand,"[4] because as average surface temperatures approach 2°C, the likelihood increases of "nonlinear responses" with "no simple proportional relation between cause and effect," such as "a dramatic reorganization of the thermohaline circulation, rapid deglaciation, or massive melting of **permafrost**."[5] If a climate Colossus now stands ready to squelch both humans and **animals**, we may have to confront it from within the unfamiliar space-time of nonlinearity, beyond the linear cause-and-effect paradigms on which we have grown to depend.

As a poet, I study nonlinear forms. A poem's primary structure is song, not story; poems accommodate fragmentation and gaps, link unlike things, and progress by rhythm and repetition. Nonlinear thinking can reframe how we challenge climate indifference and climate fear. Consider the power of nonlinear language in relation to another giant—coral, as imagined by poet Alexis Pauline Gumbs:

> *what the coral said*
>
> once. we were all singing. somewhere. we are still. moving. as something huge, vibrational, wet. we dance and keep the world in place. we shiver and know the orbit. if you let the body undulate you will remember. not all the waves are in the ocean.[6]

Gumbs's "what the coral said" concludes the eighteen-part "Anguilla" (which explores the complex, geopolitical history of that island) in Gumbs's *Dub: Finding Ceremony*. Each poem in *Dub* references an essay by **Black** feminist theorist Sylvia Wynter, in this case "Ethno? Or Socio Poetics," in which Wynter argues that, in order to justify conquest, slavery, and the exploitation of environmental resources in "the new world," settler colonials profoundly narrowed the definition of what it means to be human. For Wynter, poetry is a counterforce to the colonial ideologies pervading the economic world system.[7] In a later interview, Wynter connects poetry's emancipatory potential to nonlinearity: "I was always aware that it wasn't that I was thinking anything linearly. . . . It has to do with this beginning to question your own *consciousness* . . . a new poesis of being human."[8]

I find evidence of this new poesis of being human when I read *Dub*. Gumbs reaches "to every dawn of existence" to ask, "What if what we believe is required of humans by nature is just a story that we told ourselves about what being human is and what nature is?"[9] Via "interspecies ancestral listening,"[10] Gumbs transforms the old stories, channeling human voices among a chorus of stars, salt, seaweed, coral:

> if you let the body undulate you will remember. not all the waves are in the ocean. we don't know so much about the soloists. we don't know so much about virtue. we don't care so much about your body. it's *the* body. you are already part of it because *you* has nothing to do with it.[11]

Gumbs disrupts the hierarchical cues of punctuation and syntax, forgoing capitalization and subordination: phrases accumulate paratactically, like polyps. If it's hard to conceive of 2°C outside the rising action of deadlines and doomsday clocks that make an apocalypse feel inevitable, Gumbs unhitches time by bending form; by creating new rhythms and new avenues for thought; by remixing poetry, theory, history, and prayer. To read this work is to know that all forms—infrastructures, economies, institutions, governments—are "already part of it," and are therefore re-remixable. Set against the timescales of coral, human empires are a blip.

And this image of the coral might not be so far from Goya's giant after all; historically, the Colossus is a revolutionary symbol of the unified masses.[12] What we fear when we fear climate change could well be our own collective power. To know the true cost of fossil fuels, humans must recognize that we're part of a global organism: we are already Colossus and already coral.

Right now, rising ocean heat is inducing coral bleaching; even before we've reached 2°C, ghost-white reefs communicate a rupture, a break with the human past. As this loss compels poets to sing coral into collective memory, coral teaches us what we are and will be: ones who salvage, who regenerate, whose survival depends on becoming shelter for each other. Who are we who listen? Can we be remade?

dance into harmony now. now is already now. time has nothing to do with it. time is up. time is over. we are love in all directions. come on, sing.[13]

See also: **2040, Extinction, Settler Colonialism, Storytelling, Transitions**

Notes

1 William D. Nordhaus, "Climate Change: The Ultimate Challenge for Economics" (Nobel Prize lecture, December 8, 2018), https://www.nobelprize.org/uploads/2018/10/nordhaus-lecture.pdf.
2 Nigel Glendinning, "Goya and Arriaza's Profecia del Pirineo," *Journal of the Warburg and Courtauld Institutes* 26, no. 3/4 (1963): 363–66.
3 William D. Nordhaus, "Can We Control Carbon Dioxide? (From 1975)," reprinted in *American Economic Review* 109, no. 6 (2019): 2015–35. The original working paper can be found at https://pure.iiasa.ac.at/id/eprint/365/1/WP-75-063.pdf.
4 Carol Davenport, "After Nobel in Economics, William Nordhaus Talks about Who's Getting His Pollution-Tax Ideas Right," *New York Times*, October 13, 2018.
5 Intergovernmental Panel on Climate Change, "nonlinearity" and "rapid climate change," in "Annex B: Glossary of Terms," in *Climate Change 2001: Synthesis Report* (Cambridge: Cambridge University Press, 2001).
6 Alexis Pauline Gumbs, *Dub: Finding Ceremony* (Durham, NC: Duke University Press, 2020), 65.
7 Sylvia Wynter, "Ethno?, or Socio Poetics," *Alcheringa/Ethnopoetics* 2 (1976): 88.
8 Sylvia Wynter, "Proud Flesh Inter/views: Sylvia Wynter," interview by Greg Thomas, *ProudFlesh*, no. 4 (2006): 1–35.
9 Gumbs, *Dub*, xi.
10 Gumbs, xiii.
11 Gumbs, 65.
12 Samantha Wesner, "The Revolutionary Colossus," *Public Domain Review*, December 10, 2020, https://publicdomainreview.org/essay/revolutionary-colossus/.
13 Gumbs, *Dub*, 65.

NORMAL

What will be normal after fossil fuels?

STEPHANIE LeMENAGER

Our minds are still racing back and forth, longing for a return to "normality," trying to stitch our future to our past and refusing to acknowledge the rupture. But the rupture exists. And in the midst of this terrible despair, it offers us a chance to rethink the doomsday machine we have built for ourselves. Nothing could be worse than a return to normality.

—Arundhati Roy, "The Pandemic Is a Portal"

Most connotations of *normal* imply the prescriptive, even violent practices of the normative. Normativity does the work of biopolitical sorting; it outlaws those who fail to conform, for instance the atypical, the foreign, and the queer. Normativity enacts hegemonies. In so doing, it encourages both quietism and resistance. Since what is conceived as normal is not only normative but also ordinary, common and everyday, resistance lives—if modestly—inside the normal. The economic, infrastructural, and social forms of petronormativity become increasingly broken and surreal, as the climate crisis delivers its referendum on modern energy systems and the lifeworlds they sustained. In the era of climate change, petronormativity begins to feel provincial. More and more, everyday lives contradict the normative and hegemonic.

Across media channels in the wealthier world, variants of the question "is this the new normal?" are asked about fires, floods, and droughts hitherto conceived as 100- or 500- or 1,000-year events; political instability; social unrest; and faltering infrastructures. To elaborate Arundhati Roy's portal metaphor: in pandemic, as in climate collapse, the normal cracks open, over so many bodies.[1] *Who, if anyone, will pass through, and what worlds will they create?*

The normal indicates the location of power and, with it, epistemological dominance. Yet the plurality of everydays living under cover of the normal expresses a plurality of epistemologies, ontologies, and practices. Cultural theorists Nigel Clark and Kathryn Yusoff suggest that "artisanal pyrotechnologies" like traditional agricultural burning subtend modern cultures of energy and are more important to the aftermaths of modernity (that is, to collective futures) than the internal combustion engine.[2] Indigenous fire-tending, once criminalized by settler states, is belatedly recognized by settlers as a means of designing forests that are resistant to megafire.[3] The traditional knowledges of Indigenous Australian women who recently were called upon to predict the location of bushfire are everyday practices keyed to fundamental geologic energies like drought and fire. The Indigenous everyday of a particular nation bends toward its own future and sovereignty. The energetic practices of **Indigenous** nations are not fungible for appropriation, in other words. Let them function here as reminders of how alternative everydays persist alongside petronormativity.

So, too, Afrofuturism recognizes **Black** futures growing out of energetic practices at odds with racist and colonial norms. Historian J. T. Roane traces the ways in which Black diasporic communities in the United States have ingeniously worked the earth and sea, engaging in interstitial **commons**—gardening, gathering, aquacultures—illegible to white supremacy and its forms of property.[4] By virtue of cultural performances like code-switching, the normal can carry incommensurable modes of living the everyday, so that the normative—that prescriptive edge of the normal—becomes infiltrated by difference. Through resistant modes of the everyday, energetic practices distinct from petronormativity come into view. In time, these practices may sustain new solidarities, post-oil ways of life and forms of attachment that transgress political affiliation. *Who will pass through?*

To live in a nonnormative body day by day is to better know the present. **Greta** Thunberg recognizes the gift of insight afforded her by her neuroatypical experience on the autism spectrum as a bridge beyond provincial normativities. Thunberg voices the stakes of the climate crisis with an "uncompromising clarity," in the words of Russian American author and activist Masha Gessen, who speculates that Soviet dissidents also may have lived on the spectrum.[5] Conversely, those whose somatic experience conforms to the dictates of the normative and its cultural atmospheres of hegemony strain to see the portal. The unsustainable love for oil and coal that fueled the twentieth century—the "American century"—still flourishes in myriad cultural instantiations and infrastructures worldwide. In the United States, the white working classes are encouraged—gaslighted—to cast their lots with nihilistic industries that once served them, such as coal, oil, and Big Ag. Named for gas-fueled **electricity** and indicating a mode of deception where the dupe is encouraged to disbelieve their own experience, **gaslighting** in the era of climate collapse describes the cultural mass delusion that fossil fuels are irreplaceable. Many fear losing what they've come to expect of the everyday, even when it's no good. Billionaire CEOs entertain and distract as they circle distant stars.

How to pass through? Intellectuals tend to foreground grassroots movements and the community projects that Ashley Dawson calls "energy **commons**," over scalable planning.[6] José Luis Baerga Aguirre's and Catalina de Onís's *El Poder del Pueblo: una lucha colectiva por la vida y el medioambiente* (2021), a **documentary** about a Puerto Rican neighborhood's struggle to free themselves from the state grid, offers a compelling case study of how small communities can decolonize infrastructure.[7] (However, the shadows of corporate and U.S.-state power hover, offscreen.) For those who conceive the state as an engine of progress, Green New Deals (GNDs) in the United States, Canada, Ireland, and elsewhere seek to reinvent state power as a bulwark against corporate pillage and endless, ecocidal growth. The charisma of U.S. senator Bernie Sanders and representative Alexandria Ocasio-Cortez has contributed to global excitement about GNDs as a means by which democratic states can shepherd the world toward **renewable** energy and sustainable, working-class jobs. At varied scales, aspirations to dismantle petronormativity recognize the need to repair economic and social relations, to counter racism and the injuries of class.

Genuinely *to pass through* requires both fast technical change and complex, long-term commitments.

Shocks to normal routines and expectations of the everyday feel like—and may be lived as—disasters. The portal out of normativity, on the other hand, sounds like liberation, an affirming and deliberate transition. The imaginative work of sorting out normativity from the everyday, and loosening attachments to the normal that hold them in tension, is preparation for survival. Grids fail; the prices of fuels skyrocket; autocrats seek legitimacy in fear; climate, sickness, and war unhouse us; and, still, we each face choices about how to let the normal shatter all around us—and how to pass through.

See also: **Fire/Bushfire, Gender, Green New Deal, Local, Trans-**

Notes

1 Arundhati Roy, *Azadi: Freedom, Fascism, Fiction* (Chicago: Haymarket, 2020), 191.
2 Nigel Clark and Kathryn Yusoff, "Combustion and Society: A Fire-Centred History of Energy Use," *Theory, Culture, and Society* 31, no. 5 (2014): 203.
3 Kari Marie Norgaard, *Salmon and Acorns Feed Our People: Colonialism, Nature, and Social Action* (New Brunswick, NJ: Rutgers University Press, 2019), 72–128.
4 J. T. Roane, "Plotting the Black Commons," *Souls* 20, no. 3 (2018): 239–66, https://doi.org/10.1080/10999949.2018.1532757.
5 Masha Gessen, "The Fifteen-Year-Old Climate Activist Who Is Demanding a New Kind of Politics," *New Yorker*, October 2, 2018, https://www.newyorker.com/news/our-columnists/the-fifteen-year-old-climate-activist-who-is-demanding-a-new-kind-of-politics.
6 Ashley Dawson, *People's Power* (New York: O/R Books, 2020), 135–63.
7 Baerga Aguirre, José Luis, director. *El Poder del Pueblo: una lucha colectiva por la vida y el medioambiente*. Baerga Aguirre, José Luis and M. de Onís, Catalina, producers. 2021. 41 min. https://www.youtube.com/watch?v=n4TyYKFJGec.

NORTH DAKOTA

How does oil extraction reshape space and time?

KYLE CONWAY

It is one thing to say that the Bakken Oil Formation lies beneath a few counties in northwestern North Dakota and that this region has been the site of three oil booms (in the 1950s, the 1980s, and from 2008 to 2014, the last fueled by hydraulic fracturing). It is quite another to say that these facts define the time and space of oil extraction. The same land was and is occupied by members of the Oceti Sakowin, the Great Sioux Nation. The chronology of oil booms in North Dakota presumes a settler "chronotope," literary theorist Mikhail Bakhtin's term to describe "the intrinsic connectedness of temporal and spatial relationships."[1] The opposition to the Dakota Access Pipeline led by members of the Standing Rock Sioux Reservation in 2016 demonstrates that other chronotopes are possible, ones that call into question the conventional space-time of oil booms and the narratives underpinning them.

Let's begin with the settler chronotope. Its origins are in European Enlightenment thought and manifest in efforts to rationalize the organization of social, political, and economic life. The settler chronotope treats time as linear and space as continuous; it understands both as divisible into discrete, measurable units. History understood in this way consists in chronology: the first attempt to drill a well in North Dakota took place in 1916, but it was dry. The first successful well was drilled in 1951. By 1954, there were more than 400 active wells; by 1987, 3,500; by 2017, 15,000.[2] The effects of the 1950s boom spanned two counties, whereas the effects of the 2008–14 boom spread across the state.[3]

This chronology, however, raises an important question. Why were settlers in North Dakota in the first place? Most arrived after the 1862 Homestead Act, which divided land into 160-acre quarter-sections that settlers came to own after living on them for five years. Their legal right to mine their land (or sell or lease their mineral rights—a necessity for oil extraction) was established in three legal decisions, two by the Pennsylvania Supreme Court (*Turner v. Reynolds* in 1854, *Westmoreland v. Dewitt* in 1889) and one by the U.S. Supreme Court (*Kinney Coastal Oil v. Kieffer* in 1928).

But the laws that allowed settlers to occupy the territory were themselves the result of a long process of dispossession of Indigenous lands. In 1851, for instance, the U.S. government signed a treaty with Indigenous leaders at Fort Laramie intended to ensure settlers' safe passage across the region bounded by the Missouri River, the Rocky Mountains, and the borders of Texas and New Mexico, in exchange for annual payments of $50,000. After the U.S. government stopped its payments in 1862 (an act that precipitated the U.S.-Dakota War), it negotiated a new treaty in 1868, again at Fort Laramie. The treaty established new, smaller boundaries for the Great Sioux Reservation and designated the reservation as the "permanent home" of the Sioux, "unceded Indian territory" upon which "no white person . . . [would] be permitted to settle" (articles 15 and 16).

This narrative is marked by the settler chronotope, organized around a particular logic of cause and effect: the Fort Laramie treaties and the Homestead Act made settlement possible; court cases from 1854 to 1928 made oil extraction profitable; and the Bakken oil reserves made North Dakota's booms possible.

But what of competing notions of time and space? In 2016, the Dakota Access Pipeline (DAPL), which was to carry oil from the Bakken Oil Formation to a refinery in Illinois, garnered intense **protest** because it passed within a mile of the Standing Rock Sioux Reservation. The **pipeline** threatened Standing Rock's water supply by going under Lake Oahe, which was created in the 1950s after the U.S. government, as part of the Pick-Sloan Missouri Basin Program, flooded the river basin where the Sioux were living. In September 2014, Standing Rock representatives, including Chief David Archambault II, met with DAPL representatives to object to the proposed route, noting that it had been changed to avoid the (predominantly white) state capital, Bismarck. DAPL executives later insisted that no members of Standing Rock had participated in the consultation process, and their purported nonparticipation became a pretext for dismissing the 2016 protests.[4]

During those protests, Archambault published an op-ed in the *New York Times* outlining Standing Rock's objections. The federal government, he said, was taking "lands and resources . . . without regard for tribal interests," as it had done when "the Sioux peoples signed treaties in 1851 and 1868. The government broke them before the ink was dry."[5] The pipeline project violated these treaties, including the 1868 clause recognizing the territory as unceded. Archambault's objection reflects, in Nick Estes's description, "Indigenous notions of time [that] consider the present to be structured entirely by our past and by our ancestors"; these ideas of time read the past into the present and project the present into the future.[6] Such notions are inextricable from ideas of space that privilege proximity over borders. Archambault objected to DAPL because Standing Rock faced the greatest risk in case of a leak. DAPL executives, following a settler spatial logic, argued that Standing Rock had no standing to object because the pipeline passed outside of the reservation's borders.

The settler chronotope has both contributed to and resulted from settlement. Policies enacted by the U.S. government, such as the Homestead Act and the Fort Laramie agreements, were premised on this understanding of time and space, but these policies also helped to forge the chronotope, writing its structure into **law**. If people in North Dakota and the United States more broadly want to address the harms caused by extraction, as anti-DAPL protesters would have them do, they must work to untangle this complicated relationship.

See also: **Autonomy, Civil Disobedience, Indigenous Activism, Settler Colonialism**

Notes

1 Mikhail M. Bakhtin, *The Dialogic Imagination*, trans. Caryl Emerson and Michael Holquist (Austin: University of Texas Press, 1981), 84.
2 Kyle Conway, ed., *Sixty Years of Boom and Bust: The Impact of Oil in North Dakota, 1958–2018* (Grand Forks: Digital Press at the University of North Dakota, 2020), 21, 26.
3 David Flynn, "The Economic Consequences of Oil Development," in Conway, *Sixty Years of Boom and Bust*, 223–44.
4 Nick Estes, *Our History Is the Future: Standing Rock versus the Dakota Access Pipeline and the Long Tradition of Indigenous Resistance* (London: Verso, 2019), 42–47.
5 David Archambault II, "Taking a Stand at Standing Rock," reprinted in Nick Estes and Jaskiran Dhillon, eds., *Standing with Standing Rock: Voices from the #NoDAPL Movement* (Minneapolis: University of Minnesota Press, 2019), 39.
6 Estes, *Our History Is the Future*, 14.

ONLINE

When we are online, what lines are we on?

ANNE PASEK

The meanings of the word *online* are less narrow than we usually imagine. Though its use skyrocketed with the advent of the internet, "online" messages and practices can be found in early twentieth-century discussions of the telegraph's electric signals as well as the linear paths of the railway. To be online is to be served by and connected to fixed networks, whether they dispatch messages or material goods, and whether the line in question is optical fiber, insulated copper, or steel track. The word speaks not to a technology but to an infrastructural condition.

This history is more than just linguistic; the lines that we are on frequently run along and on top of each other. Telephone poles dot railroads, while **digital** cables run along highways. There lines are often joined with energy infrastructures that, directly or indirectly, power such efforts. High-speed fiber gets buried along the easement for a **pipeline**, while high-voltage pylons stand above them both, running to and from always-on data centers. The new gets grafted onto the old; past paths become present dependencies.[1]

These entwined spaces and definitions point to a shared logistical strategy and social ideal. Putting things on lines—whether bits of information, tons of freight, or Watts of **electricity**—serves to dramatically lessen the work of transportation. In the rolling resistance of locomotion and the electrical speed of signal traffic alike, lines undo the demands of distance by compressing space.[2] Borrowing from the immense energy potential of fossil fuels, or just the speed of light on a wire, lines connect across spaces that, by virtue of such networks, become practically and politically much less important to the sender and receiver. Lines thus work to make distinctly modern social relations: more a collection of nodes than a shared locality. It is not that lines offer linear directness, so much as they seemingly erase the need to consider path and place altogether.

These forgotten environments, however, remain vital to the functioning of such systems. In some cases, topographies are overcome through feats of technical skill: mountainsides flattened and valleys bridged to make way for the line. In others, unruly watersheds may need to be ecologically and politically coaxed to flow before electrical, data, and shipping lines can.[3] Alternatively, remoteness may serve as an insulating strategy, protecting transmission lines from human disruption just as plastic sheathes a cable.[4] The ecologies of lines are therefore a constitutive part of such systems, even if the end effect is to obscure these relations.

To be online, therefore, presents a series of paradoxes. It is to be dazzled by the new, while traversing the old. It is to be networked in place, but also unmoored. It is to overcome time, space, and environments, while also depending on the labor and maintenance of ecological engineering. Like all modern infrastructures, lines are assumed to be mostly invisible, structuring the shape and agency of polities, but rarely at the center of their demands. To be on a line is not to dwell within one, to take it as a subject and site of politics.[5]

Yet, in the face of an urgently needed energy transition, an inversion may be looming. A move toward intermittent, renewable sources of power will challenge the easy compression of communication and distribution lines. Instead of the spatially agnostic dispatch of data, goods, and energy through a network, the variable presence of sun and wind will likely contour transmission paths and dynamically shift loads, reintroducing constraints formerly overcome by fossil modernity. This is not to say that lines will be undone by such a transition. To the contrary, we may need them even more: distributed generation grids, ever-more virtual gatherings, and on-demand manufacturing imply a proliferation of new logistical lines and an expanded politics of infrastructure.

What will lines mean in such a future? This is an undecided question, and so also a promissory one. In visions of the **Green New Deal**, in experimental networks, and in **Indigenous** land and water protection movements, we find the unfolding of spaces and times that lines have long compressed. They prompt us in part to ask: what if lines were built not only for transmission, but also for living and leisure, featuring linear parks that bridge urban and rural landscapes?[6] How might community-controlled networks get more people on screens, but also outside, locating the lines of sight and solar panel placements needed to make different kinds of connectivity?[7] Could we learn to stream video while the wind blows, and pursue less energy- and data-intensive pleasures when the turbines are still?[8] And what if Indigenous data sovereignty and land sovereignty were thought and practiced together, such that consent was meaningfully negotiated for the transit of such lines, so that they in turn served to connect and benefit the nations they crossed?[9]

Lines, in this future, might invite new modes of dwelling within the spaces of interconnection. This outcome is not guaranteed and will manifest only through political struggle. But to imagine such lines, and the alter-modernities they suggest, is also to imagine the coalitions and positive political projects that might be sufficient to build the power necessary to win them. I hope to meet you on—and around—such lines.

See also: **Community, Design, Globalization, Nonlinear, Transitions**

Notes

1 Tung-Hui Hu, *A Prehistory of the Cloud* (Cambridge, MA: MIT Press, 2015), 8.
2 James W. Carey, *Communication as Culture: Essays on Media and Society* (New York; London: Routledge, 1992), 203; Nicole Starosielski, "Pipeline Ecologies: Rural Entanglements of Fiber-Optic Cables," in *Sustainable Media: Critical Approaches to Media and Environment*, ed. Nicole Starosielski and Janet Walker (Durham, NC: Duke University Press, 2016), 38–55.
3 Ashley Carse, *Beyond the Big Ditch: Politics, Ecology, and Infrastructure at the Panama Canal* (Cambridge, MA: MIT Press, 2014); Mél Hogan, "Data Flows and Water Woes: The Utah Data Center," *Big Data & Society* 2, no. 2 (2015): 1–12.
4 Nicole Starosielski, *The Undersea Network* (Durham, NC: Duke University Press, 2015), 19.
5 Tim Ingold, *Lines: A Brief History* (London: Routledge, 2007), 89.
6 Nicolas Pevzner, "The Green New Deal, Landscape, and Public Imagination," *Landscape Architecture Magazine*, July 2019, 68–81.
7 Rory Solomon, "Meshiness: Mesh Networks and the Politics of Connectivity" (PhD diss., New York University, 2020), 263, https://search.proquest.com/docview/2408892960/abstract/FF1AE2EF59B43CCPQ/1; Tega Brain, Alex Nathanson, and Benedetta Piantella, "Towards a Natural Intelligence" (Solar Protocol, 2020), http://solarprotocol.net/manifesto.html.

8 Lu Ye, "Design for Carbon-Aware Digital Experiences," *Branch Magazine*, Autumn 2020, https://branch.climateaction.tech/2020/10/11/design-for-carbon-aware-digital-experiences/.
9 First Mile Connectivity Consortium, *Stories from the First Mile: Digital Technologies in Remote and Rural Indigenous Communities* (Fredericton, PEI: FMCC, 2018).

ORGANIZE

Can the master's tools dismantle the master's house?

ZAINAB ASHRAF, RACHEL KRUEGER, GUY BRODSKY, LESLEY JOHNSTON, AND JOY HUTCHINSON

We are members of Fossil Free UW (FFUW), the student-led fossil fuel divestment chapter at the University of Waterloo (UW) in Canada. We represent just one of the more than one thousand campaigns for divestment at academic institutions worldwide, which invoke both environmental and financial reasons in calling for the withdrawal of investments in fossil fuel companies.[1] Since 2015, FFUW has urged UW to divest its endowment and pension funds from fossil fuel company holdings. After six years of campus-community organizing, we succeeded. Organizing was not without its challenges at our neoliberal, relatively apolitical institution, whose power dynamics discourage student agency. Our largest obstacle was these deeply entrenched neoliberal values, which dismiss ideas incompatible with a market approach.[2] UW agreed to divest only when they became convinced of climate-related financial (rather than environmental) risk. Our experience raises an important question for climate **action** and energy politics: how can organizers work within a neoliberal system to subvert the values of market capitalism and thereby address the existential problems of our time?

We used a combination of grassroots and insider organizing to put divestment on the agenda of the UW Board of Governors. FFUW's first grassroots campaign was mounted in November 2015. The idea was to organize those most likely to support our campaign: we urged students in the Faculty of Environment to vote to divest their student endowment fund. The decision was unanimously approved; professors across the university subsequently penned an open letter to the board urging that the university divest. By 2021, over four hundred faculty and staff had signed the letter and more than two thousand students had signed a divestment petition.

FFUW gained broader support in 2019, when many of the fifteen hundred students attending the Global Climate Strike called for divestment. As the movement grew, FFUW sought collaborators. In February 2021, a webinar on climate **justice** and the fossil fuel industry was cohosted by FFUW and RAISE (Racial Advocacy for Inclusion, Solidarity, and Equity), a student-led advocacy group that addresses racism and xenophobia on campus. Speakers highlighted the colonial violence and neoliberal capitalism endemic to the fossil fuel industry; they also identified a Eurocentric narrative underwriting the divestment movement itself. Learning from our allies, FFUW sought to affirm the intersectional values of climate justice and to make our movement inclusive.

As for insider organizing, FFUW strategically leveraged financial arguments for divestment. When the provincial government urged universities to consider the environmental, social, and governance (ESG) impacts of their investments, FFUW mounted a campaign to include climate change as a consideration under UW's ESG commitments in 2017–18. We leveraged UW's long-held title as one of Canada's "most innovative" universities to pressure the institution to be a sustainable investment leader.[3] The board relented, and UW became a signatory of the United Nations' Principles of Responsible Investment in 2020. As relationships developed, three FFUW members were invited to participate in the Finance Committee's Responsible Investment Working Group. In collaboration with committee members, FFUW helped draft a Carbon-Neutral Investment Policy in 2021.

Finally, on June 1, 2021, the board voted to divest the university's fossil fuel holdings and to adopt the Carbon-Neutral Investment Policy,[4] which aims to decrease the carbon exposure of UW's investment portfolio by 50 percent by 2030, achieving carbon neutrality by **2040**. Importantly, UW will aim to have no direct investments in companies exploring or extracting fossil fuels by 2025, effectively divesting its pension and endowment funds from the fossil fuel industry.

Divestment was adopted at UW primarily for financial reasons: climate risk and stranded assets will yield monetary losses. Though we succeeded in changing board members' stance on divestment, we recognize that change was allowed because it accorded with neoliberal logic. Our concerns are echoed by Henry Giroux, who believes that the modern university is a "neoliberal university," wherein the indicator of success is the market, rather than fostering engaged citizens committed to the public good.[5] UW's entrenchment in the neoliberal economic model will continue to impede meaningful action on the wicked problems of our time. However, our divestment win does limit UW's potential to harm the climate through their investments, even though the university still invests in associated industries.

While we could not dislodge the neoliberal perspective that holds sway at UW, we learned much from our struggle. Our pursuit of climate action through organizing created lasting change by pushing those in power to adopt meaningful policy. By engaging in this work, we take ownership over our **education**. We also prepare ourselves for thoughtful engagement with neoliberalism beyond the university. We only wish UW would do more to encourage this type of learning in the classroom, rather than requiring us to learn these lessons by confronting its own harmful policies and practices.

See also: **Civil Disobedience, Finance, Industrial Revolution**

Notes

1 Global Fossil Fuel Divestment Commitments Database, “Divestment Commitments,” accessed June 29, 2021, https://gofossilfree.org/divestment/commitments/.
2 Mathieu Blondeel, “Taking Away a ‘Social Licence’: Neo-Gramscian Perspectives on an International Fossil Fuel Divestment Norm,” *Global Transitions* 1 (2019): 200–209, https://doi.org/10.1016/j.glt.2019.10.006.
3 *Maclean's*, “Canada's Best Universities by Reputation: Rankings 2022” (last modified October 7, 2021), https://www.macleans.ca/education/canadas-best-universities-by-reputation-rankings-2022/.
4 University of Waterloo, “University of Waterloo Commits to Reduce Carbon Footprint of Its Pension and Endowment Investments 50 per Cent by 2030, Achieve Net-Neutral by 2040” (last modified June 1, 2021), https://uwaterloo.ca/news/media/university-waterloo-commits-reduce-carbon-footprint-its.
5 Henry Giroux, “Neoliberalism, Corporate Culture, and the Promise of Higher Education: The University as a Democratic Public Sphere,” *Harvard Educational Review* 72, no. 4 (2002): 425–64, https://doi.org/10.17763/haer.72.4.0515nr62324n71p1.

PARIS AGREEMENT

Does the Paris Agreement still matter?

AMY JANZWOOD

In 2015, the Paris Agreement committed its signatories to limit long-term warming to “well below 2°C above preindustrial levels,” and ideally to below 1.5°C. Instead of negotiating commitments to meet an overall goal—like a global carbon budget—the Paris Agreement shifted responsibility to countries through voluntary pledges, known as nationally determined contributions (NDCs).[1] Although the Paris breakthrough would not have been possible without this compromise, it has led to immense problems. As the 2021 UN Emissions Gap Report makes clear, total greenhouse gas emissions reductions pledged in NDCs are not enough to meet the 1.5°C target. To make matters worse, most countries are far from being on pace to meet their NDCs. Meanwhile, the all-too-familiar pattern of backsliding on government commitments continues.

Given these complications, responsibility has largely fallen on civil society to "reveal the bullshit in states' policies and practices," as political scientist Hayley Stevenson observes.[2] UN climate summits—particularly the 21st Conference of the Parties (COP21) in Paris—are opportunities for civil society groups to gain visibility and traction. COP21 attracted a record number of NGOs, many of whom linked issues such as labor, **gender**, and **Indigenous** rights to the negotiations.[3] However, COP21 also presented a dilemma for NGOs frustrated with the incremental approach of the negotiations or hoping to build a movement beyond climate summits—difficulties that compromised their effectiveness.[4] Still, a growing consensus among NGOs about the imperative of climate **justice** emerged in advance of COP21.[5] The Paris Agreement has been useful to the global climate justice movement in two ways: it is invoked by groups looking to galvanize support for government-led **action** and the 1.5°C narrative, and it is increasingly met with skepticism by groups challenging the inadequacy of governments and the international **treaty** process.

The Fridays for Future (FFF) movement, which instilled intergenerational equity in the global climate justice movement, illustrates the first tendency. **Greta** Thunberg sparked the movement in 2018 when she skipped school to protest the Swedish parliament's lack of climate action. The strikes she inspired have mobilized over 14 million people across 7,500 cities, many of whom were new to climate activism.[6] These strikes urge governments to "listen to the science" and "follow the Paris Agreement."[7] FFF has become increasingly critical of progress on climate action, calling for #NoMoreEmptyPromises, annual binding targets, and immediate emissions cuts.

Another global climate movement, Extinction Rebellion (XR), illustrates the second dynamic. It emerged around the same time as FFF but draws on a more disruptive repertoire of tactics centered around nonviolent direct action. XR expanded its concerns to biodiversity loss and the risk of social and ecological collapse. XR calls on governments to go beyond the Paris Agreement targets by reducing emissions to **net zero** by 2025. XR initially targeted the U.K. government, grinding London to a halt in its demand for accountability. However, early missteps—including not recognizing the intersections among racism, classism, and the climate crisis—and criticism of its theory of change have clouded the movement.[8] These complications surrounding XR highlight the importance of connecting struggles and systems of oppression, to which Indigenous-led movements and Black Lives Matter have brought new salience.

Other movements directly address the Paris Agreement's structural failings. Sustained, and often Indigenous-led, resistance to fossil fuel and extractive projects—termed Blockadia or **Keep It in the Ground**—have targeted fossil fuel producers directly, something the Paris Agreement does not.[9] These organizers recognize the importance of global coordination to reduce the supply of fossil fuels.[10] Out of this understanding has emerged a campaign calling on governments to phase out coal, oil, and gas—the Fossil Fuel Non-Proliferation Treaty (FFNPT). That the Paris Agreement does not mention fossil fuels has become central to this campaign. The FFNPT provides new focus for global climate action and builds on the logic of peer pressure central to the Paris Agreement while also targeting one of its critical failures.[11]

The meaning of the Paris Agreement continues to shift as the inadequacies of countries' commitments become more evident and the consequences of emissions pathways become more certain. Civil society groups and social movements have revealed the limits of state-led climate action and the necessity of globally connected movements for climate justice. They recognize that it is not enough to call on governments to implement the Paris Agreement. As evidenced by the Beyond Oil and Gas Alliance launch at COP26 in Glasgow—an international coalition of governments committed to phasing out oil and gas production—some states are responding to this mounting pressure. Though the COVID-19 pandemic suspended XR and FFF's reliance on mass mobilization, it also provided an opening for more systemic solutions and building globally connected movements vital to the ambition, action, and accountability necessary to avoid the most catastrophic effects of the climate crisis. This activism has reenergized global climate politics, connected struggles for climate justice, rapidly mobilized millions of new activists, and created a sense of urgency and collective consciousness more commensurate with the scale of the crisis we face.

See also: **Black**, **Civil Disobedience**, **Gaslighting**, **Law**, **Nonlinear**, **Scenario Planning**

Notes

1 Steven Bernstein, "The Absence of Great Power Responsibility in Global Environmental Politics," *European Journal of International Relations* 26, no. 1 (March 1, 2020): 8–32, https://doi.org/10.1177/1354066119859642.

2 Hayley Stevenson, "Reforming Global Climate Governance in an Age of Bullshit," *Globalizations* 18, no. 1 (June 12, 2020): 1–17, https://doi.org/10.1080/14747731.2020.1774315.
3 Jen Iris Allan, *The New Climate Activism: NGO Authority and Participation in Climate Change Governance* (Toronto: University of Minnesota Press, 2020).
4 Joost de Moor, "Alternative Globalities? Climatization Processes and the Climate Movement beyond COPs," *International Politics* 58 (2021): 582–99, https://doi.org/10.1057/s41311-020-00222-y; Joost de Moor, "The *Efficacy Dilemma* of Transnational Climate Activism: The Case of COP21," *Environmental Politics* 27, no. 6 (November 2, 2018): 1079–1100, https://doi.org/10.1080/09644016.2017.1410315.
5 Jen Iris Allan and Jennifer Hadden, "Exploring the Framing Power of NGOs in Global Climate Politics," *Environmental Politics* 26, no. 4 (July 4, 2017): 600–620, https://doi.org/10.1080/09644016.2017.1319017.
6 Fridays for Future, "End the Era of Fossil Fuels," accessed July 22, 2021, https://fridaysforfuture.org/; Joost de Moor et al., "New Kids on the Block: Taking Stock of the Recent Cycle of Climate Activism," *Social Movement Studies* 20 (2021): 619–25, https://doi.org/10.1080/14742837.2020.1836617.
7 De Moor et al.
8 Wretched of the Earth, "An Open Letter to Extinction Rebellion," *Journal of Global Faultlines* 6, no. 1 (2019): 109–12, https://doi.org/10.13169/jglobfaul.6.1.0109.
9 Leah Temper et al., "Movements Shaping Climate Futures: A Systematic Mapping of Protests against Fossil Fuel and Low-Carbon Energy Projects," *Environmental Research Letters* 15, no. 12 (November 26, 2020): 123004, https://doi.org/10.1088/1748-9326/abc197.
10 De Moor, "Alternative Globalities?"
11 Fergus Green, "Anti-Fossil Fuel Norms," *Climatic Change* 150, no. 1–2 (September 2018): 103–16, https://doi.org/10.1007/s10584-017-2134-6.

PERMAFROST

What do you do when the ground starts to melt beneath your feet?

BRIGT DALE

Having grown up in Norway, I remember the feeling of walking on frozen ground about to thaw, a sign of warmer spring days ahead. The sensation of softening earth now has a different valence—a sinking feeling that the ground on which one stands, upon which one's life has been built, is crumbling. Permafrost was once solid as a rock. In the Anthropocene, this solidity cannot be taken for granted.

Nowhere are the impacts of climate change felt more profoundly than in the Arctic. Here, a population of approximately four million people witness transformations of the landscape that far exceed the seasonal changes to which they are accustomed. Typically, snow-covered slopes turn muddy, then green, then the color of an autumn glow, before turning white again under the dark, sun-free polar night. This seasonal cycle requires resilience from Arctic inhabitants. While weather could be fickle, the cycle of the year had been constant: winter snow and ice will thaw in spring; in some places, the ground will remain rock solid even in summer. Now these constants have become erratic and unstable, making life more difficult.

Whether and how the Arctic exacerbates climate change is complicated. The Arctic's capacity to function as a carbon sink is at a tipping point. Subsoil depositories of carbon (as CO_2 or methane) will soon begin to release more than the fifty-eight megatons of carbon they now absorb per year.[1] Along with the thawing of permafrost comes the greening of the Arctic.[2] As ice melts, more ice-free ground allows plants to grow, with a dual effect on permafrost development: as both a carbon sink and a **local** generator of heat. Plant growth increases the capacity to store carbon. However, reductions in white surface area decrease the reflection of sunlight back into the atmosphere, increasing heat absorption. This change leads to higher local temperatures, observed by communities all over the region. As increasing temperatures make permanent frost less likely, the loss of permafrost has further knock-on effects on Arctic habitats. Frost plays an important role in holding the Arctic landscape together. When permafrost collapses, so does the soil itself, leading to abrupt, sometimes unmanageable habitat changes for flora, fauna, and human populations.

Permafrost is maintained through a balance between frost lost in summer and gained in winter. Human activities in the Arctic and elsewhere influence this balance. Tampering with soil integrity, as when establishing foundations for a building or a road, may intensify heat absorption, increasing the depth of the active layer on top of the permafrost that thaws in the summer. If this layer grows thick enough, an all-year active layer called *talik* will form between seasonal frost on top and the permafrost below, causing a constant destabilization of the ground.[3] Although some projections are more optimistic,[4] many studies anticipate that climate change and permafrost loss will continue to accelerate each other, given the potentially devastating loss of carbon sinks in a thawing Arctic and rapid changes to its landscapes. Even the most optimistic scenarios, in which the global average temperature increase is kept below 2°C, project that only two-thirds of existing permafrost will remain intact.

As permafrost melts ever deeper down, buildings and infrastructure crumble. Oil pipelines burst, roads become unstable, and homes become first unsafe, then uninhabitable. Methane gas is released, causing even more rapid climate change and increasing the possibility that viruses and bacteria that human immune systems have not encountered for millennia will be released.[5] Permafrost also acts as a sink for persistent organic pollutants and heavy metals, which, if released, would degrade the environment.[6] Even defense plans and military strategies must be rewritten as the once stable and continuous geography of permafrost areas melts away.[7] What had been secure passageways for land-based troops and equipment are no longer reliable. This predicament starkly reveals how these new and urgent threats exceed conventional notions of securitization.

The Arctic is a profoundly Anthropocenic landscape. Human desires to master the Arctic have underwritten numerous projects. It was the final frontier for exploration, to demonstrate humankind's superiority over Nature at its most harsh and demanding. Its natural resources were made accessible to extraction, and the region was made to serve geopolitical purposes. For centuries, research on permafrost has been essential to enabling this mastery. But there were costs involved, including disregard by Westerners of local knowledge about the Arctic, which could have led to a different fate for permafrost than the one now unfolding.[8]

As I write at my desk in the Lofoten Islands in Arctic Norway, autumn is arriving. Soon, the comforting sound of feet on frozen ground will accompany my morning walks and late evening runs under the blue magical winter polar light. But the freeze arrives later each year, signaling disruption to the seasonal cycle that brings snow-covered landscapes. In my journeys over frozen soil, I cannot help but think of those living farther north, whose lives and livelihoods depend on permafrost in a deep and abiding way.

See also: **Design, Extinction, Fire/Bushfire, Nonlinear, Planet, Science**

Notes

1 National Snow and Ice Data Center, "Frozen Ground & Permafrost," accessed November 23, 2021, https://nsidc.org/learn/parts-cryosphere/frozen-ground-permafrost/quick-facts-frozen-ground.

2 Isla H. Myers-Smith, Jeffrey T. Kerby, Gareth K. Phoenix, Jarle W. Bjerke, Howard E. Epstein, Jakob J. Assmann, Christian John, et al., "Complexity Revealed in the Greening of the Arctic," *Nature Climate Change* 10, no. 2 (2020): 106–17, https://doi.org/10.1038/s41558-019-0688-1.

3 Isaac Stone Simonelli, "Building on Permafrost," *Alaska Business*, September 24, 2018, https://www.akbizmag.com/industry/architecture/building-on-permafrost/.

4 Chenghai Wang, Zhilan Wang, Ying Kong, Feimin Zhang, Kai Yang, and Tingjun Zhang, "Most of the Northern Hemisphere Permafrost Remains under Climate Change," *Scientific Reports* 9, no. 1 (March 1, 2019): 3295, https://doi.org/10.1038/s41598-019-39942-4.

5 Jasmin Fox-Skelly, "There are Diseases Hidden in Ice, and They Are Waking Up." *BBC Earth*. May 4, 2017, https://web.archive.org/web/20170505151917/http://www.bbc.com/earth/story/20170504-there-are-diseases-hidden-in-ice-and-they-are-waking-up.
6 Joanna Potapowicz, Danuta Szuminska, Malgorzata Szopinska, and Zaneta Polkowska, "The Influence of Global Climate Change on the Environmental Fate of Anthropogenic Pollution Released from the Permafrost: Part I. Case Study of Antarctica," *Science of the Total Environment* 651, no. 1 (February 15, 2019): 1534–48, https://doi.org/10.1016/j.scitotenv.2018.09.168.
7 Andrew Stuhl, *Unfreezing the Arctic: Science, Colonialism and the Transformation of Inuit Lands* (Chicago: University of Chicago Press, 2016).
8 Stuhl, 108.

PIPELINE

Why is opposition to pipelines important?

KAI BOSWORTH

In July 2016, I traveled from a then-small encampment on the Standing Rock Sioux Reservation to a protest in Bismarck, **North Dakota**. About thirty **Indigenous** and non-Indigenous allies gathered with signs reading "Mni Wiconi—Water is Life" for a rally in front of the state capitol to protest Energy Transfer Partners' (ETP) Dakota Access Pipeline (DAPL). Within a month such protests would swell to many thousands, but this rally was a conventional small-town conservative-state **protest**. Though seemingly a marginal expression of power, the event was crucial for gathering us as a collective.

It's tempting to tell the story of pipeline contestation in the David versus Goliath framework: small local groups pitted against transnational infrastructure firms and their dastardly CEOs. Another narrative common in academia and activism posits that because oil is power, interrupting its flow in particular choke points is how **local** movements gain strategic power. Political theorist Timothy Mitchell argues that energy politics depends on *control of key sites*, particularly the conduits of oil transportation by rail or pipeline, and their interruption.[1] The David versus Goliath and choke point arguments yield an incomplete and misleading analysis, however, that underestimates the contradictions internal to oil economics and incorrectly presumes the scale, site, and constituent power of pipeline opposition to derive from the location or technology of the pipeline itself.

Oil does flow through central nodes: storage facilities in Cushing, Oklahoma; the switching station in Superior, Wisconsin; and the Suez and Panama canals. Yet there are so many global fields of oil production, modes of transportation (including rail, sea, and existing pipelines), refinery sites, and points of consumption that only rarely are oil flows affected seriously by any single interruption. Instead, the control of wealth, ideology, and violence has been more crucial to completing—or stopping—pipelines. Firms and states conduct media campaigns that equate oil, freedom, and life.[2] Opposition to oil pipelines from Indigenous nations and their allies is, in turn, framed as terroristic threats to such "critical infrastructure," understood as the vitality of the national **economy**.[3] And the oil industry's economic interests are not uniformly consistent: a downturn in the price of oil can spell bad news for debt-ridden oil producers heavily invested in the technologically intensive mode of oil extraction called hydrofracking. Such downturns can sometimes be good news for pipeline firms able to capture greater market share from rail shippers if desperation leads frackers to seek marginal economic savings.[4]

Many "water protectors" already understood this situation early in the **blockade** at Standing Rock, when several new weaknesses in the pipeline industry became apparent. The slogans "All Nations, All Relations, All Water" and "Standing Rock is Everywhere" indicated that *both* financial connections to the pipeline *and* nodes of resistance were transnational. For logistical and strategic reasons, pipeline opponents deemphasized the importance of the "key site" and recognized themselves as more than a local opposition group. Water protectors expanded beyond pressuring political leaders (an effective strategy in stopping the transnational Keystone XL pipeline) to making pipelines across Indigenous territory financially risky to their international investors.[5] This goal could be accomplished almost anywhere, as when urban politicians were successfully pressured to divest municipal capital from banks that were financing Warren's ETP. Though this strategy did not prevent DAPL from becoming operational, it has made financing pipelines increasingly challenging. Pipeline opponents drew attention to ongoing pipeline opposition in Mexico, Chad/Cameroon, Russia, and India, and showed that the oil industry is less of a coordinated Goliath than it seemed.

From where does the power of pipeline opposition emerge? Not through any particular territorial defense or interruption of energy at a choke point. Instead, power derives from constructing political collectives around an understanding of anticapitalism and anti-imperialism as entwined with responsibility to multispecies relations.[6] Tactically, capacious land-based political organizing led to anticolonial and socialist alliances by way of what Kul Wicasa scholar Nick Estes calls "Indigenous internationalism."[7] Back in Bismarck, the American Indian Movement flag flying overhead reminded me that even this small collective was already transnationally constituted. More than the control of key sites or technologies, such organized political power can extinguish the glee of pipeline CEOs in dark times.

See also: **Finance, Globalization, Indigenous Activism, Organize**

Notes

1 Timothy Mitchell, *Carbon Democracy: Political Power in the Age of Oil*, 2nd ed. (London: Verso, 2013), 40, 47, 67, 103.
2 Matthew T. Huber, *Lifeblood: Oil, Freedom, and the Forces of Capital* (Minneapolis: University of Minnesota Press, 2013).
3 Andrew Crosby and Jeffrey Monaghan, *Policing Indigenous Movements: Dissent and the Security State* (Halifax: Fernwood, 2018); Kai Bosworth and Charmaine Chua, "The Countersovereignty of Critical Infrastructure Security: Settler-State Anxiety versus the Pipeline Blockade," *Antipode* 55 (2023): 1345–67, https://doi.org/10.1111/anti.12794.
4 Brian Gruley, "Pipeline Billionaire Kelcy Warren Is Having Fun in the Oil Bust," *Bloomberg*, May 19, 2015, https://www.bloomberg.com/news/features/2015-05-19/pipeline-billionaire-kelcy-warren-is-having-fun-in-the-oil-bust; Robert Wright, "US Rail Groups Track Shale Boom," *Financial Times*, September 11, 2012, https://www.ft.com/content/808f9fac-f93f-11e1-945b-00144feabdc0; Neil Munshi, "US Shale Boom Times Look to Be Over for Railcar Makers," *Financial Times*, January 7, 2014, https://www.ft.com/content/446a4b52-6906-11e3-bb3e-00144feabdc0.
5 Shiri Pasternak, Katie Mazer, and D. T. Cochrane, "The Financing Problem of Colonialism: How Indigenous Jurisdiction Is Valued in Pipeline Politics," in *Standing with Standing Rock: Voices from the #NoDAPL Movement*, ed. Nick Estes and Jaskiran Dhillon (Minneapolis: University of Minnesota Press, 2019), 222–34; Kylie Benton-Connell and D. T. Cochrane, "'Canada Has a Pipeline Problem': Valuation and Vulnerability of Extractive Infrastructure," *South Atlantic Quarterly* 119, no. 2 (2020): 325–52.
6 Glen Sean Coulthard, *Red Skin, White Masks: Rejecting the Colonial Politics of Recognition* (Minneapolis: University of Minnesota Press, 2014); Anne Spice, "Fighting Invasive Infrastructures: Indigenous Relations against Pipelines," *Environment and Society* 9, no. 1 (2018): 40–56; Winona LaDuke and Deborah Cowen, "Beyond Wiindigo Infrastructure," *South Atlantic Quarterly* 119, no. 2 (2020): 243–68.
7 Nick Estes, *Our History Is the Future: Standing Rock versus the Dakota Access Pipeline, and the Long Tradition of Indigenous Resistance* (London: Verso, 2019).

PLANET

What happens to politics on an increasingly unlivable planet?

LEAH ARONOWSKY

The planet is not what it used to be. For much of human history, the Earth lurked in the background. It existed as a timeless form, a silent backdrop to the drama of human affairs. From time to time, the environment may have imposed itself in the form of droughts, earthquakes, typhoons, or little ice ages. But the Earth itself—its status as a habitable planet—was taken for granted. Indeed, as Dipesh Chakrabarty observes, this notion of a stable, enduring Earth was a precondition for a tradition of Western political thought concerned with human freedom and flourishing: on a planet seemingly free from constraints, humans were free to mold their destinies.[1]

In recent decades, the planet has impinged on everyday life. As the effects of global warming materialize in unprecedented weather events, rising sea levels, and melting ice sheets, the lived experience of climate change has become reality. What happens to politics on an increasingly unlivable planet?

This is not the first time that humans have confronted their planetary bounds. In the late 1960s, as neo-Malthusian alarmism about *population bombs* and *limits to growth* took hold in the cultural consciousness, economists in Europe and the United States began to promote a program of international politics organized around the notion of Spaceship Earth. Borrowing from the iconography of the space cabin, Spaceship Earth construed the planet as a closed world whose finite natural resources required expert management. This image evinced a profound confidence in technological mastery over Earth, leveraging cutting-edge technologies like computer simulations and global environmental monitoring systems for a science of world planning.

In theory, Spaceship Earth represented a rejection of traditional politics in favor of a purportedly apolitical technocracy, placing decision-making power in the hands of "scholars, scientists, experts . . . true international servants, bound by no national or political mandate," as George Kennan, architect of the Cold War's foreign policy of containment, put it.[2] In practice, Spaceship Earth became a shorthand among world elites for the need to address **resource** scarcity within existing international governance structures. At a moment when **decolonization** and Cold War geopolitics threatened to undermine the authority of institutions like the United Nations, Spaceship Earth promised to revitalize the postwar internationalist project of cultivating world **community**. Spaceship Earth at the UN would link international **development** and foreign aid to environmental conservation while providing developing nations a framework for holding the industrial North to account for the world's most egregious pollution problems.[3]

Traces of Spaceship Earth and its high modernist approach to planetary crisis persist today in the form of geoengineering evangelism and Elon Musk's settler-colonial dream of pioneer towns on Mars. But a growing consensus holds that if a politics founded on humanity's exceptional separation from nature led us to this point, then politics on a warming planet will require something fundamentally new. For Bruno Latour, this predicament means looking to Gaia as a new political trope for our times. Gaia, the theory that life maintains the environmental conditions necessary for its existence, rewrites our understanding of planetary history. It frames the earthly features that humans rely on for survival—the air we breathe, the temperature, the nutrients we ingest—not as inert resources to be managed (à la Spaceship Earth) but as the products of a host of dynamic biological processes. (As I have shown elsewhere, the Gaia hypothesis emerged in a collaboration between James Lovelock and Royal Dutch Shell.)[4] In Gaia, Latour argues, humans can no longer ignore "how many other beings they need in order to subsist"; this necessity is where the theory derives its political potency because Gaia amounts to a political philosophy that foregrounds dependency rather than freedom.[5] It expands understandings of "self-interest" by emphasizing that it is in humans' own interest to speak up for creatures on whom they depend. For Latour, a politics that takes the materiality of the world seriously will generate alliances that scramble traditional class and national boundaries.

But if the populist realignments of the past decade are any indication, politics organized around common interests no longer have a unifying effect. And Latour's vision for what a Gaian politics would look like in practice—with delegates representing the oceans, atmosphere, and forests participating in negotiations at a UN-like body—looks not so dissimilar from the One World internationalism of the 1970s.[6] But Latour is right about one thing: the tropes we use to represent the planet in our politics matter. They condition and constrain the limits of the possible. "It matters what thoughts think thoughts," Donna Haraway observes.[7] The tropes we adopt to ground our politics in the material forces of the planet, to bring politics "down to earth" as it were, can work against the normalizing grain of everyday life and help envision alternative ways of inhabiting the world. Whether tropes alone can compel meaningful action, however, remains to be seen.

See also: **2040, Abandoned, Globalization, Permafrost, Populism**

Notes

1. Dipesh Chakrabarty, "The Global Reveals the Planetary: A Conversation with Bruno Latour," in *The Climate of History in a Planetary Age* (Chicago: University of Chicago Press, 2021), 205–18.
2. George Kennan, "To Prevent a World Wasteland: A Proposal," *Foreign Affairs* 48, no. 3 (1970): 401–13, 409–10.
3. Perrin Selcer, *The Postwar Origins of the Global Environment: How the United Nations Built Spaceship Earth* (New York: Columbia University Press, 2018); Stephen J. Macekura, *Of Limits and Growth: The Rise of Global Sustainable Development in the Twentieth Century* (Cambridge: Cambridge University Press, 2015); Sabine Höhler, *Spaceship Earth in the Environmental Age, 1960–1990* (London: Pickering & Chatto, 2015).
4. Leah Aronowsky, "Gas Guzzling Gaia, or: A Prehistory of Climate Change Denialism," *Critical Inquiry* 48, no. 2 (2021), 306–27.
5. Bruno Latour, *Down to Earth: Politics in the New Climatic Regime*, trans. Catherine Porter (Medford, MA: Polity, 2018), 87.
6. Bruno Latour, *Facing Gaia: Eight Lectures on the New Climatic Regime*, trans. Catherine Porter (Medford, MA: Polity, 2017).
7. Donna Haraway, *Staying with the Trouble: Making Kin in the Chthulucene* (Durham, NC: Duke University Press, 2016), 35.

POPULISM

What can energy politics learn from populism?

RODRIGO NUNES

How are populism and climate politics connected? One likely answer would involve politicians rallying their base in support of fossil fuel extraction by invoking jobs for **local** communities and defending "our way of life." Another is the specter of ecofascism, which invokes climate crisis to promote stringent border regimes, ethnonationalism, and subordinating democratic procedure to the necessity of turning the nation-state into a lifeboat. In both cases, the association is negative and belies a fear that populism lends itself more to the goals of the right than to a just energy transition. Is this necessarily so?

Despite—or perhaps because of—how casually it is used, *populism* is a slippery term, as many commentators observe.[1] Yet scholars generally agree about one thing: populist rhetoric always divides society into two camps, the people and the elite, the latter of which usurps power to its own ends, thwarting the sovereignty of popular will. Purporting to give voice to popular will, populism demands that this injustice be redressed. Since it is constituted by this antagonism, populism is "not so much a coherent ideological tradition as a set of ideas that, in the real world, appears in combination with quite different, and sometimes contradictory, ideologies."[2] Or, as the Argentinean political philosopher Ernesto Laclau remarks, not so much a content as a form.

This ideological plasticity is one of several factors that inspire distrust among progressives. Historically, populists have often improved the lot of the poor while limiting their political **autonomy** through clientelistic ties and conciliation with elites. The claim to channel the popular will risks concentrating power in the hands of charismatic leaders and the potential for abuse. Finally, appeals to the people are compatible with right-wing tropes about national identity, cultural homogeneity, and an imagined organic past.

No one has done more to claim populism for the left than Ernesto Laclau, who challenges such objections. His account of populism as an *operation* complicates ideas of the people as inherently conservative. True, notions of something like a popular essence sound dangerously regressive. But what populism does when it professes to speak for the people, says Laclau, is constitute what it claims to represent: *asserting* popular unity is how it is *created*. That there is no essence is the central question. For Laclau, the people can be constituted in different ways at different times because political identities do not follow automatically from individuals' social positions; rather, they are politically constructed. Although we might say that someone has an objective interest dictated by their class, race, gender, age, or post code, they will not necessarily recognize or prioritize this interest as such. Populism names the operation through which hegemonic political identities are built.

At any given time, a society will harbor numerous unsatisfied demands. When these accumulate to the point that they can no longer be absorbed piecemeal, the potential emerges for populist politics—an operation that gathers together those demands by equating them. This operation works metonymically, according to Laclau: one demand, slogan, image, or leader is made to stand for the whole. This synecdochic operation creates an image of society split into two antagonistic camps: these things (represented by a particular one among them) are what the people want; the force preventing them from being achieved is the elite.

This logic might appeal to climate activists: is the climate not implicated in all other issues, and thus the perfect basis on which to build a social majority of this type? Here Laclau's thought raises further questions. Populism does work through already existing demands, but people do not necessarily see the environmental dimension of their demands. One can have an interest without recognizing it; the task of politics is to catalyze that recognition. Laclau goes further: in constructing a common identity, "we are dealing not with a conceptual operation of *finding* an abstract common feature underlying all social grievances, but with a performative operation constituting [equivalences] as such."[3] In other words, the fact that many issues have an environmental dimension does not mean it will be the most effective way of connecting them, nor that connecting them entails merely explaining their interconnection. The point is rather to discover what demands, slogans, and images best embody those connections and to make them vivid and intensely felt, while simultaneously ensuring that this operation produces an identity compatible with climate **justice** and energy transition in other respects (by not differentiating between clean and fossil fuel jobs, for instance).

Laclau's theory of populism could teach energy politics the following lessons: to start from issues that people already recognize rather than focusing on abstract notions of the environment; to focus less on logical connections among issues than on making them understandable as on the same side; and to build equivalences in ways that exclude demands (or solutions) inimical to the longer-term goals of social and environmental justice.

See also: **Communication, Nationalism, Neoextractivism, Organize, Storytelling**

Notes

1 Ernesto Laclau, *On Populist Reason* (London: Verso, 2006); Jan-Werner Müller, *What Is Populism?* (Philadelphia: University of Pennsylvania Press, 2016); John B. Judis, *The Populist Explosion: How the Great Recession Transformed American and European Politics* (New York: Columbia Global Reports, 2016); Cas Mudde and Cristóbal Rovira Kaltwasser, *Populism: A Very Short Introduction* (Oxford: Oxford University Press, 2017).
2 Mudde and Kaltwasser, *Populism*, 6.
3 Laclau, *On Populist Reason*, 97.

PROTEST

Why is protection more important than protest in Indigenous struggles?

KIMBERLY SKYE RICHARDS

"Occupy inaugurated the end of protest, a period in which activism must be reinvented—realigned with its spiritual calling—to be effective once again," writes Micah White, cocreator of Occupy Wall Street.[1] The Occupy experiment—to "get money out of politics" and solve income inequality—tested the hypothesis that a mass, urban, unified, nonviolent public occupation could compel a democratic government to capitulate to the people's demands. White concludes that despite being the "strongest, most sophisticated, and broadly based social movement in fifty years," Occupy Wall Street "failed to live up to its revolutionary potential"; nonetheless, its "constructive failure" revealed that social change won't happen through the contemporary repertoire of protest tactics—public spectacles, marches, or mass media frenzy.[2] If protest has become, at least in Western democracies, "an accepted, and therefore ignored, by-product of business-as-usual," then what alternative paradigms of revolutionary social change might effect social change?[3]

This question is urgent in relation to the politics of energy, particularly on **Indigenous** land, where *protest* risks obscuring mechanisms of power. David Graeber usefully distinguishes protest—"an appeal to the authorities to behave differently"—from direct action—"acting as if the existing structure of power does not exist."[4] Something like this "defiant insistence on acting as if one is already free" is arguably at work in movements to protect Indigenous land and water.[5] How can we understand protection as an alternative to protest? The gathering of water protectors at Standing Rock in 2016 against the construction of the Dakota Access Pipeline (DAPL) epitomizes protectionist movements that prevent ecological destruction, manifest Indigenous sovereignty, and build compelling alliances.

DAPL was designed to transport crude oil from the Bakken shale in western **North Dakota** (where fracking had released billions of gallons of oil) to southern Illinois for refining, before being shipped to the Gulf Coast, where it could be dispatched to market. The $3.8 billion pipeline was originally routed to cross the Missouri River north of Bismarck (the population of which is 92.4 percent white), but the planned route was moved in September 2014 to less than a mile north of the Standing Rock Sioux reservation. Enraged that DAPL would cross beneath Lake Sakakewea, Lake Oahe, and the Mni Sose (the Missouri River, the main water source of the tribe and eighteen million people living downstream), thousands of water protectors assembled to halt construction. Numbering between ten and fifteen thousand, these protectors posed a collective corporeal challenge to the state's legitimacy and its authority to act on Indigenous land.

Protectionism embodies a compelling theory of social change that recognizes no separation between past and present conflicts involving those who recognize and respect the self-determination of Indigenous nations, on the one hand, and those who ignore those rights and attempt to destroy those people for land and capital accumulation, on the other.[6] As Nick Estes (Lower Brule Sioux Tribe) puts it, the long tradition of Indigenous resistance to the trespass of settlers, dams, and pipelines across land and water bodies, now reanimated in movements like #NoDAPL, is "the future." Indeed, such conflicts will only proliferate as risky technologies (like fracking and deepwater drilling) enable countries like the United States and Canada to access hard-to-reach fossil fuel deposits on domestic territory, hoping to retain energy independence and favorable market position.[7]

In general terms, water protectors use their bodies to protect the land and water by standing in the way of the machinery of development—be it plows or police. The direct action of water protectors enacts a participatory politics that prefigures the social and spiritual relationships they desire, performatively (re)producing Indigenous anticolonial practices with ethical and political dimensions that are deeply embedded in place. Protectionist practices are based in "grounded normativity," which Glen Coulthard (Yellowknives Dene) describes as "the modalities of Indigenous land-connected practices and longstanding experimental knowledge that inform and structure our ethical entanglements with the world and our relationships with human and nonhuman others over time."[8] Water protection is, then, a practice to prevent colonialism from further strangulating grounded normativities.

Despite the recurring refrain "we are protectors, not protestors," mainstream media reporters repeatedly characterized water protectors as protestors, a fundamental political *mis*recognition.[9] Negative associations of protest with violence also helped legitimate the deployment of police and to posit Indigenous demands as antithetical to the public interest. Drawing on reports by Fairness and Accuracy in Reporting (FAIR), Sandy Grande (Quechua) observes that "until and outside of the widely-circulated images of armored vehicles, riot police, water cannons, war bonnets, teepees and painted ponies, the Lakota peoples hardly existed, virtually erased from public consciousness."[10] And yet "the non-spectacular reality was that the overwhelming majority of the time at the Oceti Sakowin encampment was spent in prayer, cooking, training, eating, laughing, building, teaching, working, washing, cleaning, singing, listening, reading, and tending."[11]

The assembly of bodies at sites of land and water defense is distinct from protest because of the Indigenous ontologies and epistemologies that ground the practice.[12] It may be tempting to understand assemblies of water protectors as enacting the forms of political power through collective action that Judith Butler argues that subjugated actors can seize to resist precarity amid neoliberal governmentality. But water protection is about more than the assertion of sovereignty: it is an enactment of responsibility.[13] The popular refrain "water is life" is a translation of the Lakota slogan *Mni Wiconi*, which more precisely means "water is alive." *Mni Sose* is a relative, and protecting her is a practice of *Wotakuye*, "being a good relative."[14]

Our constitution *as* **water** and our dependence *on* water entangle us with all life at the same time that they render us vulnerable. This vulnerability fosters new allegiances premised on the recognition that we are all related—the belief encapsulated by the Lakota concept *Mitakuye Oyasin* and practiced as a discipline of what Lakotas call *Wicekiye*, meaning "honouring relations."[15] As Robin Wall Kimmerer (Potawatomi) and Kathleen Dean Moore claim, "Everyone can join the people of Standing Rock and say *No*. No more wrecked land. No more oil spills. No more poisoned wells. We don't have to surrender the well-being of communities to the profit of a few. We can say *Yes*. Yes, we are all in this together. Yes, we can all stand on moral ground. Yes, we can all be protectors of the water and protectors of the silently watching future."[16] Thus, even when extractive development rams through, social change occurs through the performative politics of protection as it unsettles the colonial order. Distinct from appeals to established power that are inherent to protest, protectionism provides an ethical, relational foundation for invigorated anticolonial politics and practices of everyday life.

See also: **Blockade**, **Civil Disobedience**, **Justice**, **Settler Colonialism**, **Solidarity**

Notes

1 Micah White, *The End of Protest: A New Playbook for Revolution* (Toronto: Knopf, 2016), 40.
2 White, 35, 26.
3 White, 27.
4 David Graeber, "Occupy Wall Street's Anarchist Roots," *Al Jazeera*, November 30, 2011.
5 Graeber.
6 Nick Estes, *Our History Is the Future: Standing Rock versus the Dakota Access Pipeline, and the Long Tradition of Indigenous Resistance* (New York: Verso, 2019), 14.
7 The Red Nation, *The Red Deal: Indigenous Action to Save Our Earth* (Brooklyn: Common Notions, 2021), 14.
8 Glen Sean Coulthard, *Red Skin White Masks: Rejecting the Colonial Politics of Recognition* (Minneapolis: University of Minnesota Press, 2014), 27.
9 Iyuskin American Horse, "'We Are Protectors, Not Protestors': Why I'm Fighting the North Dakota Pipeline," *Guardian*, August 18, 2016, https://www.theguardian.com/us-news/2016/aug/18/north-dakota-pipeline-activists-bakken-oil-fields.
10 Sandy Grande, "Refusing the Settler Society of the Spectacle," in *Handbook of Indigenous Education*, ed. Elizabeth McKinley and Linda Tuhiwai-Smith (Singapore: Springer, 2019), 1–17, 4.
11 Grande, 4.
12 Elizabeth Ellis, "Centering Sovereignty: How Standing Rock Changed the Conversation," 172–97, and Andrew Curley, "Beyond Environmentalism: #NoDAPL as Assertion of Tribal Sovereignty," 158–68, both in *Standing with Standing Rock: Voices from the #NoDAPL Movement*, ed. Nick Estes and Jaskiran Dhillon (Minneapolis: University of Minnesota Press, 2019).
13 Judith Butler, *Notes toward a Performative Theory of Assembly* (Cambridge, MA: Harvard University Press, 2018).
14 Craig Howe and Tyler Young, "Mni Sose," in Estes and Dhillon, *Standing with Standing Rock*, 59.
15 Estes, *Our History Is the Future*, 15.
16 Robin Wall Kimmerer and Kathleen Dean Moore, "The White Horse and the Humvees—Standing Rock Is Offering a Choice," *Yes! Magazine*, November 3, 2016, https://www.yesmagazine.org/democracy/2016/11/05/the-humvees-and-the-white-horse2014two-futures.

RENEWABLE

What do renewables renew?

MARIANNA DUDLEY

Harnessing wind, water, and sun for energy is nothing new. For millennia, humans have converted solar energy into calories by growing food and have used water and wind to power machinery. The ability to generate electrical power from renewable sources, developed in the nineteenth and twentieth centuries, is relatively recent but firmly established.[1] Yet notice how renewable energy discourse emphasizes, and has always emphasized, the new—new technological capabilities, new sites of production, and newly emergent markets. **Wind** power was presented as "the energy of the future" alongside nuclear power at the 1951 Exhibition of Industrial Power (United Kingdom). In the twenty-first century, headlines proclaiming renewables as "the future of energy" proliferate daily. On its website, the World Economic Forum explains "why hydrogen could be the future of green energy," predicts that "the future looks bright for solar power," and ponders "how to build public trust in a sustainable energy future." Such future-focused techno-optimism does renewable energy a disservice. Renewable energy is *already here.* It has been *for some time.* These depictions of renewables as new, or as fuels of the future, exemplify some problems with how energy transitions are understood in relation to time, particularly the temporal horizons of past and future. What, exactly, do renewables renew? What futures do they promise?

In the most literal sense, renewables are fuels that renew themselves, or can be renewed through human action. Renewable energy is produced from sources that are "naturally replenishing, but flow-limited."[2] The thermal energy of the Sun and Earth's core as well as the kinetic energy of wind and water can be harnessed to generate **electricity**: these sources are, in effect, unlimited, even if not always constant or consistent. Many fuels commonly dubbed *renewable* are actually inexhaustible: they do not have to be replenished, even if they are not always available for capture. Unlike fossil fuel deposits, winds, waves, and solar rays are not depleted by human use. And unlike the vast scales of time involved in fossilization, fuels derived from biomass can be harvested and reharvested within timescales amenable to human action. *Renewable*, then, names the physical characteristics of these energy sources and contrasts them with finite ones.

Renewables might also be understood to remake the landscapes in which they are captured for human use. We often think of renewables as being not only organic and natural (as if **natural gas** is not natural) but also intangible since the kinetic energy of wind and waves becomes visible only as it works upon other objects. As Christina Rossetti wrote,

> Who has seen the wind?
> Neither I or you:
> But when the leaves hang trembling,
> The wind is passing through.[3]

We often see renewable energy through the medium of technologies that harness it: the turning blades of turbines on the horizon, or panels silently soaking up solar rays, for example. But some technologies rearrange entire landscapes and ecosystems, as historians of hydropower have noted.[4] The process of generating energy enmeshes technology together with renewable fuels and forces; different renewable energy sources require distinct technologies that reshape topography in various ways. All of them require humans to install, maintain, and decommission infrastructure. Renewable energy, although transient by nature, is mediated through technology that can reshape place as well as the people who inhabit it.

This capacity to reshape landscape and the built environment is part of the appeal to newness when renewables are described as future fuels. Gleaming fields of solar panels and wind turbines replace slag heaps and oil spills in the optimistic imaginary of energy transition. Given the urgent imperative for industrialized societies to wean themselves from fossil fuel dependency to reduce levels of carbon dioxide in the atmosphere and thereby limit global warming, renewable energy not only generates electrical power without carbon emissions but also fuels and renews hope. Renewables appear to break with destructive fossil-fueled energy systems without having to give up access to the massive amounts of electricity to which many humans have grown accustomed. In 2020, renewable energy produced 29 percent of global electricity, a proportion increasing every year. This trend is cause for hope: renewable energy technologies work, limits and barriers to use are known,[5] and costs continue to fall. *Renewable* combines the hopefulness of *renew* with the practicality of *able*—the capabilities actually exist. Renewables renew a sense of possibility: if energy sources can be **clean**, then societies can use their outputs without consequence.

But renewables also renew some of the limitations and harms of the fossil-fueled energy regime, not least because they have been added to the energy mix to meet an ever-increasing demand for energy, rather than replacing fossil fuels or reducing overall energy use. Coal and gas still produce 63.3 percent of global electricity and, along with oil, provide 84.3 percent of total energy (factoring in transport and heating).[6] While it may seem counterintuitive, renewables can help to prolong the era of fossil fuels, rather than hastening its demise. But even in a completely renewable future, some of the harms of the present will remain, and new ones are already emerging. Claims for the cleanness of renewables are tarnished by the metals and minerals needed for the technologies to function. Wind turbines require concrete, steel, iron, fiberglass, polymers, aluminum, copper, zinc, and rare earth elements.[7] Batteries use lithium, cobalt, and graphite. Solar power runs on sunshine, but also chromium, copper, manganese, and nickel. Extraction of these materials is a global enterprise, with mines in countries such as Australia, Brazil, Chile, China, Madagascar, Myanmar, Portugal, South Africa, and the United States that entail shocking environmental and social costs.[8] Renewables are thus creating new landscapes of extraction, exploitation, and environmental harm, in the name of clean energy. Renewables thus threaten to renew and repeat the extractive colonialisms of fossil fuel systems, whereby some societies and regions benefit from energy access at the cost of others. If *renewable* tends to evoke the clean natural forces that propel the technologies, we must also reckon with the dirtier, less visible flows of materials that constitute and mediate those technologies.

On a warming planet, hope is itself a precious resource. As national governments and international organizations look to renewable energy to simultaneously keep the lights on and hit zero-carbon targets, we must acknowledge all of the things that renewables can renew. If the expanded use of these new energy sources merely fosters the continuation of old systems and behaviors premised on fossil fuels, then the hope pinned to renewables will be false. Hope may be better placed in cultivating critical thinking around renewables and forging more equitable and sustainable relationships with energy.

See also: **Africa, Mining, Net Zero, Solar Farm, Sustainability**

Notes

1 In Britain, where I focus my research, the first wind turbine connected to a public supply grid was built in 1951. Calder Hall, the first civil nuclear power station, opened in 1956. Denmark, the United States, and the Soviet Union were developing wind turbine technology even earlier.

2 U.S. Energy Information Administration, "Renewable Energy Explained" (last modified May 20, 2021), https://www.eia.gov/energyexplained/renewable-sources/.

3 Christina Rossetti, *Sing Song: A Nursery Rhyme Book* (London: George Routledge, 1872).

4 Richard White, Sara B. Pritchard, and others developed the concept of the "envirotechnical system" by analyzing the impact of hydropower on rivers.

5 One important barrier to scaling up renewables is the technological challenge of storing electricity in order to ensure consistent load capacities. See Sean Kheraj, "More: Energy History and Energy Futures," *Historical Climatology*, April 10, 2019, http://www.historicalclimatology.com/features/more-energy-history-and-energy-futures.

6 Hannah Ritchie and Max Roser, "Electricity Mix" (Our World in Data, 2020), https://ourworldindata.org/electricity-mix.

7 International Energy Agency, *The Role of Critical Minerals in Clean Energy Transitions* (Paris: IEA, 2021).

8 Thea Riofrancos, "The Rush to 'Go Electric' Comes with a Hidden Cost: Destructive Lithium Mining," *Guardian*, June 14, 2021, https://www.theguardian.com/commentisfree/2021/jun/14/electric-cost-lithium-mining-decarbonasation-salt-flats-chile.

RESOURCE

Can waste serve as a resource for energy transition?

KESHA FEVRIER AND ANNA ZALIK

A shift from fossil fuels to green energy powered by critical minerals has been widely embraced, but ideas about how best to achieve this transition vary widely. Mainstream policymakers advocate capital-friendly solutions like the circular economy, which encourages the sustainable use of natural resources by designing smarter products and minimizing waste.[1] This approach includes transforming global waste streams of discarded electronics and plastics into tradeable commodities—that is, into secondary *resources* like scrap metal and post-consumer recycled resin—that are reinjected into circuits of production. However, actual examples of resource recycling under the banner of the circular economy risk reproducing historical legacies of uneven **development**. The people and places involved in waste recycling are often economically marginalized, racialized, and predominantly located in the Global South. Reserves of critical minerals necessary for **renewable** energy technologies and infrastructures are, to a large degree, located in low-income countries. **Mining** practices are not only replicating the harms long associated with mining but also reliant on fossil fuels, which contradicts the goal of reducing emissions. Therefore, the imperatives of climate action and energy **justice** require something more radical: fundamentally redirecting the global **economy** away from consumption, profit generation, and hydrocarbon dependency.

We argue that understanding the role of resources in an energy transition requires greater attention to **waste**. We define *resource* as commodified nature, "biophysical materials, and ecosystems services that are disentangled from existing socio-ecological relations and rearticulated into new relations to . . . realize a goal or state."[2] *Waste* is generally understood as the discarded end or downpipe residuals of production systems, materials also associated with hazard, filth, risk, and disorder.[3] In conventional economics, resources emerge through the purposeful (de)construction and transformation of nature and its integration into larger circuits of capital. This process, while serving the growth imperative, inevitably makes waste. Yet, under circular economy models, waste is unmade, conceptualized as a valuable commodity, and reconstructed as a resource. Resource and waste are thus socially and technologically defined categories that are increasingly determined by collective socioecological, spatial, and economic pressures and changing norms. The conceptual boundaries between waste and resource have become porous, given their co-constitution and the parallels between them.[4] The integration of resources and waste into human systems facilitates processes of worldmaking through which the use of resource enables or creates different kinds of socioenvironmental and spatial organization.

In the current pursuit of a cleaner, greener energy system, a circular resource-waste-resource relationship has been embedded within a capitalist growth agenda. Yet societies predating industrial capitalism also employed resources in circular fashion. The organic nature of waste and the cost of raw materials facilitated material recirculation through practices of reuse, repair, and recycling. While not extensively commodified or monetized, the reintegration of waste as resource—as raw material and feedstock—was essential to ecologically sound agricultural systems and for sustaining basic forms of social reproduction. The nineteenth century, however, marked a dramatic shift in this relationship, when the resource-waste-resource circuit was decoupled from noncapitalist relations. In the twentieth century, global Fordism, dependent on cheap and abundant fossil fuels, ushered in a shift toward globalized consumer capitalism, generating excessive volumes of consumables and waste alike.

Today, under the proposed transition from hydrocarbons to green energy, waste is being commodified within a circular model, taking on the characteristics of a resource through intensified industrial-scale recycling. Proponents argue that waste recycling prevents further climate change by reducing both carbon emissions and energy consumption that would otherwise occur from extracting and mining virgin materials. Unsurprisingly, industrial-scale recycling also creates new sociospatial avenues for extended value extraction under capitalism. Similarly, current circular economy models depoliticize the socioeconomic and political conditions through which waste is created, treated, and disposed of by suggesting that waste is always-already manageable and can be perpetually recycled into a resource.[5] Indeed, these models rely on a promise of ecological modernization while ignoring the frequently racist and classist exploitation of humans and more-than-human nature under which waste is recycled and via which consumption, profit, and growth are generated.

Finding viable alternatives to this resource-waste-resource value chain embedded in global socioeconomic inequality will require confronting and dismantling neoliberal capitalist common sense, which assumes that waste can be efficiently managed only when revalued as an economic good and integrated into circular economy models. However, the idea that intensifying recycling can both meet the increasing demand for materials like critical minerals and reduce reliance on mining is flawed. Transitioning to a non-fossil energy system in a growth-oriented or even a steady-state economy requires an immense quantity of physical materials, including nature and waste, treated as a resource.[6] Moving forward with an energy transition under a capitalist order would thus entail another round of the destruction of nature and labor—what James O'Connor called capitalism's second contradiction—regardless of whether waste-to-resource recycling is intensified through circular economy models.[7] Thus, some critical geographers insist that the "epistemological and ontological limits of the World of Resources must be *unbounded* to allow for other ways of organizing socio-ecologies."[8] It is important to recognize that the neoliberal capitalist circular economy is nothing less than a project of worldmaking, one that threatens to repeat and re-entrench the earlier worldmaking projects of capitalism and colonialism. In this context, we can learn from Adom Getachew's vision of worldmaking that emphasizes the repressed political aspirations central to anti-imperialist thought and the **Black** Radical Tradition.[9] The resource-waste-resource nexus located within imperial, capitalist logics must be unmade and remade in order to foster socioenergetic systems based on collective needs rather than economic growth.

See also: **Clean, Decolonization, Degrowth, Globalization, Transitions**

Notes

1 Jouni Korhonen, Antero Honkasalo, and Jyri Seppälä, "Circular Economy: The Concept and Its Limitations," *Ecological Economics* 143 (2018): 37–46.
2 Gabriela Valdivia, Matthew Himley, and Elizabeth Havice, "Critical Resource Geography: An Introduction," in *The Routledge Handbook of Critical Resource Geography*, ed. Valdivia, Himley, and Havice (New York: Routledge, 2021), 1–20.
3 Sarah A. Moore, "Garbage Matters Concepts in New Geographies of Waste," *Progress in Human Geography* 36, no. 6 (2012): 780–99.
4 Michael Thompson, *Rubbish Theory: The Creation and Destruction of Value* (London: Pluto Press, 2017), 50; Gavin Bridge, "Resource Geographies 1: Making Carbon Economies, Old and New," *Progress in Human Geography* 35, no. 6 (2011): 820–34.
5 Francisco Valenzuela and Steffen Böhm, "Against Wasted Politics: A Critique of the Circular Economy," *Ephemera* 17, no. 1 (2017): 42.
6 Peter A. Victor and Martin Sers, "The Limits to Green Growth," in *Handbook on Green Growth*, ed. Roger Fouquet (Cheltenham: Edward Elgar, 2019), 30–51.
7 James O'Connor, "The Second Contradiction of Capitalism: The Material/Communal Conditions of Life," *Capitalism Nature Socialism* 5, no. 4 (1994): 105–14.
8 Valdivia, Himley, and Havice, "Critical Resource Geography," 6.
9 Adom Getachew, *Worldmaking after Empire: The Rise and Fall of Self-Determination* (Princeton, NJ: Princeton University Press, 2019).

RETROFIT

What does it mean to breathe life into buildings?

FALLON SAMUELS AIDOO AND DANIEL A. BARBER

The catchphrase *reduce, reuse, recycle* helps environmental activists encourage publics in the United States to adopt new relationships to production and consumption. The same sentiment is evident in the construction industry, with recent efforts to reduce new construction, reuse existing building stock, and recycle building materials. Yet awareness of the need to adapt the built environment to the challenges of climate change is hardly new. Architecture is a cultural practice imbricated with energy and resources, and the turn to retrofit suggests that this relationship is changing dramatically. Practiced globally in diverse disciplines and political and cultural contexts, retrofit registers historical and contemporary responsiveness to unsustainable building conventions and endangered built environments. *Retrofit*, however, has many meanings: to mitigate weathering, abate decay, reduce **waste**, limit emissions, extend livability, and redress inefficiency, and even to contest colonialism, mobilize Indigeneity, and materialize local ingenuity.

Until recently, academic, cultural, philanthropic, and governmental sponsors treated environmental **design** and hazard response as irrelevant to redesigning or preserving built environments for the future. Today, research on retrofits can be found across the disciplines and practices of architecture, preservation, planning, and engineering, as well as history, archaeology, anthropology, geography, and the material sciences. Retrofit is part of a shift away from practices that consume energy and materials thoughtlessly for economic gain, and toward a rigorous and radical engagement with questions of demand management and social relations more broadly. In this context, retrofit is essential to many attempts to reimagine approaches to practice, from the #RetroFirst movement in the United Kingdom to the cool infrastructures discourse in Southeast Asia or the global efforts of the Architecture Lobby to reform labor practices and design priorities around a new set of principles.

Unfortunately, much contemporary retrofit practice only superficially engages **sustainability** discourse and practices generally accepted in architecture.

Rather than use projects to address social and environmental injustice, architects specify new materials for renovations and new construction that increase (rather than decrease) hydrocarbon fuel extraction and consumption. As the 2022 IPCC report on mitigation observes, one driver of increasing carbon emissions in the building sector is the "low ambition level . . . when existing buildings are renovated," especially in the overindustrialized North.[1] Such projects offer rich opportunities for renovating and realigning mechanical systems to correlate with emissions reduction goals, but buildings of the past also offer complex and varied examples of how the structure of buildings, their program, and the urban fabric in which they sit can reduce or eliminate reliance on fossil-fueled mechanical systems. In other words, retrofit strategies are best understood in relation to broader urban and infrastructural adjustments to economies, practices, and ways of life that could be organized as sufficiency interventions (that is, doing *more* with less) rather than as continued marginal improvements in efficiency.

These approaches that attend to the history, context, and energetic position of a building offer a compelling horizon for research, both for transforming the building industry and for producing knowledge about technological trajectories and their social effects. Research on retrofits could generate new conceptual frameworks to reckon with three ideas of long-standing concern to preservation and conservation stakeholders worldwide: integrity, authenticity, and appropriateness. Research in the private sector and at public universities, as well as public-private ventures that center **Indigenous** building practices and materials such as mass timber, can cultivate new incentives and practices for designers and developers. These new perspectives move away from the prevalent bias toward stability, sustainability, and consistency that characterizes recent architecture focused on energy efficiency; in prioritizing these goals, capital markets have made them more feasible than other strategies.

Retrofit, in other words, is about more than targeted design interventions. It can catalyze more profound transformations: forging new economic and social relations, investing in circular economies and material reuse, embedding a radical approach to maintenance and care in Indigenous epistemologies and nonhegemonic worldviews. Retrofit might foster vastly different relationships between social and ecological systems. At stake in retrofit and reuse is the imperative of confronting the divergent approaches that designers of the Global South and Colonial North, as well as settlers and stewards of settled land, can bring to the movement for decarbonizing the world's monuments, national historic sites, and other tangible heritage.

The subaltern theories of being in the world that underwrote **decolonization** movements throughout Asia, Africa, and South America offer a framework for reexamining the ethical dimensions of hydrocarbon fuel-efficient practices in designing the built environment. This situated approach to building knowledge and practices calls into question models of knowledge production and change making that emphasize the migration of starchitects or sustainability experts between World Monument sites and UNESCO summits. Opposition to LEED building in the Global North reflects a different spatial epistemology of energy and ecology than the organization's global standards promote.[2] To achieve LEED benchmarks in the Global North, designers across Europe and America have prescribed (via architectural specifications) the extraction of rare earth minerals from the Global South. Likewise, research exchanges that consume fossil energy and emit carbon not only are an ironic pursuit for retrofit scholars and practitioners but also tend to reinscribe colonial knowledge into building practices. Retrofit is a call not merely to reimagine supply chains in the building industry but to reimagine the principles, priorities, and processes of breathing life into buildings. Making things anew calls for an economy of care for built, social, and natural worlds.

See also: **Resource, Settler Colonialism, Sustainability, Transitions**

Notes

1 Intergovernmental Panel on Climate Change, "Technical Summary," in *Climate Change 2022: Mitigation of Climate Change* (2022), 71, https://www.ipcc.ch/report/ar6/wg3/.
2 LEED is Leadership in Energy and Environmental Design, a widely used system for rating the sustainability of a building.

SABOTAGE

Can sabotage ever be more than a romantic gesture?

TED HAMILTON

In the late 1700s and early 1800s, laborers in the British textile industry reacted against low wages and declining working conditions by smashing their owners' machines. The owners shifted their operations from far-flung sources of waterpower to centralized mills run on steam, where they had the upper hand in arranging hours, wages, and conditions.[1] The capitalists' new steam was powered by burning coal, and fossil-fuel-powered industrialization began in earnest.

One July 24, 2017, two women held an impromptu press conference in front of the Iowa Utilities Board. They claimed credit for a months-long series of arson and welding torch attacks against the Dakota Access Pipeline, which they argued threatened Indigenous sovereignty and environmental health. They closed their statement by saying, "Water is Life, oil is death."[2]

From one resistance campaign to another, across the entire time span of anthropogenic warming, in the natal mills of the **Industrial Revolution** and on the plains of the great North American refining network, several questions emerge: What is sabotage for? What does it accomplish? And can it ever be more than a romantic gesture against the onslaught of history?

"Sabotage means primarily: *the withdrawal of efficiency*," Elizabeth Gurley Flynn, a militant with the Industrial Workers of the World, declared in 1917.[3] The efficiency she spoke of was productive; sabotage was among the workers' means of controlling the rate and quality of capitalist production: "Sabotage is an unfair day's work for an unfair day's wage."[4] Through the mid-twentieth century, sabotage retained this labor-specific connotation. It included tactics such as strikes, slowdowns, work-to-rule, and, in its classic form, the destruction of machines. Of the paradigmatic saboteurs, the nineteenth-century mill-wrecking Luddites, Gavin Mueller writes that "their revolt was not against machines in themselves, but against the industrial society that threatened their established ways of life."[5] At the other end of the cotton economy, slaves in the United States regularly destroyed their owners' tools and equipment.[6]

Whatever the means, sabotage primarily involved a conflict over a specific form of energy: labor power. Owners wanted more of it, workers wanted to give less, and sabotage was a means of impeding its extraction. To accomplish this goal, and to impede it, both sides tried to assert control over flows of water and fossil fuels. By moving mills, controlling coal trains, and building and dynamiting dams, they adjusted the ratios of production in their favor.

After World War II, the definition of sabotage both constricted and expanded, coming to refer specifically to interference with tools or materials even as its possible applications moved beyond the owner-worker nexus to include world powers bombing munitions factories and infecting centrifuges with computer viruses. (More flows, more efficiencies: water, nuclear, electric.) But sabotage became increasingly associated with vigilante actions against symbols of technological progress. When motivated by environmental concerns, it became known as *ecotage*.

Sabotage is not necessarily technophobic—but it is conservative. Although the folk etymology ascribing the first act of sabotage to French weavers who stuck their *sabots* (wooden shoes) in a loom is fanciful, the term's craft-age ploddingness remains. Thorstein Veblen, thinking of the shoe, defined sabotage as "any manoeuvre of slowing-down, inefficiency, bungling, obstruction."[7] Increasingly, what is to be slowed down is not a single site of productive exploitation, not a single instance of unjust energy transfer, but technological change itself. By 1975, this turn of the Luddite legacy had grown so dominant that Edward Abbey could write of the mythical loom breaker Ned Ludd: "They called him a lunatic but he saw the enemy clearly. Saw what was coming and acted directly."[8]

For Abbey, that enemy was industrial civilization writ large. Its signature crimes (and the targets of *monkeywrenching*, the version of ecotage to which his novel gave its name) were the highways, power lines, billboards, and dams marring the landscape of the western United States. Abbey's hero George Hayduke—who burns, saws, and detonates his way through these targets while eschewing any definite class or political identity—embraces an explicitly conservative program of "counter-industrial revolution," whose aim is to "keep it like it was."[9]

Keep what like it was? Some modern-day ecoteurs—such as the Indian farmers who burned down a Cargill factory producing genetically modified seeds—seek to preserve modes of subsistence against capitalist encroachment, in much the same fashion as the Luddites. Others, like the radical Earth First! movement, operate from an explicitly biocentric rationale, seeking to impede incursions against wildlife and wild spaces. The efficiency that these saboteurs hope to withdraw belongs not to a labor-powered machine on a shop floor but to an entire economic and ideological system that appears to have an inhuman momentum of its own, an inertia of production through destruction. By the turn of the century, a feeling of alienation from this system had so pervaded the ecotage scene that Kirkpatrick Sale could call his history of the Luddites' influence on radical environmentalism *Rebels Against the Future*.[10]

Jessica Reznicek, one of the women who sabotaged the Dakota Access Pipeline, claims that her actions were not property destruction but rather "property improvement."[11] Understood with reference to efficiency—the ratio of useful work performed to total energy expended—one might say that this modern variety of sabotage aims, through intimate interference with the tools of energy transfer, to enact a new-old idea of the useful: the maintenance of social and natural balance. Destruction as improvement, then, and the saboteur as agent of a lost, or dreamed-of, stability.

See also: **Bankrupt, Black, Blockade, Indigenous Activism, North Dakota, Protest**

Notes

1 Andreas Malm, *Fossil Capital: The Rise of Steam Power and the Roots of Global Warming* (New York: Verso, 2016).
2 Ruby Montoya and Jessica Reznicek, "Statement of Jessica Reznicek & Ruby Montoya on DAPL 7-24-17," *Bill Quigley: Social Justice Advocacy*, July 24, 2017, https://billquigley.wordpress.com/2017/07/25/statement-of-jessica-reznicek-ruby-montoya-on-dapl-7-24-17/.
3 Elizabeth Gurley Flynn, *Sabotage: The Conscious Withdrawal of the Workers' Industrial Efficiency* (Chicago: I.W.W. Publishing Bureau, 1917), 5.
4 Flynn, 8.
5 Gavin Mueller, *Breaking Things at Work: The Luddites Are Right about Why You Hate Your Job* (New York: Verso, 2021), 38.
6 John Hope Franklin and Lee Schweninger, *Runaway Slaves: Rebels on the Plantation* (New York: Oxford University Press, 1999).
7 Thorstein Veblen, *The Engineers and the Price System* (New York: B.W. Huebsch, 1922), 1.
8 Edward Abbey, *The Monkey Wrench Gang* (New York: Avon, 1985), 57.
9 Abbey, 190, 68.
10 Kirkpatrick Sale, *Rebels Against the Future: The Luddites and Their War on the Industrial Revolution: Lessons for the Computer Age* (Reading, MA: Addison-Wesley, 1995).
11 Jessica Reznicek, "Uncomfortable," *Via Pacis* 41, no. 1 (April 2017): 1, https://frankcordaro.wordpress.com/2021/01/26/uncomfortable-by-jessica-reznicek-april-2017-via-pacis-p-1/.

SCALE

What if scale was about equity rather than magnitude?

CYMENE HOWE

Questions about scale almost inevitably arise in conversations about energy. For instance, what is the scale of energy needs in a community? What is the scale of output for an energy system—whether a hydropower station, a solar array, or a liquid natural gas plant? All energy systems can be scalable—made larger or smaller in their design and intent. A tiny set of solar panels might be the proper scale for an off-grid household that uses minimal power, while an average household in Canada would need about four and a half tons of coal for a year's worth of **electricity**.[1] The Gansu Wind Farm in China produces enough electricity every day to meet the needs of ten million people in **India**, but not quite enough for one million consumers in the United States—a striking commentary on the variable **scales** of consumption itself.[2]

Folded into these calculations is the scale of monetary investment for construction costs, ongoing maintenance, and the eventual decommissioning that is the fate of all energy infrastructures. The fiscal calibration of energy scaling asks, is it worth it to get *x* quantity of energy for *y* number of people? Scale is therefore tightly bound up with value—both monetary value and the social value of providing energy for people. In conventional thinking about upscaling or downscaling energy systems, scale is tied to economic and energetic values: a metricized, transactional equation between resource investment and the benefits expected from the generation of power.

In the scientific terms of energy physics, scale is about magnitude. Orders of magnitude illustrate correlations between energetic outputs and responses, large and small, across our world. The energy of a photon with a frequency of 1 hertz is truly slight, at ten to the negative thirty-fourth power (10^{-34}) joules (J). The kinetic energy of one red human blood cell is, by scalar accounting, quite a bit larger at 10^{-15}J. Imagine the sound energy (also known as vibration) that flows to your eardrums as you listen to a whisper for one second (10^{-14}J). A "tera-electronvolt" (10^{-7}J) is equivalent to the kinetic energy of a mosquito in flight. This amount is quite small compared to the one joule of kinetic energy produced when a little apple drops, as it did on Isaac Newton, one meter through Earth's gravity. Catapult up in scale and note that it takes 3.3×10^{2}J to melt one gram of ice. Scale up much, much further (1.5×10^{22}J) to the total energy of the Sun that strikes the Earth each day. That amount is a bit less energy than what is estimated to be contained in the world's coal reserves (2.4×10^{22}J), which is a little shy of the total energy captured in the world's fossil fuels (3.9×10^{22}J). Don't forget the whisper.

Energy physics understands scale as proportional to work achieved. By this logic, the physics of scaling energy is related to what *can* be accomplished rather than what *ought* to be accomplished. There is a transactional quantification of joules in the service of transformation. But certain perspectives are woven into, and also left out of, this logic. When we think about energy and scale, we think of equivalences—of work done and bodies moved, air warmed and velocities achieved. But we could think of energy and scale quite differently, not as magnitude but as equity. Scale is not merely a measurement; it is also a mode of orientation, comparison, connection, and positioning. Scale is relational, always a view from somewhere. And by someone. Since the early twentieth century, the largest power plants have not been fueled by coal, nuclear fission, or fuel oil; hydroelectric dams have been the kings of scale. At Three Gorges in China or Itaipu on the border between Brazil and Paraguay, the energy produced by these enormous structures is measured by the number of kilowatt hours or horse power produced, rather than by the fair distribution of that energy to nearby populations, or by the numbers of people and other creatures displaced.

In many places in the world, energy has its equivalence in wealth. The more money you have, the more energy you can have. If you are the billionaire owner of a multiroom mansion that you want brightly lit day and night, and you are willing to pay for all those watts, so be it. An electricity provider will be at your service. In these places, seemingly infinite energy is for sale. In some places in the world, even with money to spend, you might not be able to secure the energy you desire. In places where there is "energy poverty," transmission systems falter, power plants fail to produce, lines are not laid, deliveries are stalled, and energy dissipates. The fair, just, and equal distribution of energy across the world—both within and among nation-states—is uneven at best.

Energy scales are a process and a site ripe for remediation. Rather than ensuring that the entitled few have all the energy they desire, we could calibrate energy systems to reach the many in more egalitarian ways. In that scaling, energy would be linked to ethics and not only to economics. Survivals of colonialism would be brought into account, including processes like the toxic extraction of uranium, mined rare earth minerals, and other dispossessions based on the extractive capitalism of energy regimes always hungering for profit, even and especially in the current transition toward renewables.

In *Friction*, Anna Lowenhaupt Tsing observes that "scale has become a verb that requires precision." It is a term and a process vulnerable to the urges of technoscientific growth (at all cost). Tsing warns that the quality called *scalability* risks becoming synonymous with the capacity to expand toward the goal of increased magnitude without ever "rethinking basic elements" and instead falling into "ideologies of scale."[3] In fetishizing the size of energy systems, therefore, we may lose track of an essential element: that while energy directed toward human needs has been scaled for magnitude, energy scaled for equity is still a work in progress.

See also: **Class, Clean, Degrowth, Digital, Finance, Renewable, Sustainability**

Notes

1 Energyrates.ca, "Residential Electricity and Natural Gas Plans" (n.d.), https://energyrates.ca/residential-electricity-natural-gas/.

2 Hannah Ritchie and Max Roser, "A Sense of Units and Scale for Electrical Energy Production and Consumption" (Our World in Data, November 22, 2017), https://ourworldindata.org/scale-for-electricity.
3 Anna Lowenhaupt Tsing, *Friction: An Ethnography of Global Connection* (Princeton, NJ: Princeton University Press, 2012), 505, 347.

SCALES

What can the human voice tell us about the scale of climate change?

LISA MOORE

In 2009, I published a novel called *February* about the *Ocean Ranger*, an oil rig that sank off the coast of Newfoundland in 1982.[1] All eighty-four men on board died. The story, like all stories, has morphed with the passage of time; the meaning of this history changes, significances accrue. The urgency of climate crisis has recalibrated everything in the novel. When I wrote it, what mattered most to me in the story was the unnecessary loss of life. It's still what matters most, but the **scale** has changed. Or the scale is the same, but our understanding of the scale has changed.

In 2019, I was asked to be a co-librettist for an adaptation of the novel into an opera. Even as the artistic director of *Opera on the Avalon*, Cheryl Hickman, called to ask if I was interested in the working on the project, I was Googling *librettist*. It means writing the words. I'd never been to the opera. I know nothing about **music**.

I said, yes, absolutely, I'd love to.

My co-librettist Laura Kaminsky was an outrageously talented composer. Laura lives in New York and I am in Newfoundland. We worked together for over two years, she sending me MIDI files and showing me the score by sharing her screen on Zoom. We worked as people died in northern Italy from the **coronavirus**; we worked as they died in New York. We worked as they died all over the globe. We worked as they stormed the Capitol. We worked as Newfoundland, where I was writing from, was hit by the snowstorm of the century and people were buried in their houses, able to escape only by tobogganing on their bums from second-story windows. Grocery stores closed for seven days; the city shut down, power lost. The elderly and vulnerable cut off from outside help. Power restored and lost again.

Laura and I talked about **art** and music and light and sound. We talked about oil. Seismic testing, gardening. We talked about people out of work and in poverty. We talked about shutting down, floods and wildfires, the production of oil. We talked about the convoy of truckers in Canada, spreading to France, and other places; and the unleashing of new kinds of ugliness, of fascism; and what music could do, might do. We talked about the nature of opera. We listened to certain phrases of music, over and over; we chopped a word. Sometimes it took fifteen minutes to chop a single word, or syllable, or add a note; sometimes we put the word back.

In the novel, time shifts from the present to the past, mimicking the nature of memory as it is altered by trauma. There are scenes in *February* when Helen and Cal, a married couple, are together, though the reader knows already that Cal has died on the *Ocean Ranger*, has already read the scene where the rig goes down. Helen longs for him, grieves for him; and because of the nature of a novel, its peculiar freedoms, how it can lasso time, rein it in, sandwich different times together cheek by jowl, splice the present and the past, glue them together, Cal can be present, even as he is already dead.

For the first workshop with singers, as we rehearsed, I realized with a shock that time behaves differently onstage than in a novel. With a literal chill, I saw a difference between a scene in the novel where Cal appears to be in the present and already dead simultaneously (the reader able to hold both conflicting realities at once, at the same time), and the same scene onstage. Onstage, the presence of Cal appears to be the presence of a ghost. This is perhaps because singing is so visceral, so undeniably corporeal and present and *there*, that the scene doesn't read as a flashback to an earlier time, but rather, Cal could be understood *only* as a spirit. He is already lost. What we see and hear is a shade or stain of something gone forever: *already* gone.

The singers had come from New York and other parts of the United States. And there were singers from Newfoundland, too. We met wearing masks and were socially distanced. The story of the *Ocean Ranger* was always unbearably sad, but I've found, over time, it has become even more sad. The sadness has also changed in scale. The more we understand about climate change (about hubris, about inevitability), the sadness engulfs, overwhelms.

In one part of the opera, Helen, the widow, and her grown-up son, Johnny, sing a duet. The two singers did a thing in rehearsal when they sang the duet for the first time, feeling their way for the notes and rhythm and cadence and emotion. As the notes ascended, requiring what seemed like impossible amounts of breath, lungsful of oxygen, the singers each slowly raised one arm as their voices went higher and higher. They each formed a kind of upside-down cup with their rising hand, as if the notes were a rising water fountain and they were cupping the very top of the note or fountain, physically holding onto it, as their voices climbed. As the notes rose, higher and higher, so did their bodies, following their rising arm until they were on tiptoe. But, unbelievably, reaching higher and higher still.

And at the very highest note their voices became stronger, more voluminous, more powerful, and also eerie, uncanny and charged with—what? *Grace* is the best word here, or perhaps *beauty*, undiluted beauty. Counter to expectation, as they reached up with their voices, both of their faces broke open wide with smiles of astonishment at what they had achieved together. And unmitigated joy. At what the human body could do and make.

At the same moment (the hairs on my arms standing up, my nose tingling as it does before tears, especially tears in public, which are always a kind of ambush), there was an upside-down cascade of profound grief (their voices now running back down the scale, going lower and lower, under the water), because caught in that moment was the ephemerality of the human body, the human voice, our world, the sense of time running out, breath running out, all of us drowning, and the hope that goes into singing anyway. Singing anyway.

See also: **2040, Abandoned, Shipping, Storytelling**

Note

1 Lisa Moore, *February* (Toronto: House of Anansi Press, 2009).

SCENARIO PLANNING

Is the future a waste of time?

JOSEFIN WANGEL

Scenario planning has been linked to energy for more than half a century. A quick Google search reveals the story of Royal Dutch Shell's use of scenario planning to adeptly anticipate and navigate the 1973 energy crisis.[1] While Shell didn't invent scenario planning, its specific use of this technique highlighted the shortcomings of the traditional "predict and provide" method in energy production, popularizing a future-focused approach that embraces uncertainties.[2] This shift significantly impacted how we perceive and engage with the future, moving away from an approached premised on an "elevation of trend to destiny" and dominated by a mindset (and accompanying practices) that assumes "more of the same," to an approach that focuses on risk and rupture.[3]

The energy crises of the seventies also highlighted how energy production's possibilities are intertwined with the societal context of its production and usage. Despite seeming obvious to some, energy systems are commonly still developed and managed as if they were purely or primarily a matter of physics, technology, and energy resources. By contrast, scenario planning is a tool with which to engage with future uncertainties resulting from uncontrollable choices and contingency. Since such uncertainties tend to be found outside the energy system as technology, scenario planning aims to avoid reductionist approaches and to widen the understanding of what an energy system is, or at least to diversify the understanding of what needs to be taken into consideration when developing and managing energy systems.

The most common and well-known approach for scenario planning is the scenario matrix, in which two key uncertainties are selected and combined to form four different scenarios. But there are also approaches that focus on a single key uncertainty and its implications across a continuum of concerns (such as the IPCC's RCP scenarios),[4] as well as approaches that engage with more than two uncertainties (such as the Manoa method developed by Wendy Schultz to allow for more complexity than the scenario matrix approach typically affords).[5]

Scenario planning has many merits. But when it comes to achieving a just and sustainable post-carbon future, this way of engaging with the future is not enough. Scenario planning was developed as a tool for exploring futures that are possible—not those that are preferable. It can help us identify uncertainties and know how to manage them. It cannot help us to imagine what a desired future might look like, nor how to get there.

Fortunately, the seventies energy crises spurred the development of not only scenario planning but also a visionary scenario approach that would become known as backcasting. In "Energy Strategy: The Road Not Taken?," a 1976 essay published in *Foreign Affairs*, Amory Lovins describes two possible paths for the U.S. energy system. The first is a "hard" path that "is essentially an extrapolation of the recent past," with a continued expansion of the current energy system as well as a continued dependence on fossil fuels. The second, "soft" path is based on substantial energy efficiency and decentralized **renewable** energy, which "diverges radically from incremental past practices to pursue long-term goals."[6] Lovins sought to turn energy planning on its head, by suggesting that the energy system should be developed not as a path-dependent reaction to perceived threats but instead as a long-term vision of what a desirable system might look like.

Yet, did this approach succeed? While the use of backcasting has gained traction in energy policy, particularly for climate change mitigation, the energy system's structure, dominated by vested interests, has been resistant to change. In contemporary energy planning, centralized and fossil-fuel-dependent "predict and provide" approaches continue to dominate. Even if scenario planning and visionary approaches had been widely adopted, it is unclear whether they would have made any substantial difference. Most scenario planning and visions literature and practices are premised on an understanding of change as mobilized primarily by altering discourse—that is, the way we think and talk about things. Consequently, in times of crises, calls for alternative narratives and new stories tend to proliferate. There's no doubt that the way we think and talk about things is important. But if this talk is not translated into **action** in the present, this focus on future possibility risks becoming little more than escapism.[7] This predicament raises the question: can we invigorate scenario planning and other future engagement methods to move beyond "strategic conversation,"[8] or is scenario planning essentially a waste of time?

See also: **2040, IPCC, Paris Agreement, Science, Storytelling**

Notes

1 Angela Wilkinson and Roland Kupers, "Living in the Futures," *Harvard Business Review*, May 2013.
2 Thomas Chermack, Susan Lynham, and Wendy Rouna, "A Review of Scenario Planning Literature," *Futures Research Quarterly* (Summer 2001): 7–31; David Hounshell, "The Cold War, RAND, and the Generation of Knowledge, 1946–1962" (Santa Monica, CA: RAND, 1998).
3 Rene Dubos, "Trend Is Not Destiny," *New York Times*, November 10, 1975.
4 The Representative Concentration Pathway (RCP) scenarios describe possible futures that differ in terms of their radiative forcing in the year 2100. Radiative forcing is a concept used to quantify the energy balance of Earth. A positive radiative forcing means a net gain in energy, which causes global warming. The first four RCPs (RCP2.6, RCP4.5, RCP6, and RCP8.5) were published as part of the IPCC's "Fifth Assessment Report" in 2014.
5 Wendy Schultz, "Manoa: The Future Is Not Binary," *APF Compass* 4 (2015): 4–8.
6 Amory Lovins, "Energy Strategy: The Road Not Taken?," *Foreign Affairs*, October 1976.
7 Josefin Wangel, "Troubling Speculation," in *Beyond Efficiency*, ed. Josefin Wangel and Eléonore Fauré (Baunach: AADR, 2021), 191–98.
8 Kees van der Heijden, *Scenarios: The Art of Strategic Conversation* (Chichester: Wiley, 1996).

SCIENCE

Do scientists believe that science matters?

ADAM SOBEL

As a child in New York City in the 1970s, I loved *Star Trek*. My favorite character was the science officer, Mr. Spock. Some things this fact says about me are obvious, I suppose. Less obvious is what the show said *to* me, and what I believed, about the real world in which I was growing up. *Star Trek* articulated a vision of science's role in a pluralistic, democratic, and fundamentally decent society: the United Federation of Planets, a utopian, interplanetary United States, in which science serves **democracy** by informing the decisions of enlightened leaders and creating technical solutions to problems.

Late in a typical episode, as the crew of the *Enterprise* became desperate in the face of a dangerous situation, Spock would suddenly make sense of the data and develop a science-based proposal to save the day. Captain Kirk would trust his first officer and order it implemented. This standard plot line articulates the linear model of the role of science in policy: science indicates a need for **action** in the public interest, and government takes that action.[1] Implicit in the linear model is the understanding—as on the bridge of the *Enterprise*—that science makes the world better not just because it has the right answers but because there is a benevolent, just power structure that listens to it.

I entered what we now call climate science as a new graduate student in meteorology in 1993. I went to grad school to improve my career prospects, without fully giving in to the mundane and mercenary aspects of the rest of the capitalist **economy**. Academic science seemed the most practical route to something like a life of the mind. But I also wanted to work in an area that held some potential to be a positive force in human society. Atmospheric science was such an area: a discipline where the linear model of science seemed to work. In 1974, Mario Molina and Sherwood Rowland had predicted that chloroflourocarbons (CFCs) would pose a threat to the ozone layer.[2] That theoretical prediction began a regulatory process to restrict production of CFCs. In 1985, the dramatic loss of ozone over Antarctica was documented from ground-based measurements and confirmed shortly after by satellite observations. By 1987, the first Montreal Protocol, which banned production of CFCs internationally, was signed.

That this unprecedented feat of global environmental regulation occurred during the presidency of Ronald Reagan was astonishing, given Reagan's otherwise strong opposition to government regulation of the private sector. If one were inclined to believe that the U.S. political system had some inherent rationality at its core, and that science was an essential element of that rationality, this course of events might have reinforced that belief.

When I was a graduate student in the nineties (writing my thesis on the stratospheric dynamics of the ozone hole, though after graduation I would switch to studying tropical meteorology, and later climate change and extreme weather events) it was much easier than it is today to believe that the linear model might also still work on climate. That Spock might yet convince management to cut greenhouse gas emissions. The science on global warming—though more than solid enough to justify action—was less advanced, atmospheric CO_2 lower, and the planet cooler than today. The signing of the Montreal Protocol was still very recent. And while the political situation was disturbing, it was easier to see it as transient and curable. There was still a semblance of a shared reality embodied by the national media. The internet was new and social media was a decade away. The federal election process was not under attack.

But global warming would prove more difficult to solve than ozone had been. Fossil fuels were more deeply entrenched and harder to replace than CFCs were, and the fossil fuel industry more deeply committed to a strategy of denial—and more successful at it—than the chemical companies had been regarding ozone.

And the Republican takeover of the House of Representatives under Newt Gingrich in 1994 signaled a new era of ugliness and dysfunction in U.S. politics that has in the intervening years moved only closer to its logical conclusion. Gingrich and his movement portrayed the opposing party as inherently unworthy to rule—a stance typical of authoritarian governments—and attacked the media and other sources. In *How Fascism Works*, philosopher Jason Stanley quotes right-wing radio host Rush Limbaugh, who identifies "the four corners of deceit: government, academia, science and media. Those institutions are now corrupt and exist by virtue of deceit."[3] Limbaugh became a right-wing media star in the Gingrich era; in 2020, he was awarded a Presidential Medal of Freedom by Donald Trump.

My faith in the national myth, and with it the linear model of science in the service of democracy, wasn't fully conscious. If you had asked me back in the nineties, I would have said that global warming was a serious problem and our national and global inaction on it terribly worrying. I would have expressed despair and concern about the direction of U.S. politics. Nonetheless, in a way that I can see only now, I still believed.

I don't believe anymore. So now what? What is the right course of action for a scientist?

Keep the faith, seek understanding for its own sake, and train the young, in preparation for a better day? Find people to whom one's science might still make a difference, and ask what they need? Turn to political action, looking for ways to do so *as a scientist*, somehow harmonizing the passion activism requires with the dispassion that science does? Or do something else entirely?

After many conversations with colleagues and friends, I've come to think that this crisis is not mine alone, nor can the young escape it. I don't know if we can find the right ways to engage with the current reality or find them quickly enough to matter. But we can't keep kidding ourselves.

Science is a set of processes and methods for understanding the external world, allowing us to predict and sometimes to control some aspects of it. But that control is exerted through the systems of power that exist in the larger society. We are not on the bridge of the *Enterprise*. If we want our science to fulfill its promise to make the world better, we can start by telling ourselves a more believable, grown-up story about how and why it can be made so.

See also: **Abandoned, Communication, Evidence, Expert, Planet, Storytelling**

Notes

Thanks to Deborah Coen and Leah Aronowsky for insightful suggestions.

1 Sheila Jasanoff and Brian Wynne, "Science and Decisionmaking: Human Choice and Climate Change," in *Human Choice and Climate Change*, vol. 1: *The Societal Framework*, ed. Steve Rayner and Elizabeth L. Malone (Washington, DC: Batelle Press, 1998), 1–87.

2 Mario Molina and F. Sherwood Rowland, "Stratospheric Sink for Chlorofluoromethanes: Chlorine Atom-Catalysed Destruction of Ozone," *Nature* 249 (1974): 810–12.

3 Jason Stanley, *How Fascism Works* (New York: Random House, 2018), 52.

SETTLER COLONIALISM

How can modern energy systems be Indigenized?

SĀKIHITOWIN AWĀSIS

Following the raid by Royal Canadian Mounted Police on the Wet'suwet'en anti-pipeline camps in February 2020, roving rail and highway blockades took place across Turtle Island (North America) in **solidarity**.[1] In the mainstream media, Indigenous resistance to energy projects is often portrayed as protests that inconvenience settler society. Meanwhile, Indigenous community members often refer to themselves as **water** protectors and understand their actions as upholding Indigenous laws. Settler colonialism dispossesses Indigenous ways of knowing, legal systems, and Indigenous peoples themselves from the landscape. On Turtle Island, settler colonialism is a set of ongoing processes characterized by interrelated class, racial, gender, and state powers that compose hierarchical structures imposed on Indigenous communities. Indigenous resistance to settler colonialism and destructive energy projects is not merely activism, but a conflict between fundamentally different legal systems and worldviews. I cannot summarize the vast diversity of Indigenous knowledge systems here, nor do I have the knowledge or authority to do so. Instead, I map some of the ways that Indigenous communities are advancing the futurity of (non)human communities through resistance to and development of energy projects.

As a result of settler-colonial structures of power, **Indigenous** communities are often disproportionately burdened by the social and environmental impacts of energy projects. Under colonial capitalism, lands and bodies are commodified and framed as objects for exploitation.[2] Contamination from industrial processes dispossesses Indigenous peoples from the land and disrupts place-based relationships that constitute the foundation of Indigenous health and ways of living.[3] For many Indigenous people engaged in resistance movements, land is not a **resource** over which to exert control, but a source of identity, kinship, knowledge, **law**, decision-making power, and sustenance. Indigenous ways of knowing reframe settler-colonial notions of "rights" to energy resources as intergenerational reciprocal responsibilities to (non)human relatives.

The opposite of dispossession is consensual relationship building with land.[4] Indigenous peoples' refusal of dispossession is a generative process of becoming more deeply attached to the land. Indigenous-led blockades are vibrant, life-affirming places that advance futurity through the sharing of culture, knowledge, and ceremonies that foster intimacy with the land and waters.[5] Linking the reciprocal responsibilities of humans and beavers, Nishnaabeg scholar Leanne Betasamosake Simpson compares blockades to dams, which share disruptive and life-giving capacities:

> One can stand beside the pile of sticks blocking the flow of the river, and complain about inconveniences, or one can sit beside the pond and witness the beavers' life-giving brilliance—deep pools that don't freeze for their fish relatives, making wetlands full of moose, deer and elk food and cooling spots, places to hide calves and muck to keep the flies away, open spaces in the canopy so sunlight increases creating warm and shallow aquatic habitat around the edges of the pond for amphibians and insects, plunge pools on the downstream side of dams for juvenile fish, gravel for spawning, home and food for birds. Blockades are both a negation destruction and an affirmation of life.[6]

Nonhumans have a lot to teach about building more equitable and sustainable futures. Indigenous legal systems that center specific place-based responsibilities are central to Indigenous resistance movements. Along the proposed Trans Mountain **pipeline** route, the Secwepemc Tiny House Warriors have built ten tiny homes to assert Secwepemc law and jurisdiction. An alliance of Anna's hummingbirds, Tsleil-Waututh community members, and allies halted construction for four months during the summer 2021 nesting season. In resistance to the proposed Line 3 pipeline expansion, White Earth Anishinaabeg unanimously enacted the Rights of Manoomin (wild rice) in December 2018, recognizing that wild rice in their territory has the inherent right to exist, flourish, and regenerate.[7] Manoomin has not only legal rights but also a distinct role in decision making, with the right to intervene in any action concerning their rights. Indigenous resistance movements are regenerating mechanisms for shared decision making with nonhumans by embodying governance practices and upholding responsibilities to land and waters.

Questions that remain include these: How are Indigenous place-based responsibilities applied to energy development? Can modern energy systems can be Indigenized? Indigenous peoples are developing energy systems in adherence to community-specific laws and ethical systems. The Amisk (Beaver) Solar

Project of the Beaver Lake Cree Nation is directly pushing back against tar sands extraction on their territory. The Nigig (Otter) Power Wind Farm of the Henvey Inlet Anishinabek is helping to foster economic self-determination. M'Chigeeng First Nation has established several wind turbines on their territory, through the Mother Earth Renewable Energy project. The deep respect for nonhumans and consensual basis of this project were evident at the groundbreaking ceremony when elder Alma Jean Migwans shared, "We acknowledge the birds and ask them to change their course to avoid the turbines and we ask the winds to use their power."[8] Indigenizing energy systems relies on consensuality between humans and nonhumans.

Indigenous place-based ways of knowing call for understanding energy as a gift, rather than a commodity.[9] Indigenizing energy systems encourages understanding local energy sources on an intimate, personal level: who they are, where they come from, who their relatives are, and how lands and waters are impacted when energy is gifted, generated, processed, transported, and consumed. More attentive relationships with nonhumans will entail more sensitive impact assessments and more responsible action when land-based relationships are disrupted.

See also: **Alberta, Animals, Blockade, Fire/Bushfire, North Dakota, Protest**

Notes

1. Wet'suwet'en solidarity rail blockades occurred in Tyendinaga, Toronto, Vancouver, Winnipeg, Hazelton, Hamilton, and London.
2. Leanne Betasamosake Simpson, *As We Have Always Done: Indigenous Freedom through Radical Resistance* (Minneapolis: University of Minnesota Press, 2017), 41–44.
3. Chantelle Richmond and Nancy Ross, "The Determinants of First Nation and Inuit Health: A Critical Population Health Approach," *Health and Place* 15, no. 2 (June 2009): 403–11.
4. Simpson, *As We Have Always Done*, 43.
5. Leanne Betasamosake Simpson, *A Short History of the Blockade: Giant Beavers, Diplomacy, and Regeneration in Nishnaabewin* (Edmonton: University of Alberta Press, 2021).
6. Leanne Betasamosake Simpson, "Being with the Land, Protects the Land," *Abolition Journal* (February 2020).
7. White Earth Band of Minnesota Chippewa Tribe, *Rights of Manoomin*, Resolution no. 001-19-009, White Earth Reservation Business Committee (December 2018).
8. Robin Burridge, "M'Chigeeng Launches Wind Farm, First for an Ontario First Nation," *Manitoulin Expositor*, June 29, 2011, https://www.manitoulin.com/mchigeeng-launches-wind-farm-first-for-an-ontario-first-nation/.
9. Tyler McCreary, "Between the Commodity and the Gift: The Coastal GasLink Pipeline and the Contested Temporalities of Canadian and Witsuwit'en Law," *Journal of Human Rights and the Environment* 11, no. 3 (December 2020): 122–45.

SHIPPING

What would actual free shipping look like?

CHRISTIAAN DE BEUKELAER

On July 24, 2020, the *Avontuur* arrived in the port of Hamburg carrying sixty-five tons of cargo. The sailing ship, a hundred-year-old traditionally rigged two-masted schooner, had left her home port in northern Germany on January 17. It took the crew of fifteen, of which I was one for most of the voyage, six months to sail more than fifteen thousand nautical miles and to return with a hold full of coffee, cacao, rum, and gin.[1]

The *Avontuur* is among a handful of sail cargo vessels. Others include the *Tres Hombres* and *De Gallant*. While varying in size and philosophy, they share a common goal: decarbonizing the shipping industry by returning to sail as a primary means of zero-emission propulsion.[2] After all, until the Promethean invention of the steam engine, every ship was emission-free.

Roughly a year after the *Avontuur*'s arrival in Hamburg, the Intergovernmental Panel on Climate Change warned in August 2021 about "widespread, rapid, and intensifying" climate change. Barring immediate and drastic action, "limiting warming to close to 1.5°C or even 2°C will be beyond reach."[3] The only real solution, they argued, is to reduce emissions to net zero as soon as possible. Embracing wind propulsion would surely help the shipping industry bring its emissions down to zero by 2050, right? It seems like common sense that a ship propelled by the wind creates no emissions.

While this idea may seem appealing, there are at least three reasons for caution. First, shipping is already the least carbon-intensive mode of cargo transport, even if the sheer volume of maritime shipping means that the industry emits more (2.89 percent of global emissions) than the aviation industry (2.4 percent).[4] With emissions ranging from 5 to 90 gCO_2/tkm, shipping compares favorably with every other mode of cargo transport, including train (10–60 gCO_2/tkm), truck (70–190 gCO_2/tkm), and especially **airplane** (375–2900 gCO_2/tkm).[5] While slow steaming and wind-assisted propulsion may further reduce these emissions, far more is needed to turn shipping into a zero-emission industry.[6]

Second, a great deal of carbon is embedded in the ship, the portside infrastructure, and the operations that make the shipping industry work. One might argue that internalizing emissions that occur across the industry throughout the lifetime of vessels and infrastructure is unfair. But systematic externalization and creative carbon accounting are precisely why we face climate collapse. In preparing to build the *EcoClipper*, a newly designed sailing cargo vessel with traditional rigging, the company commissioned a lifecycle emissions analysis, which found that while both the carbon emissions (2 gCO_2/tkm) and the broader environmental footprint are significantly lower than those of a container ship, they do not attain zero emissions. As neither slow steaming nor **wind** propulsion have been widely embraced, aggregate emissions from shipping are set to grow significantly over the next decades.[7]

Third, given the enormous projected growth of demand for maritime cargo transport (tripling between 2015 and 2050) and the limited progress in lowering shipping's carbon intensity, the complete decarbonization of the industry remains highly uncertain.[8] Since this goal would require massive decreases in either demand for cargo or the carbon intensity of ships, shipping volumes must decrease immediately.[9] This change would lower the industry's aggregate emissions in order to buy time for technological innovation without depleting the remaining carbon budget at the current pace. As long as emissions are not zero, aggregate emissions will pose a problem. At a billion tons of cargo per year, the industry can't afford complacency or inertia.

Newly designed ships that incorporate advances in engineering and manufacturing promise to eliminate the need for fossil fuels.[10] While some zero-emission fuels are promising, they do not eliminate the emissions in shipbuilding, operations, and portside activity.[11] Until the entire chain is fully decarbonized, shipping will never be fully emission-free. And no sailing ship will change that.

In sum, maritime cargo transport is highly efficient, thanks to its low carbon intensity and enormous scale. But going from low to zero carbon intensity proves difficult, particularly when internalizing environmental impacts of shipping beyond propulsion. To make matters worse, there is little evidence that the industry is, in fact, taking the radical climate action necessary to avoid the worst. So, whether or not fully emission-free ships exist, what will be most necessary for determining whether shipping can be emission-free is rethinking the political economy of the industry.

See also: **Design, Globalization, IPCC, Net Zero, Nonlinear**

Notes

1 Christiaan De Beukelaer, "COVID-19 at Sea: 'The World as You Know It No Longer Exists,'" *Cultural Studies* 35, no. 2–3 (May 4, 2021): 572–84, https://doi.org/10.1080/09502386.2021.1898020.

2 Christiaan De Beukelaer, "Sail Cargo: Charting a New Path for Emission-Free Shipping," *UNCTAD Transport and Trade Facilitation Newsletter*, November 5, 2020, https://unctad.org/news/sail-cargo-charting-new-path-emission-free-shipping.

3 Intergovernmental Panel on Climate Change, "Climate Change Widespread, Rapid, and Intensifying—IPCC," *IPCC Newsroom*, August 9, 2021, https://www.ipcc.ch/2021/08/09/ar6-wg1-20210809-pr/.

4 International Maritime Organisation, "Fourth IMO Greenhouse Gas Study" (London: IMO, 2020), 4; International Air Transport Association, "Fact Sheet: Climate Change & CORSIA" (Geneva: IATA, 2021).

5 R. Sims et al., "Transport," in *Climate Change 2014: Mitigation of Climate Change. Contribution of Working Group III to the Fifth Assessment Report of the IPCC*, ed. O. Edenhofer et al. (Cambridge: Cambridge University Press, 2014), 610.

6 Sarah Mander, "Slow Steaming and a New Dawn for Wind Propulsion: A Multi-level Analysis of Two Low Carbon Shipping Transitions," *Marine Policy* 75 (January 2017): 210–16, https://doi.org/10.1016/j.marpol.2016.03.018; Nishatabbas Rehmatulla et al., "Wind Technologies: Opportunities and Barriers to a Low Carbon Shipping Industry," *Marine Policy* 75 (January 2017): 217–26, https://doi.org/10.1016/j.marpol.2015.12.021; Wind Assisted Ship Propulsion, "New Wind Propulsion Technology A Literature Review of Recent Adoptions" (Wind Assisted Ship Propulsion, 2020).

7 European Parliament, "Emission Reduction Targets for International Aviation and Shipping" (Brussels: Directorate General for Internal Policies, 2015), http://www.europarl.europa.eu/RegData/etudes/STUD/2015/569964/IPOL_STU(2015)569964_EN.pdf.

8 International Transport Forum, "ITF Transport Outlook 2019" (Paris: Organisation for Economic Co-operation and Development, ITF, 2019).

9 Christiaan De Beukelaer, "Tack to the Future: Is Wind Propulsion an Ecomodernist or Degrowth Way to Decarbonise Maritime Cargo Transport?," *Climate Policy*, October 21, 2021, 1–10, https://doi.org/10.1080/14693062.2021.1989362; Simon Bullock, James Mason, and Alice Larkin, "The Urgent Case for Stronger Climate Targets for International Shipping," *Climate Policy*, October 27, 2021, 1–9, https://doi.org/10.1080/14693062.2021.1991876.

10 Nishan Degnarain, "A New Golden Age of Sailing Is Here: Where Is the Leadership?," *Forbes*, December 1, 2020, https://www.forbes.com/sites/nishandegnarain/2020/12/01/a-new-golden-age-of-sailing-is-here-where-is-the-leadership/?sh=22fa91aa1504.

11 Lloyd's Register and UMAS, *Zero-Emission Vessels 2030: How Do We Get There?* (London: Lloyd's Register & University Maritime Advisory Services, 2017); Alice Bows-Larkin et al., "High Seas, High Stakes: High Seas Final Report" (Manchester: Tyndall Centre for Climate Change Research, 2014), https://www.research.manchester.ac.uk/portal/files/40102807/High_Seas_High_Stakes_High_Seas_Project_Final_Report.pdf.

SOLAR FARM

Can solar farms foster energy justice, instead of obstructing it?

DUSTIN MULVANEY

A solar farm is a colloquial term for a utility-scale power plant that uses the sun as the source of energy. Sometimes called *solar parks*, solar farms used to generate electricity include photovoltaics—which convert the sun's photons to electricity—and concentrated solar power plants—which concentrate the heat from the sun using mirrors to make steam that turns a steam turbine. Solar farms are widely heralded as a key technology to decarbonize electricity; they emit twenty times fewer greenhouse gas emissions per unit of energy than fossil fuels.

Solar farms are also often sites of socioecological conflict. Among the issues that make them controversial are their size and placement. Projects can occupy thousands of acres, have considerable impacts on landscapes, and restructure social relations when sited in areas of ecological, agronomic, or cultural significance. As solar farms have become more widespread, so have conflict and social resistance, with some projects exacerbating existing community vulnerabilities. Research from rural **India** documents how one of the largest solar farms in the country displaced vulnerable landless peoples, who lost access to the places where they make their sustenance.[1] The centrality of land to solar means that such "green grabbing"—the privatization of public or common land for **clean** energy—is likely to continue in arid places with high-quality solar resources. In Morocco, a large-scale solar farm called the Ouarzazate (Noor) solar power station dispossessed local land users of three thousand hectares. With an eye toward the European electricity market, the project was framed as a benefit to the country's position as an energy exporter and echoed strategies the state had used to expropriate common lands under colonial rule.[2]

While solar farms promise economic and community benefits such as jobs, tax revenues, and associated economic activity, these promised benefits can be elusive. Studying a large solar farm on the Yucatán Peninsula, researchers found that the social benefits offered by developers did not resolve existing conflicts on the ground, partly because they failed to recognize the historic colonial dispossession of Indigenous peoples from common lands.[3] In such contexts, potential community benefits can seem relatively small.

Another reason solar farms cause social friction is a lack of opportunities for public participation or community consent over land use decisions. In Gujarat, India, the siting process for the Charanka Solar Park, at the time India's largest solar park, raised questions about whose voice counts in decision-making processes.[4] Local farmers and pastoralists had little input into the development process and lost important land access once the project was approved for construction. Many infrastructures are developed without meaningful public input in part because of an adversarial "decide-announce-defend" permitting process. Companies decide where a project will be located, purchase or acquire the rights to develop the land, announce it to the public, and then staunchly defend the project during public review.

The large footprint of solar farms can also displace wildlife and otherwise impact ecosystems, especially when sited in or near protected areas.[5] Large-scale solar installations can fragment habitat by impeding the movement of species across the landscape. Some studies suggest that a lake effect cue or polarized light mechanism may attract aquatic birds to solar farms in desert regions.[6] At times, "green versus green" arguments over land use sometimes pit groups working toward climate change mitigation and renewable energy against groups advocating the protection of wildlife habitat and biodiversity conservation—another unexpected outcome of energy transition.[7]

Not all solar farms cause negative changes in land use. Brownfields, industrialized sites, salt-contaminated cropland, **abandoned** mines, and retired landfills have all been developed as solar farms; these sites generate less social conflict and less additional land damage than "quality" solar habitat or open spaces.[8] Some solar farms aim for ecological restoration of their sites by enhancing pollination and other ecosystem services or by providing habitat and forage.[9] In the future, solar farms might further pursue synergies between energy and agricultural production. Agrivoltaics are a type of solar farm integrated into an agricultural production system with photovoltaic arrays arranged and spaced to allow for crop rows; others are co-sited with apiaries and honeybees. Ranchvoltaics integrate solar with places for livestock grazing. Floating solar farms—floatovoltaics—position solar over water and are sometimes integrated with aquaculture or wastewater treatment.

The shift to renewable energy will substantially increase the number of solar farms across the planet. Without adequate foresight and planning, solar farms will inevitably spur land use conflicts owing to the diffuse nature and land intensiveness of solar power. Solar farms built with community input and designed to cause minimal ecological, agronomic, or cultural resource damage are less likely to encounter frictions in development and to restructure social relations in ways that leave some communities vulnerable. The lesson of solar farms is not that the search for new energy cultures is over but that producing just energy futures will involve ongoing effort and contestation.

See also: **Animals, Commons, Design, Farm, Justice, Sustainability**

Notes

1 Ryan Stock and Trevor Birkenholtz, "The Sun and the Scythe: Energy Dispossessions and the Agrarian Question of Labor in Solar Parks," *Journal of Peasant Studies* 48, no. 5 (2021): 984–1007.
2 Karen Eugenie Rignal, "Solar Power, State Power, and the Politics of Energy Transition in Pre-Saharan Morocco," *Environment and Planning A: Economy and Space* 48, no. 3 (2016): 540–57.
3 Ivet Reyes Maturano, "Social Dispossession, the Real 'Benefit' of Green Projects in Yucatan," *Development* 65 (2022): 63–70.
4 Komali Yenneti and Rosie Day, "Procedural (In)justice in the Implementation of Solar Energy: The Case of Charanaka Solar Park, Gujarat, India," *Energy Policy* 86 (2015): 664–73.
5 Rebecca R. Hernandez, Madison K. Hoffacker, Michelle L. Murphy-Mariscal, Grace C. Wu, and Michael F. Allen, "Solar Energy Development Impacts on Land Cover Change and Protected Areas," *Proceedings of the National Academy of Sciences* 112, no. 44 (2015): 13579–84.
6 Karl Kosciuch, Daniel Riser-Espinoza, Michael Gerringer, and Wallace Erickson, "A Summary of Bird Mortality at Photovoltaic Utility Scale Solar Facilities in the Southwestern US," *PLOS One* 15, no. 4 (2020): 1–21.
7 Dustin Mulvaney, "Identifying the Roots of Green Civil War over Utility-Scale Solar Energy Projects on Public Lands across the American Southwest," *Journal of Land Use Science* 12, no. 6 (2017): 493–515.
8 Rebecca R. Hernandez, Alona Armstrong, Jennifer Burney, Greer Ryan, Kara Moore-O'Leary, Ibrahima Diedhiou, Steven M. Grodsky, Leslie Saul-Gershenz, Rob Davis, Jordan Macknick, Dustin Mulvaney, Garvin A. Heath, Shane B. Easter, Brenda Beatty, Michael F. Allen, and Daniel M. Kammen, "Techno-ecological Synergies of Solar Energy for Global Sustainability," *Nature Sustainability* 2, no. 7 (2019): 560–68.
9 Kara A. Moore-O'Leary, Rebecca R. Hernandez, Dave S. Johnston, Scott R. Abella, Karen E. Tanner, Amanda C. Swanson, Jason Kreitler, and Jeffrey E. Lovich, "Sustainability of Utility-Scale Solar Energy—Critical Ecological Concepts," *Frontiers in Ecology and the Environment* 15, no. 7 (2017): 385–94.

SOLIDARITY

How can solidarity catalyze energy democracy?

ANA ISABEL BAPTISTA

The pace and scale of energy **transitions** are catalyzing new social and political arrangements in communities around the world. These changes are breaking open possibilities for transforming not only how we produce and consume energy but also how we relate to each other. Marginalized communities are at the front lines of the intersecting crises of climate change, rapacious inequality, and ecological collapse. To find hope amid destruction, frontline communities in powerful alliances must lead transformative change to construct a new world. These alliances must emerge from what Paulo Freire calls the radical posture of solidarity: "True solidarity is found only in the plenitude of this act of love, in its existentiality, in its praxis."[1]

We often conceive of solidarity as making common cause with those who differ from us in instrumental ways. Instrumental solidarity can be motivated by shared interests or identities, but radical solidarity is driven by intersectional and collective bonds among individuals, communities, or movements. These bonds reflect a deep and abiding love. Intersectional solidarity can be an expression of radical activism; Fernando Tormos argues that "fighting oppression out of love for the other is not foreign to radical activism. . . . Activists conceptualize their own agency as emerging out of love, spirituality, and an intersectional consciousness."[2] This form of radical solidarity could enable the construction of new energy systems that link frontline communities with those who share a commitment to socially just, regenerative, reciprocal, and liberatory relationships.

How can solidarity help to shift exploitative energy systems toward more just energy futures? One approach is the social and solidarity energy economy (SSE), in which people actively take over and transform energy systems. The SSE model of energy transitions requires more decentralized, renewable energy sources; participatory democratic governance; and economic arrangements grounded in cooperation, fairness, mutuality, and reciprocity over private, concentrated profit. SSE can take various forms, including cooperative associations, participatory public utilities, or translocal associations.[3] SSEs are an expression of energy **democracy** and energy justice. Energy justice implies a fairer, more equitable distribution of the benefits and burdens of energy production and prosumption,[4] a shared approach to governing energy **commons**, and a centering of concerns on marginalized communities, as well as a recognition of their agency and the importance of their leadership.[5] Without solidarity among and between differently situated communities and social movements, the goals of energy **justice** will remain elusive.

Radical forms of energy solidarity reflect decentralized energy production, storage, and transmission systems that are democratically controlled. These energy systems are based on the production of **local** clean and renewable forms of energy, utilized in more cooperative and less consumptive ways.[6] Examples include community renewable energy (CRE) systems and mutual aid projects.[7] The Red de Apoyo Mutuo de Puerto Rico, or mutual aid network of Puerto Rico, helped to distribute, build, and train residents in generating sources of **renewable** power after Hurricane Maria.[8] A foundational principle of this network is solidarity, which they define as "a consistent and daily action based on respect, empathy and the understanding that we need each other. . . . We believe in the importance of social ties that unite people and in our ability to act as a whole for the good of all."[9] Unlike instrumental solidarity motivated by transactional interests among differently situated strangers, solidarity in this definition is grounded in a relationship of equals directly impacted by exploitative energy systems who share bonds of love of their **community**, mutual respect, and a stake in collective well-being. These types of mutual aid efforts can be the most powerful forms of resurgence for frontline communities.[10] To build more just, democratic, and decentralized energy systems will require such collective efforts motivated by radical energy solidarities.

See also: **Economy, Electricity, Finance, Solar Farm**

Notes

1 Paulo Freire, *Pedagogy of the Oppressed*, trans. Myra Bergman Ramos (New York: Seabury Press, 1970), 35.
2 Fernando Tormos, "Intersectional Solidarity," *Politics, Groups, and Identities* 5, no. 4 (2017): 713.
3 Jon Morandeira-Arca, Enekoitz Etxezarreta-Etxarri, Olatz Azurza-Zubizarreta, and Julen Izagirre-Olaizola, "Social Innovation for a New Energy Model, from Theory to

Action: Contributions from the Social and Solidarity Economy in the Basque Country," *Innovation* 37 (2024): 33–59.
4 Prosumption involves thinking about production and consumption together.
5 Shalanda Baker, Subin DeVar, and Shiva Prakash, "The Energy Justice Workbook" (Initiative for Energy Justice, December 2019), https://iejusa.org/workbook/.
6 Marieke Oteman, Mark Wiering, and Jan-Kees Helderman, "The Institutional Space of Community Initiatives for Renewable Energy: A Comparative Case Study of the Netherlands, Germany and Denmark," *Energy, Sustainability and Society* 4, no. 1 (2014): 1–17.
7 Ankit Kumar and Gerald Taylor Aiken, "A Postcolonial Critique of Community Energy: Searching for Community as Solidarity in India and Scotland," *Antipode* 53, no. 1 (2021): 200–221.
8 Red de Apoyo Mutuo de Puerto Rico, "Principios," accessed May 17, 2021, https://redapoyomutuo.com/principios.
9 Red de Apoyo Mutuo de Puerto Rico.
10 Dean Spade, "Solidarity Not Charity: Mutual Aid for Mobilization and Survival," *Social Text* 38, no. 1 (2020): 142.

SPORT

Can sport be untangled from oil?

GRAEME MACDONALD

In October 2013, a soccer match in UEFA's Champions League between FC Basel (Switzerland) and FC Schalke (Germany) was disrupted by an audacious environmental **protest**. Abseiling Greenpeace activists swooped down over the main stand and unfurled a huge banner bearing the slogan "Don't Foul the Arctic." Their spectacular stunt, beamed to a large international television audience, protested the detention by Russian authorities of the Arctic 30 group of Greenpeace members who were arrested while attempting to board an oil platform set for exploratory drilling in the Arctic.[1]

The platform was owned by Gazprom, the Russian majority-state-owned energy company that, in a controversial 2012 deal, became the official lead sponsor of UEFA's most prominent and lucrative competition. That it took a sanctions campaign enacted after the invasion of Ukraine in 2022 to convince European football's governing body to terminate a decade-long, recently extended contract spoke volumes about the peculiar, entangled relations between soccer (and other sports, including track and field, American football, basketball, ice hockey, rugby, golf, cycling, and cricket) and carbon-intensive industries.

Climate change is projected to have significant impacts on sporting activities, including athletic performance and tactics, the resilience of stadium infrastructure, playing surfaces, and equipment, fan and team travel, game times, league schedules, and media coverage.[2] As an activity with a range of participants and audiences at local and global scales, sport has the potential not only to raise climate awareness but also (especially at the elite level) to propose, mediate, and install behavioral and structural changes necessary for mitigation. The sporting industry has not been shy about advocating **sustainability** and declaring an environmental ethics of the type demanded by **net-zero** culture.[3] Some players and teams have been particularly environmentally conscious, including clubs where an authentic transition ethos is evidenced by the promotion of decarbonized operations. But such commitments remain exceptions rather than the rule, particularly in light of the tight relationship that has evolved between sport and extractive industries since the turn of the century. The link between sport and extraction is far larger than the oil industry's controversial sponsorship of art and museum institutions. This entanglement has turned sport into a battleground over who gets to command and control the transition and its ethical or ideological shape. Sport can be a medium for the continuity of petroculture, or its demise.

That Gazprom has been an official partner of FIFA and other UEFA competitions only confirms the insidious, profitable nature of this toxic relation between sport and oil. The most prominent investment outside Europe has been from the Middle East and the Gulf States, notably Qatar, Abu Dhabi, and Saudi Arabia, where the lines among private **corporation**, financing, and state power blur, as sovereign wealth funds built from oil and gas have moved beyond sponsorship to outright ownership of clubs in Europe and around the globe.[4] Some of the most historic and recognizable cultural institutions in the world—a long list of top teams including Paris Saint-Germain, Real Madrid, Barcelona, Chelsea, Manchester City, Manchester United, AC Milan, and Newcastle, with an emotionally invested audience of billions—either are owned by or have sponsorship or financing deals with the oil industry or the investment arms and extended institutions of oil states. World football, from center to periphery, from the elite level to kids' clubs, has become saturated with petroleum.

A generous reading of this expansive pattern of funding and acquisition of cultural assets would understand it as a signal—however weak—of an energy transition in action: that petroleum-wrought interests, anxious about the longevity of their fuel source and fossil-based economies, may mean what they say in their stated objectives to diversify and fund renewable and regenerative social projects at home and abroad. But the fact is that the soft power of sponsorship and the cultural acquisition of sporting symbols is usually accompanied by a form of international diplomacy that seeks opportunity for states and corporations to maintain and extend their oil operations. Indeed, soccer was essential to a diplomatic push for new energic relations cemented between Russia and Iran in 2023.[5] This fact follows a retentive geopolitical pattern where sporting ties and financial aid create pathways for core-periphery energy infrastructure and flows in many territories, from the EU to China in **Africa** and South Asia. Time and again, sport serves as beneficial cultural grease for constructing new energic chains of **development**, new markets, and new relations in the fossil energy regime.

The 2023 FIFA World Cup in Qatar—among the most premium events in world popular culture—was the apotheosis of petrosourced **finance**: the crowning achievement of fossil-fueled sporting influence and strategic planning.[6] But the previous tournament was held in Russia, with Gazprom as official partner. The next will be held across the vast North American territories of Mexico, the United States, and Canada, three historical and contemporary petrostates with significant interests in continuing oil. The link between sport and energy is set to continue unabated.

See also: **Action, Civil Disobedience, Corporation, Neoextractivism**

Notes

1 Cassady Craighill, "Greenpeace Shows Gazprom the Red Card at the Champions League Game in Switzerland" (Greenpeace, October 1, 2013), https://www.greenpeace.org/usa/greenpeace-shows-gazprom-the-red-card-at-the-champions-leagues-game-in-switzerland/.

2 See Leslie Mabon, "Football and Climate Change: What Do We Know and What Is Needed for an Evidence-Informed Response?," *Climate Policy* 23 (2023): 314–28, https://doi.org/10.1080/14693062.2022.2147895.

3 See Francis Taiyo, Joanne Norris, and Robert Brinkman, "Sustainability Initiatives in Professional Soccer," *Soccer & Society* 18, nos. 2–3 (2018): 396–406, http://dx.doi.org/10.1080/14660970.2016.1166769.

4 See Salma Thani and Tom Heenan, "The Ball May Be Round, but Football Is Becoming Increasingly Arabic: Oil Money and the Rise of the New Football Order," *Soccer & Society* 18, no. 7 (2017): 1012–26, https://doi.org/10.1080/14660970.2015.1133416.

5 Russian teams and athletes are being banned from most international sporting events and competitions. Jonathan Liew, "Russia Has Seamlessly Returned to Football—and Nobody Seems Overly Perturbed," *Observer*, March 26, 2023.

6 See Danyel Reiche, "Investing in Sporting Success as a Domestic and Foreign Policy Tool: The Case of Qatar," *International Journal of Sport Policy and Politics* 7, no. 4 (2015): 489–504, https://doi.org/10.1080/19406940.2014.966135; see also David Conn, *The Fall of the House of FIFA: The Multimillion-Dollar Corruption at the Heart of Global Soccer* (London: Yellow Jersey Press, 2017); Emma Hughes and James Marriott, *All That Glitters: Sport, BP and Repression in Azerbaijan* (London: Platform, 2015).

STORYTELLING

What can storytelling really do?

MISTY MATTHEWS-ROPER

For a research project on climate **communication**, I organized and facilitated four cli-fi book clubs. The novels we discussed led to conversations about climate anxiety, climate **science**, tentative hope for the future, and the role of stories in our lives. Some striking comments arose. One was the notion that "stories might not always be the truth, as we know it. But: they become people's truths."[1] Another concerned mobilizing people for change: "It's really stories that get people to change anything, or to do anything."[2] Story as truth and story as mobilization—both ideas can help us understand why stories matter in the face of climate collapse.

Story as truth. Our perception of ecological crisis is guided by stories. Pop culture creates climate change stories, and governmental reports offer climate narratives. The sixth Intergovernmental Panel on Climate Change (**IPCC**) report shattered any remaining illusions that climate change isn't happening or is solely a future problem.[3] The report narrates a story in which scientists' grim predictions from the 1980s have come to pass; people around the world are already feeling, physically and mentally, the catastrophic effects of a changing climate.

The story offered by the IPCC is framed in scientific language on the premise that more information will lead to **action**. Yet over the past thirty years, environmental communication researchers have shown that information alone is insufficient to inspire social change.[4] We have not an information deficit but rather a narrative deficit.[5]

Social scientists have demonstrated that human behavior is influenced more by emotions and worldviews than by scientific findings,[6] and have traced how stories reinforce ways of knowing and being in the world.[7] Many narratives tell us that our future will be defined by uncertain doom and gloom, and that only techno-scientific solutions can save us.[8] With these stories continually in the news, it is no wonder many people despair for the future.[9] If we let them, these stories will become self-fulfilling prophecies. But they don't need to become our truth; we can change our story.[10]

Story as mobilization. When talking about how stories help motivate humans, my book club participants echoed insights from literary theory: stories cultivate empathy,[11] teach us how to deal with uncertain futures,[12] and enable readers to explore their feelings (in the case of cli-fi, feelings specifically about climate change).[13] What unites these theories is their focus on affect. They contend that to deal with phenomena that are new or different, like climate change, humans require more than facts. Contemplating a climate-changed future often leads to negative thoughts and can be emotionally overwhelming.[14] In order to mobilize into action, we need to be able to talk about and engage with our feelings about this crisis.[15]

The abundance of negative climate change narratives has given rise to new genres that focus on hope. These range from optimistic solar-punk stories that imagine life after a just energy transition,[16] to hopepunk stories that show how life is worth living even after the worst comes to pass.[17] The climate is undeniably changing. But humans, too, can change and adapt. Stories can provide a space to share feelings of guilt, fear, anger, and hope.[18] They can help us redefine what it means to be human and to live a good life.

Do stories matter on a dying **planet**? One might say that capitalist and individualist stories helped to create this crisis. To move beyond these planet-killing ways, we need to create new stories. These stories should contain hopeful truths to mobilize us into action. What future do you want? Start imagining it. Talk about it with your friends and family.[19] Stories help us understand reality. They can also become our reality.

See also: **2040, Art, Community, Documentary, Permafrost, Scales**

Notes

1 P51, book club participant, Zoom meeting with the author, May 9, 2022, BC3—*Blackfish City*.

2 P26, book club participant, Zoom meeting with the author, May 9, 2022, BC3—*Blackfish City*.

3 Hans-Otto Pörtner, Debra Cynthia Roberts, Melinda M. B. Tignor, Elvira S. Poloczanska, Katja Mintenbeck, Andrès Alegría, Marlies Craig, et al., "Climate Change 2022: Impacts, Adaptation and Vulnerability: Contribution of Working Group II to the Sixth Assessment Report of the Intergovernmental Panel on Climate Change" (IPCC, 2022), https://www.ipcc.ch/report/ar6/wg2/.

4 Susanne C. Moser, "Reflections on Climate Change Communication Research and Practice in the Second Decade of the 21st Century: What More Is There to Say?," *Wiley Interdisciplinary Reviews: Climate Change* 7, no. 3 (2016): 345–69, https://doi.org/10.1002/wcc.403.

5 Siri Veland, M. Scoville-Simonds, I. Gram-Hanssen, A. K. Schorre, A. El Khoury, M. J. Nordbø, A. H. Lynch, G. Hochachka, and M. Bjørkan, "Narrative Matters for

Sustainability: The Transformative Role of Storytelling in Realizing 1.5°C Futures," *Current Opinion in Environmental Sustainability, Sustainability Governance and Transformation* 31 (April 1, 2018): 41–47, https://doi.org/10.1016/j.cosust.2017.12.005.

6 Sarah E. Wolfe and Amit Tubi, "Terror Management Theory and Mortality Awareness: A Missing Link in Climate Response Studies?," *Wiley Interdisciplinary Reviews: Climate Change* 10, no. 2 (2019): e566, https://doi.org/10.1002/wcc.566.

7 Francesca Polletta, Pang Ching Bobby Chen, Beth Gharrity Gardner, and Alice Motes, "The Sociology of Storytelling," *Annual Review of Sociology* 37, no. 1 (2011): 109–30, https://doi.org/10.1146/annurev-soc-081309-150106.

8 Cheryl Hall, "Beyond 'Gloom and Doom' or 'Hope and Possibility': Making Room for Both Sacrifice and Reward in Our Visions of a Low-Carbon Future," in *Culture, Politics and Climate Change*, ed. Deserai Crow and Maxwell Boykoff (New York: Routledge, 2014), 23–38.

9 Britt Wray, *Generation Dread: Finding Purpose in an Age of Climate Crisis* (Toronto: Knopf, 2022).

10 Andreas Malm, *How to Blow Up a Pipeline* (New York: Verso, 2021).

11 Alexa Weik von Mossner, "Why We Care about (Non) Fictional Places: Empathy, Character, and Narrative Environment," *Poetics Today* 40, no. 3 (September 1, 2019): 559–77, https://doi.org/10.1215/03335372-7558150.

12 Marco Caracciolo, *Contemporary Fiction and Climate Uncertainty: Narrating Unstable Futures* (London: Bloomsbury, 2022).

13 Matthew Schneider-Mayerson, "The Influence of Climate Fiction: An Empirical Survey of Readers," *Environmental Humanities* 10, no. 2 (November 1, 2018): 473–500, https://doi.org/10.1215/22011919-7156848.

14 Wray, *Generation Dread.*

15 Blanche Verlie, *Learning to Live with Climate Change: From Anxiety to Transformation* (London: Routledge, 2021), https://doi.org/10.4324/9780367441265.

16 Rhys Williams, "'This Shining Confluence of Magic and Technology': Solarpunk, Energy Imaginaries, and the Infrastructures of Solarity," *Open Library of Humanities* 5, no. 1 (September 13, 2019): 60, https://doi.org/10.16995/olh.329.

17 Alyssa Hull, "Hopepunk and Solarpunk: On Climate Narratives That Go Beyond the Apocalypse," *Literary Hub*, November 22, 2019, https://lithub.com/hopepunk-and-solarpunk-on-climate-narratives-that-go-beyond-the-apocalypse/.

18 Verlie, *Learning to Live with Climate Change.*

19 Manjana Milkoreit, "Imaginary Politics: Climate Change and Making the Future," *Elementa: Science of the Anthropocene* 5 (November 6, 2017): 62, https://doi.org/10.1525/elementa.249.

SUSTAINABILITY

Does sustainability's popularity indicate the failure of environmentalism?

STEPHANIE FOOTE

Everywhere I look, I see the word *sustainability* grinning smugly at me. It lounges on billboards and in magazines. It preens itself in blog posts and news articles. It's in the mouths of school administrators and blazoned on neighborhood block party invitations. It glides through car commercials and alcohol ads. It's a key term for university presidents and corporate CEOs, who set up offices of sustainability as quickly as they can get them staffed. Its ubiquity is rivalled only by its cheerful piety. What could be wrong with sustainability, with its vague but earnest promises of moral leadership and a better planetary future? It's the ideal prayer to eulogize a **planet** on life support: no one knows how it works but everyone seems to believe in it.

Because the word *sustainability* is so pervasive, it's hard to remember how recently it entered general discussions about ecological or environmental matters. The environmental movement has, over its various histories, been guided by key terms: *conservation* in the late nineteenth century, *recycling* in the late twentieth. Such terms tell us something about the focus of environmental concerns at different historical moments and reveal how provisional terms like *nature* and *the natural world*—ideas always under debate—once appeared stable enough to ground appeals to the public and investment from the government. Those terms mark a moment when debates about how to define the **scale** of environmental problems solidified around a central, often anodyne, issue; they became convenient and effective marketing tools designed to elicit state and public interest in, say, the National Parks project or the bottle return bill. If they reveal something about how the general public was enlisted to engage environmental issues, they also paper over debates about other ways to approach and define those issues, other histories in which the future could have been otherwise.

What, then, does *sustainability* tell us about the debates that have shaped environmental thinking in the late twentieth and early twenty-first centuries?

Unlike *recycling*, *sustainability* was not an invention of the adman or the marketer. It began its current celebrity life in *Our Common Future*, the 1987 Brundtland Report published by the World Commission on Environment and Development. That report asserted that if economic development continued on its present course, it would outstrip the planet's carrying capacity, intensify the wealth gap among and within nations, accelerate resource wars, and increase catastrophic weather events. The report's recommendations for addressing the unequal impact of environmental harms, balanced against the need to create economic and social opportunities for people in the developing world, relied on the argument that the measure of successful industrial or economic development should not be the accrual of wealth but instead its effects on the planet's future inhabitants.[1]

The Brundtland Report used *sustainable* as a modifier: sustainable development, sustainable agriculture, sustainable growth. *Sustainable* was a way to rethink overlapping environmental crises and solutions spanning economics, law, politics, culture, food systems, transportation, and a host of other concerns. Since 1987, sustainable growth and **development** have become important factors in how international coalitions, NGOs, and other climate change governmental task forces have framed their commitment to mitigating climate change while also signaling that they understand the inequities caused by the history of capitalist depredation. In this sense, sustainability seems to promise a set of strategies that can mitigate past harms while imagining a shared future. The U.S. government's EPA website, for example, notes that "the National Environmental Policy Act of 1969 committed the United States to sustainability, declaring it a national policy 'to create and maintain conditions under which humans and nature can exist in productive harmony, that permit fulfilling the social, economic and other requirements of present and future generations.'"[2]

And here, of course, is the rub. What economic requirements do present and future generations need? How can sustainability possibly promote harmony and balance when the system to which it is linked is the root cause of climate change and the immiseration and poverty of most of the world's people? What is sustainable for global capital has not historically proven all that sustainable for many of the people who encountered it. Sheered from its status as a modifier, *sustainability* has become an empty signifier, a word that cannot even claim the barest political charge. Now used by the very governments and industries that it was meant to change, *sustainability* is a synonym for efficiency, for shitty business as usual but with recycled toilet paper.

Yet the term endures because we have yet to find a better one to grip the popular imagination. Diffused and defused, *sustainability* has become a free-floating promise that global capitalism will somehow not destroy the earth. It seems to promise that the future is just around the corner, that it can be meticulously managed and wildly imagined at the same time.

What if the real trouble with sustainability is not that it promises to usher in a better future, but that it has no real conception of a future in which some lifeworlds don't have to be extinguished to make room for the uncontrolled growth of others? What if sustainability relieves citizens of the urgent political task of recognizing the cascading extinctions that will surely follow from the present moment's inability to fully understand how ravenously hungry is the extractive machine of capital expansion? As a promise from corporations to subjects, sustainability assures us that resources can be replaced if we use them or invent them or manage them wisely. But its diagnosis of the failure of that promise is that subjects can no longer believe in a future that looks just like the present. What, after all, can be replaced if places themselves and the lifeworlds that inhabit them have disappeared?

See also: **Action, Commons, Economy, Extinction, Finance, Gaslighting, Lifestyle, Waste**

Notes

1 See World Commission on Environment and Development, "Our Common Future" (1987), https://sustainabledevelopment.un.org/content/documents/5987our-common-future.pdf.
2 See U.S. Environmental Protection Agency, "Learn about Sustainability" (October 16, 2023), https://www.epa.gov/sustainability/learn-about-sustainability.

TRANS-

What is trans- about transition?

EMERSON CRAM

The prefix *trans-* denotes movement through, over, and across boundaries, states, and categories of all kinds, both human and nonhuman.[1] In petromodernity, *trans-* organizes a constellation of relations and malleabilities in this sense of movement across. Crude oil moves through pipelines named "TransCanada," "Transcontinental Pipeline," or "Energy Transfer Partners," across vast swaths of transcontinental or regional space, often in sovereign

Indigenous lands and waters. Refineries transform crude oil into a catalogue of petroleum products, some powering modes of transportation, others on their way to become asphalt, fertilizer, plastics, lipstick, or vinyl records. These modes of building life depend on the relationality of *trans-*, even as they efface what this prefix could offer in imagining futures beyond fossil fuels. What work could *trans-* do in imagining and working toward a just transition? How could the categorical ruptures or reconfigurations associated with *trans-* contribute to processes of worldmaking and regenerative practices of anticolonialism and anti-extractivism?[2]

In the context of petromodernity, *trans-* centers the multiple categorical crossings necessary to produce petroculture as "normal life." This normative aspect of petroculture involves a particular story of the human and the human body, in relation to extractive processes that produce capitalist value.[3] The disciplinary mechanisms that constitute this normativity foster affective attachments to those regimes of value and the technologies that enable them. Petroculture's normative condition is what Stephanie LeMenager has called "living infrastructure," or modes of living within the vast infrastructures enabled by petroleum since the mid-twentieth century.[4] The **normal** life of petromodernity encompasses embodied memory and emotional attachments to its infrastructural aesthetics. But the task of imagining different energy futures provokes questions about who has the authority to imagine what or how bodies *should* be and how value *should* be generated. The normate constitution of the human form is among the many inheritances of making the modern world through extractive and multiscalar violence.[5] As Mary Annaïse Heglar reminds us, for example, enslaved persons and plantation labor constituted the dominant mode of energy production in the Americas before the consolidation of modern energy infrastructures, deepening the entanglement of fossil fuels and anti-Blackness.[6] The *trans*-ition in *just transition* may open up ethical questions about what bodies *could be*, beyond regimes that demand the endless accumulation of value.

Stories about extraction in the energy humanities might open new sites of political inquiry if we imagined life in this conjuncture as a normative entanglement of bodies and technologies bound up in regimes of value. Considering what extraction stories tell us about knowledge practices and regimes of value, Imre Szeman and Jennifer Wenzel call for more precise differentiations between the *act* of extraction and extractivism as a name for the logics, ideologies, and cultural rationales structuring and legitimating such **action**.[7] The hyphen attached to *trans* calls attention to the underlying logics and technologies that create energy relations, sometimes by force. That hyphen "marks the difference between the implied nominalism of *trans* and the explicit relationality of *trans-*," the latter of which resists object specificity.[8] The unruliness of such a frame has much to offer thinkers, practitioners, and artists at a moment when we urgently need a regenerative energy regime and different modes of imagining and materializing bodily difference and social relationships.

Not all boundary crossings entail violence. Not all attachments tend toward the normative. The energy humanities have much to learn from trans and disability studies in relation to normative conditions and possibilities of resistance. Trans-analytics would ask what kinds of boundary crossings become normalized and taken for granted and what kinds are criminalized, deemed unnatural, or rendered socially and economically disposable. The queer and trans of color disability collective Sins Invalid details ten principles of disability justice, three of which intersect with the project of transforming energy politics:

> Anti-Capitalist Politics: In an **economy** that sees land and humans as components of profit, we are anti-capitalists by the nature of having non-conforming body/minds.
>
> Recognizing Wholeness: People have inherent worth outside of commodity relations and capitalist notions of productivity. Each person is full of history and life experience.
>
> Sustainability: We pace ourselves, individually and collectively, to be sustained long term. Our embodied experiences guide us toward ongoing **justice** and liberation.[9]

These principles of disability justice, like those articulated by the Just Transition Alliance, challenge regimes of value premised on **resource** logics, and they look toward a more regenerative horizon for environmental futures. They do so by privileging interdependence and care over and above the exploitation, disablement, and disposability of human and nonhuman life. These principles make it possible to imagine transition as transformation, a crossing toward a more just future for all.

See also: **Black, Nonlinear, Pipeline, Settler Colonialism, Transitions**

Notes

1 Oxford English Dictionary, s.v. "trans- (prefix), sense 1," accessed Sept. 3, 2024, https://doi.org/10.1093/OED/9149015485.

2 Just Transition Alliance, "What Is a Just Transition" (n.d.), http://jtalliance.org/what-is-just-transition/; Catalina de Onís, *Energy Islands: Metaphors of Power, Extractivism, and Justice in Puerto Rico* (Berkeley: University of California Press, 2021).
3 E Cram, *Violent Inheritance: Sexuality, Land, and Energy in Making the North American West* (Berkeley: University of California Press, 2022).
4 Stephanie LeMenager, *Living Oil: Petroleum Culture in the American Century* (New York: Oxford University Press, 2014).
5 Cram, *Violent Inheritance*; Armond R. Towns, *On Black Media Philosophy* (Berkeley: University of California Press, 2022).
6 Mary Annaïse Heglar, "Forgive Humans, Not Oil Companies," *The Nation*, November 16, 2021, https://www.thenation.com/article/environment/fossil-fuel-prison-abolition/; C. Riley Snorton, *Black on Both Sides: A Racial History of Trans Identity* (Minneapolis: University of Minnesota Press, 2017).
7 Imre Szeman and Jennifer Wenzel, "What Do We Talk about When We Talk about Extractivism," *Textual Practice* 35, no. 3 (2021): 506.
8 Susan Stryker, Paisley Currah, and Lisa Jean Moore, "Introduction: Trans-, Trans, or Transgender?," *Women's Studies Quarterly* 36 (2008): 11.
9 Sins Invalid, "10 Principles of Disability Justice" (September 17, 2015), https://www.sinsinvalid.org/blog/10-principles-of-disability-justice.

TRANSITIONS

If transition is inevitable, will it be just?

MATTHEW S. HENRY

A common mantra for climate justice activists is "transition is inevitable, but **justice** is not." This idea can teach us something about how we think about energy transitions. First, let's consider inevitability. Many energy transitions are occurring simultaneously, with decreasing costs of **wind**, solar, and **battery** storage; declining coal markets and a rise in stranded fossil fuel assets; the growing popularity of electric vehicles; and increasing preference for utility-scale low-carbon energy. But not all transitions are alike. While the technology necessary for decarbonizing the U.S. economy exists, significant political challenges remain. In Wyoming, opposition to renewables is accompanied by strenuous efforts to preserve the coal industry, a longtime primary source of state revenue. But Colorado, a major natural gas producer, established an Office of Just Transition and unveiled in 2019 an expansive plan to transition to 100 percent **renewable** energy by **2040**. In the Global South, questions of climate colonialism have inflected transition debates, including the phaseout versus phasedown controversy at COP26, where **India** resisted calls from Global North countries to abandon a fuel crucial to economic development. It is perhaps more useful to discuss energy transitions in the plural, as dependent on an array of technological, economic, political, and social factors that interact across spatial and temporal scales.

There are many possible paths to prevent a global average temperature rise of 2°C, per the 2015 **Paris Agreement**. Historically, however, most energy transitions are protracted processes. It took seventy years for petroleum to capture 25 percent of the U.S. market. It took coal nearly 130 years to reach that level of market share. To be sure, there exist notable exceptions in the adoption of end-use technologies—cookstoves in China and air conditioners in the United States—in national-scale shifts in energy supply, like natural gas in the Netherlands in the sixties and nuclear in France in the seventies and eighties. Rapid transitions are indeed possible and are more likely when politically prioritized in response to resource scarcity or crises like climate change.[1]

Questions of timescale are questions of justice. Who pays for energy transitions and when? The concept of *just transition* seeks to reconcile the urgency of climate action with economic equity in regions historically dependent on the fossil energy economy. For coal and power plant communities in the United States and Europe, *just transition* refers to worker retraining and economic diversification initiatives to facilitate approaches to decarbonization that prioritize **community** reinvestment. Financing and implementation of such initiatives vary, from multigovernmental funding initiatives like the Structural Reform Support Service in the European Union to a combination of governmental, philanthropic, and private industry funding in the United States, such as the Just Transition Fund and Appalachian Regional Commission.

While emerging initiatives focus on economic metrics and technological innovation, the key remaining hurdles for decarbonization are sociopolitical. Despite consensus that fossil fuels are primary drivers of climate change, they remain central to the global political imaginary. Moreover, cultural ties to extractive industries are well-documented in fossil-fuel-producing regions. In Appalachia and the U.S. Intermountain West, pride in a cultural heritage of fossil fuel labor exists in tension with a professed affinity for public lands, outdoor recreation, and place-based identity, complicating discussions about transition. Speaking against the fossil fuel industry, or even mentioning transition, can lead to social ostracization and political censorship. At their most extreme, such dynamics dovetail with the growing threat of militant climate denialism, "fossil fascism," and white supremacist militia movements on the far right.[2] Many energy transition communities face significant obstacles to obtaining structural support.

It is also important to account for the historically racialized dimensions of energy regimes, including **clean** and renewable energy. Writing of solar transitions, Myles Lennon evokes the Black Lives Matter movement to emphasize that "colonialism transformed energy—the ability to change matter—into a commoditized form that made certain lives not matter."[3] For example, the Obama-era Genesis Solar Energy Project in California caused friction between the Colorado River Indian Tribes and the federal government over the destruction of Mohave cultural artifacts. New Mexico's 2019 Energy Transition Act set ambitious goals for decarbonization and economic relief for tribal communities impacted by the closure of coal-fired power plants. But Indigenous advocates were disappointed by the bill's failure to address the legacies of **settler colonialism**, including the impacts of coal mining on public health and traditional lifeways.[4] Additionally, supply chain issues have enormous implications for end-use technologies in the Global North. The extraction of nickel necessary for cathode batteries in electric vehicles has negatively impacted the lands of the Indigenous Dolgan peoples of the Russian Arctic.[5] Allegations of forced labor and coal use to mine Chinese silicon necessary for solar panels raise issues about the **sustainability** and ethics of the solar boom.[6]

The inevitability of *just* energy transitions is far from assured. An overemphasis on economic indicators in transition planning can reproduce social hierarchies that further marginalize Black, Indigenous, low-income, and Global South communities. Climate justice frameworks, with their capacious view of equity amid change, offer one way forward. The Climate Justice Alliance defines just transition as the "shift from an extractive economy to a regenerative economy,"[7] which includes but extends beyond economic justice for fossil fuel communities to include racial justice, food sovereignty, and a commitment to antiracist and anticolonial futures. Within this framework, energy transitions must be place-based, community-driven, and responsive to systemic inequities. Transition policy must be characterized by a commitment to energy democracy, a shift from a centralized, corporate energy economy to one governed by and reflective of the values of **local** communities and conducive to health and well-being for all.

See also: **Black, Net Zero, Nonlinear, Retrofit, Trans-**

Notes

1 Benjamin Sovacool, "How Long Will It Take? Conceptualizing the Temporal Dynamics of Energy Transitions," *Energy Research & Social Science* 13 (2016): 202–15.

2 Cara Daggett, "Petro-Masculinity: Fossil Fuels and Authoritarian Desire," *Millennium: Journal of International Studies* 47 (2018): 32.
3 Myles Lennon, "Decolonizing Energy: Black Lives Matter and Technoscientific Expertise amidst Solar Transitions," *Energy Research & Social Science* 30 (2017): 18–27.
4 Emma Foeheringer Merchant, "New Mexico's 100% Clean Energy Law Praised for Worker Retraining, but Community Concerns Remain," *Green Tech Media*, March 29, 2019, https://www.greentechmedia.com/articles/read/new-mexico-100-clean-energy-law-community-concerns.
5 Maddie Stone, "Russian Indigenous Communities Are Begging Tesla Not to Get Its Nickel from This Major Polluter," *Grist*, September 21, 2020, https://grist.org/justice/russian-indigenous-communities-are-begging-tesla-not-to-get-its-nickel-from-this-major-polluter/.
6 Morgan Bazilian and Dustin Mulvaney, "To Lead the Green Energy Future, Solar Must Clean Up Its Supply Chains," *World Economic Forum*, September 28, 2021, https://www.weforum.org/agenda/2021/09/to-lead-the-green-energy-future-solar-must-clean-up-its-supply-chains/.
7 Climate Justice Alliance, "Just Transition: A Framework for Change" (n.d.), https://climatejusticealliance.org/just-transition.

TREATY

How can treaties help states transition away from fossil fuels?

MOHAMED ADOW, REBECCA BYRNES, CARLOS LARREA, AND BHUSHAN TULADHAR

Efforts to limit demand for fossil fuels have failed to effect measurable reductions in greenhouse gas emissions, which has spurred growing calls to limit or end the supply of fossil fuels. A proposed Fossil Fuel Non-Proliferation Treaty aims to do just this: to end expansion of new fossil fuel production, phase out existing production, and support an equitable transition for workers, communities, and economies.[1]

Why a treaty? International cooperation would help align fossil fuel production plans with climate goals and create obligations for which governments can be held to account. By committing all countries to end production (with different timelines based on each country's wealth and fossil fuel dependence), a treaty could counter arguments that one country reducing its fossil fuel production will simply cause another to increase their production by a corresponding amount (known as "carbon leakage").

A treaty could also include mechanisms through which wealthy countries can subsidize fossil-fuel-dependent countries in the Global South. Some countries are locked into fossil fuel dependency and rely on production revenues to feed their debt (Ecuador); others are considering exploiting newly discovered fossil deposits as a revenue source (Kenya), just as countries in the Global North have done for centuries; still other countries are highly dependent on fossil fuel imports for energy use (Nepal) and need assistance to transition to clean energy.[2]

Such support could be provided through a Global Just Transition Fund and other innovative mechanisms such as debt for nature swaps, an idea that Ecuador tried with the Yasuni-ITT Initiative, which pledged to leave oil reserves untapped in return for programs to reduce dependence on oil extraction.[3] Support could be directed to ensuring 100 percent renewable energy access, diversifying economies to reduce reliance on revenues from fossil fuel production, and providing a just transition for workers and communities.

Negotiating a new treaty is not without challenges. Fossil-fuel-producing countries are unlikely to sign onto an instrument that restricts their ability to produce fossil fuels, at least initially. However, the 2017 Treaty on the Prohibition of Nuclear Weapons demonstrated that a critical mass of non-nuclear-armed states could agree upon obligations that were stronger—including a ban on nuclear weapons—than if nuclear-armed states had a seat at the table. Enshrining this ban in international **law** created normative and political pressure on nuclear-armed states, even if they are not parties to the treaty.[4] A critical mass of non-, small, and midsize fossil fuel producers establishing a fossil fuel treaty could generate similar political pressure, with innovative mechanisms to influence major producers, such as a commitment to purchase fossil fuels only from treaty signatory countries committed to phasing out production.

Some critics have argued that a fossil fuel treaty can't come fast enough to reduce fossil fuel production at the scale and speed needed. Fortunately, **evidence** shows that once a group of countries agrees to negotiate a treaty, negotiations can be concluded in as little as a year.[5] Even were the process to take longer, social movements can create significant political change when they coincide with political windows of opportunity and frame their demands compellingly.[6] Significant public momentum is already behind the call for a treaty, with cities and subnational governments, parliamentarians, scientists and academics, and many other organizations supporting the public campaign.[7] This support has been driven partly by the persuasive framing of fossil fuels as weapons of mass destruction, analogous to the threat posed by nuclear weapons. The Fossil Fuel Non-Proliferation Treaty is also working with local partners to adapt the framing to country-specific contexts. For example, in India health and development tend to be the most effective issues in communicating about the need to end fossil fuel extraction.[8]

And political windows are beginning to emerge. In 2021, the International Energy Agency recognized that in order to limit global warming to less than 1.5°C, no new fossil fuel projects can be built. The 2021 **IPCC** report led the UN secretary-general to announce a "death knell for fossil fuels," which were named as a key component of climate mitigation in the decision text of the UN Framework Convention on Climate Change negotiations for the first time since 1992.[9] In November 2021, fourteen countries and provinces jointly launched the Beyond Oil and Gas Alliance, a soft-law coalition of countries committed to phasing out production.[10]

The public campaign and changing political landscape bring the adoption of a fossil fuel treaty—and a global, equitable fossil fuel phaseout—within closer reach. While a treaty is no silver bullet and may take time to establish, the significant buy-ins from civil society, subnational governments, parliamentarians, and experts demonstrate that it is a compelling idea, commensurate with the scale of the threat posed by climate change. A treaty will not only help prevent catastrophic climate change but also address many structural challenges of the fossil fuel era by facilitating a transition to a new global energy and economic system.

See also: **Action, Democracy, Justice, Keep It in the Ground, Organize, Populism**

Notes

1. Peter Newell and Andrew Simms, "Towards a Fossil Fuel Non-Proliferation Treaty," *Climate Policy* 20, no. 8 (July 8, 2019): 1–12, https://doi.org/10.1080/14693062.2019.1636759; Fergus Green and Richard Denniss, "Cutting with Both Arms of the Scissors: The Economic and Political Case for Restrictive Supply-Side Climate Policies," *Climatic Change* 150, no. 1–2 (September 2018): 73–87, https://doi.org/10.1007/s10584-018-2162-x.
2. Greg Muttitt and Sivan Kartha, "Equity, Climate Justice and Fossil Fuel Extraction: Principles for a Managed Phase Out," *Climate Policy* 20, no. 8 (May 31, 2020): 1–19, https://doi.org/10.1080/14693062.2020.1763900.
3. Science Panel for the Amazon, "The Amazon We Want," accessed August 31, 2021, https://www.theamazonwewant.org/.
4. Nick Ritchie and Ambassador Alexander Kmentt, "Universalising the TPNW: Challenges and Opportunities," *Journal for Peace and Nuclear Disarmament* 4, no. 1 (January 2, 2021): 70–93, https://doi.org/10.1080/25751654.2021.1935673.
5. International Campaign to Ban Landmines, "Frequently Asked Questions: The Treaty," accessed February 21, 2022, http://www.icbl.org/en-gb/the-treaty/treaty-in-detail/frequently-asked-questions.aspx.

6 Georgia Piggot, "The Influence of Social Movements on Policies That Constrain Fossil Fuel Supply," *Climate Policy* 18, no. 7 (August 9, 2018): 942–54, https://doi.org/10.1080/14693062.2017.1394255.
7 Personal communication with Poland Michael, campaign director, Fossil Fuel Non-Proliferation Treaty, 2022.
8 Gurprasad Gurumurthy et al., "No Future for Fossil Fuels" (white paper, Climate Action Network South Asia, March 2021).
9 UNFCCC, "Decision 1/CP.26" (November 12, 2021), https://unfccc.int/sites/default/files/resource/Overarching_decision_1-CP-26_0.pdf.
10 Beyond Oil and Gas Alliance, "Redefining Climate Leadership," https://beyondoilandgasalliance.org/.

VEGAN

What does transition look like on your plate?

ALLISA ALI AND JORDAN B. KINDER

It's become a commonplace that the widespread adoption of plant-based diets could be integral to combating climate change. Studies continue to show that key drivers of climate change include industrial and animal agriculture. Meat and dairy production accounts for approximately 78 percent of all agricultural greenhouse gas emissions (GHGs),[1] and contributes to at least 14.5 percent of total global GHGs.[2] Worldwide meat production is on track to increase 73 percent by 2050, and milk production will likely increase 58 percent.[3] Given these trends, a widespread adoption of veganism, in which animal products are excluded from one's diet and use, may be an essential component of a greener, more equitable future.

Energy transition is as much a cultural pursuit as it is a technological or political one. This axiom also speaks to the political imagination that underwrites our collective relationship to food. What promises and challenges arise when situating diet in transition discourse? Shifts in relationships to food are shaped by a tension between individual ethical positions expressed through consumer choice, on the one hand, and a need for broader structural transformation, on the other. For an energy transition, individually opting out of a fossil-powered **lifestyle** is not sufficient. After all, those in the Global North reside in a petroculture in which our everyday lives are conditioned by the combustion of fossil fuels. The same may be said of the Western or standard American diet, heavy on meat and dairy. Does veganism offer a lifestylist micro-solution available only to a privileged **class**, or can it inform a larger transformative politics that insists upon structural changes beyond the individual?

An answer to this question must begin with an important recognition: there is not one single veganism, but many, each with differing motivations in particular historical contexts. Early forms of vegetarianism were informed by religioethical concerns. Numerous religions—from Buddhism to Jainism—follow a vegetarian diet in order to avoid inflicting pain on **animals**. In 1944, British woodworker Donald Watson coined the term *vegan* to denote more rigid vegetarian diets that also exclude dairy and eggs. Since then, veganism has spread widely across the planet, adopted by those motivated by health concerns or moral considerations.[4] These strands of ethical vegetarianism have historically been informed by concerns surrounding the well-being of nonhuman animals, but recent turns to veganism are often inspired by environmental commitments.

Veganism has often been understood as a privileged pursuit accessible only to those with the resources to pattern lives upon norms that counter dominant ones. And within the animal rights movement, to which much of Western veganism is historically bound, the whiteness is blinding. As Jessica Hart has argued, activist-oriented veganism in the Global North has become a "colonised space" where Indigenous rights and sovereignty are devalued and oppressed in pursuit of animal rights and environmental goals.[5] Targeting Indigenous seal hunting among Inuit peoples in the eighties and nineties, for instance, vegan activists undermined self-determination and autonomy, in ways consonant with practices of **settler colonialism**.[6] Despite increasing BIPOC representation in vegan communities, such as **Black** veganism in the United States, many contemporary approaches to veganism remain politically troubling. Historian Troy Vettese calls for a regime of compulsory veganism in order to meet pressing environmental goals, in which all citizens must adhere to a plant-based diet. These sorts of arguments tend to homogenize diverse populations and ignore cultural and social implications rather than incorporate diverse voices and perspectives.[7]

An inspiring vision for the place of veganism in a lived climate politics is offered by what Jason Hannan calls "climate veganism," that is, a veganism whose tenets emerge as much out of concern for climate change as for animal welfare.[8] Climate veganism speaks to a potential role of veganism beyond lifestyle choices. As Matt Huber describes it, lifestyle environmentalism "sees modern lifestyles—or what is sometimes called *our way of life*—as the primary driver of ecological problems," and posits solutions to these problems in these terms.[9] Such lifestyle environmentalisms posit the individual as the site of transformation, an orientation that avoids recognizing the larger power relations that produce a problem like climate change in the first place.

Whither veganism in the pursuit of a more socially, ecologically, and energetically just future? So long as veganism remains shackled to lifestyle environmentalism, the possibilities for its widespread adoption seem both unlikely and politically questionable. Yet, as the Intergovernmental Panel on Climate Change's vice chair pointed out in a press call on the publication of Working Group I's contribution to the IPCC's Sixth Assessment Report, to avoid catastrophic warming "requires unprecedented transformational change."[10] Alongside calls for a just energy transition, we need to advocate for a just food and agricultural transition that adopts climate veganism.

See also: **Behavior, Cycling, Farm, Habit, Justice, Transitions**

Notes

1. Christina Sewell, "Removing the Meat Subsidy: Our Cognitive Dissonance around Animal Agriculture," *Journal of International Affairs* 73, no. 1 (Fall 2019 / Winter 2020): 310.
2. Oliver Lazarus, Sonali McDermid, and Jennifer Jacquet, "The Climate Responsibilities of Industrial Meat and Dairy Producers," *Climatic Change* 165, no. 1 (2021): 30.
3. Bingli Clark Chai et al., "Which Diet Has the Least Environmental Impact on Our Planet? A Systematic Review of Vegan, Vegetarian and Omnivorous Diets," *Sustainability* 11, no. 15 (2019): 4110.
4. Claire Suddath, "A Brief History of Veganism," *Time*, October 30, 2008, https://time.com/3958070/history-of-veganism/.
5. Jessica Hart, "A Critically Reflective Approach to Veganism: Implications for Indigenous Rights and Green Social Work," *Social Alternatives* 39, no. 3 (2020): 29.
6. Alethea Arnaquq-Baril and Aaju Peter, dirs., *Angry Inuk* (Eye Steel Film, National Film Board of Canada, Unikkaat Studios, 2016).
7. Troy Vettese, "To Freeze the Thames: Natural Geo-Engineering and Biodiversity," *New Left Review*, no. 111 (2018): 63–86.
8. Jason Hannan, "Introduction," in *Meatsplaining: The Animal Agriculture Industry and the Rhetoric of Denial*, ed. Jason Hannan (Sydney: Sydney University Press, 2020), 16.
9. Matt Huber, "Ecological Politics for the Working Class," *Catalyst* 3, no. 1 (Spring 2019), https://catalyst-journal.com/2019/07/ecological-politics-for-the-working-class.
10. Brian Kahn and Dharna Noor, "The IPCC Warns This Is a Make-or-Break Decade for Humanity," *Gizmodo*, August 9, 2021, https://gizmodo.com/the-ipcc-warns-this-is-a-make-or-break-decade-for-human-1847444573.

WASTE

What is at stake in using waste for energy?

AARTI LATKAR AND AALOK KHANDEKAR

In **India**, municipal waste management has become a matter of urgent concern. Alongside the increasing amount and complexity of waste, the limits of conventional landfill-oriented models of waste management have also become more apparent. It is no longer sufficient to envision waste management as merely dumping degradable organic waste at sites relatively removed from population centers where it decomposes on its own. Municipal waste today can be extremely heterogenous in its composition, comprising everything from organic matter to various kinds of plastics, (unclaimed) human and animal carcasses, and biomedical waste. Such waste doesn't easily degrade on its own; it tends to generate a strong stench that can carry over large distances; and it produces leachates and other pollutants that contaminate soil, **water**, and air, presenting potential health hazards for local populations. The siting of new landfills thus almost inevitably elicits pushback from communities living in the vicinity of proposed landfill sites. Waste, therefore, often becomes socially and politically contentious, in addition to being environmentally harmful.

Responding to these issues, waste governance in India has received significant attention over the past two decades, resulting in a series of high-profile policy initiatives such as the Swachh Bharat Mission (Clean India Mission, 2014) and regulatory interventions such as the Solid Waste Management Rules and Regulations, first released in 2000 and revised in 2016. These regulations discourage open dumping and burning of waste but continue to promote centralized waste processing and management strategies. Such solutions often take the form of universalized, top-down, and expert-driven impositions that rarely align with the sociopolitical contexts in which they are to be implemented. They also demonstrate a general trend toward privatizing waste management, which introduces a new imperative of profitability for owners and shareholders. Waste has always been a source of economic value, heretofore supporting a rich informal economy of recycling (though itself fractured along axes of **class**, caste, and **gender**).[1] New arrangements for waste management call this informal economy into question and raise new questions about labor and livelihoods.

Administrative, popular, and scholarly attention to waste has often focused on the **scale** of cities. Shifting attention to other scales, such as that of a town, foregrounds other, newer challenges to waste management. The volumes of waste generated in towns are more modest, with small profit potentials unlikely to entice private investors. Towns are also politically subordinate, with less decision-making authority than cities, making it more difficult to invest in waste management infrastructures and arrangements sensitive to preexisting practices and other contextual particularities. Universalized approaches to waste management are often championed without regard to local conditions. Consider, for example, the deployment in towns of high-tech dump trucks with such advanced features as separators to segregate dry and wet waste, shutters to keep the waste covered, and a hydraulic jack to facilitate easy dumping. In practice, the narrow roads found in many town neighborhoods will require new tasks for waste workers operating these trucks, as they have to haul waste manually from houses inaccessible to them. And shutters limit the amount of waste that can be accommodated per trip, resulting in an increased number of trips to dump sites, more fuel usage, and greater transport costs. Other challenges, including the absence of at-source segregation or reliable roads leading to dump sites, also add to workers' woes. Thus, in effect, the seemingly upgraded waste management technology of the dump truck worsens labor relations and makes waste collection less rather than more efficient.[2]

Similarly, while government-aided schemes enable acquisition of new trucks for towns, the accompanying imaginaries of privatized waste management often assume that waste collection firms (rather than the town) will own these more sophisticated dump trucks. However, the lack of an extant, reliable infrastructure of waste management in many towns often means that there are few takers for these new arrangements. Often, the result is practices of waste management that are uncertain and undetermined over long periods of time,[3] so that it becomes difficult to identify the locus of responsibility—and accountability—in the (mis)management of waste.

One potential solution to these waste management challenges that has recently garnered support and funding from government agencies is new facilities that convert waste to energy. For larger urban centers, waste-to-energy conversion plants continue to be a preferred solution, even though nearly half of already-established plants have failed to function as envisioned, in part because of the lack of adequate amounts of waste with sufficient calorific value for effective plant operation.[4] Although many technologies can be employed to harness energy from waste, an important prerequisite for such technologies to yield results is consistency in the quality and composition of waste.[5] Achieving this consistency might be possible in the context of industrial waste, but usually proves impossible with municipal waste: while national policies champion the production of energy from waste and mandate the segregation of waste at source, such measures are seldom implemented effectively in practice. One reason for this limitation is that policy mandates rarely account for the labor associated with waste segregation, which relies on and reproduces preexisting gender, caste, and class hierarchies.[6]

In the context of the Global South, attention to waste thus foregrounds multiple **transitions:** of energy, **development**, and urbanization. And even as waste comes to be reimagined as a resource and fuel source in the development of clean technologies, questions remain: Under what conditions, and for whom, can waste become a **resource**? What relations of sociopolitical power are at stake in treating waste as a source of energetic power?

See also: **Commons, Finance, Industrial Revolution, Justice, Renewable**

Notes

1 Vinay Gidwani, "The Work of Waste: Inside India's Infra-Economy," *Transactions of the Institute of British Geographers* 40, no. 4 (2015): 576, https://doi.org/10.1111/tran.12094; Barbara Harriss-White, "Formality and Informality in an Indian Urban Waste Economy," *International Journal of Sociology and Social Policy* 37, no. 7/8 (January 2017): 417–34, https://doi.org/10.1108/IJSSP-07-2016-0084.

2 Kaveri Gill, "The Environment as Disingenuous Trope: Tracing Waste Policy and Practice in a Medium Hill Town of the Himalayas, India," *Journal of Developing Societies* 37, no. 2 (June 2021): 184–200, https://doi.org/10.1177/0169796X211001246.

3 Aarti Latkar, "Infrastructural Limbo: Management of Municipal Waste in a Town Setting" (MA thesis, IIT Hyderabad, 2021), 42.

4 Sunita Narain and Swati Singh Sambyal, *Not in My Backyard: Solid Waste Management in Indian Cities* (New Delhi: Centre for Science and Environment, 2016), 28–33.

5 Lal Chand Malav, Krishna Kumar Yadav, Neha Gupta, Sandeep Kumar, Gulshan Kumar Sharma, Santhana Krishnan, Shahabaldin Rezania, et al., "A Review on Municipal Solid Waste as a Renewable Source for Waste-to-Energy Project in India: Current Practices, Challenges, and Future Opportunities," *Journal of Cleaner Production* 277 (December 2020): 13, https://doi.org/10.1016/j.jclepro.2020.123227.

6 Aman Luthra, "Housewives and Maids: The Labor of Household Recycling in Urban India," *Environment and Planning E: Nature and Space* 4, no. 2 (March 2020): 475–87, https://doi.org/10.1177/2514848620914219.

WATER

How can microbe metabolism foster sustainable infrastructures?

MELODY JUE

In 2019, I brought my graduate students to observe a gently bubbling pool of shit—or rather, several pools. We had arrived at the El Estero Water Resource Center to observe a key part of Santa Barbara's city infrastructure, the treatment of wastewater.[1] Making our way through verdant gardens with a faint whiff of sewer, we followed our guide up a staircase to the top metal deck of a series of deep, cement-rimmed pools. As we gazed over the metal railing, we observed a brown, simmering liquid with white froth, reminiscent of a large jacuzzi. Our guide explained that new renovations to the facility had expanded its pool space in order to better utilize the natural capacity of microbes to break down organic matter. While this process conscripts microbes as a technology in the circuits of wastewater treatment, other technologies are enlisted for microbes themselves: 49 percent of the wastewater treatment's electricity costs go toward aeration, producing a steady supply of oxygen bubbles that keep the microbes active and metabolically thriving. As Somak Mukherjee observed, the entire infrastructure of wastewater treatment was dedicated to "care for the microbe," facilitating an externalized process of digestion.[2]

The El Estero Water Resource Center showcases one of the intimacies of water and energy, although perhaps not the most obvious one. Hydropower utilizes the flow of water to generate energy; nuclear power plants require water for the generation of electricity as well as cooling; hydraulic fracking uses (up) between 1.5 and 16 million gallons of water per well; and oil pipelines are known to contaminate freshwater sources (such was the concern at Standing Rock: "Water is life. Mní Wičóni. This is all we have left—our river, and the lands you didn't take last time").[3] In these varied cases, water mediates the extraction of energy, elements enlisted in the smooth flow of capitalism. This smoothness extends to shared metaphors: like water, energy is often figured as flow or current; as utilities, forms of energy and water are carried by cables or pipelines. Yet key differences remain. Unlike water, many forms of energy (oil or electrical) are not meant to be touched; they are buffered from human contact via forms of technical mediation. Further, unlike energy forms, water is a serial **resource**: it is not consumed through use, although it may undergo contamination that makes it unusable. Following the movements of water through and beyond bodies—from drinking to **waste**—is a transcorporeal process that can help identify forms of harm or environmental injustice.[4]

The ferment of microbes at El Estero Water Resource Center offers a multispecies perspective on the intimacies of water and energy. Energy enters the picture not only in terms of the **electricity** required for the aerators, but also through the microbial bodies' decomposing different elements of wastewater (sulfur, nitrogen, ammonia) into forms of energy that they can use metabolically. These microbial metabolic processes also generate their own byproducts that can be used as other forms of energy. During one step of the wastewater treatment process, solids are removed from liquid waste, and later used to produce methane gas. This process of cogeneration produces 51 percent of the plant's electrical power.[5] Following Natasha Myers and Timothy Choy, we might consider wastewater microbes "co-conspirators," in the etymological sense of "breathing together" (we aerate them to support respiration) and the sense of sharing a common purpose: water treatment.[6]

Wastewater treatment is a useful site to think about water and energy within large-scale metabolic processes that move across scales of relation. Historian of science Hannah Landecker writes that metabolism "is the interface between inside and outside, the space of conversion of one to another, of matter to energy, of substrate to waste, of synthesis and breakdown. A process-thing, it is always in time."[7] Yet metabolism also takes time. While the aeration technologies at El Estero speed up microbial digestion to nine days, Surojit Kayal observes that the East Kolkata Wetlands in **India** model a form of wastewater treatment that obviates the need for expensive infrastructure, but takes about a month.[8] Instead of aerators, it involves the intentional shaping of wetlands as a metabolic infrastructure for wastewater purification, directing canals toward large fishponds separated by small strips of land. These ponds generate large amounts of fish and vegetables, an example of multispecies worldmaking through the careful management of water. If Rob Nixon's "slow violence" has been invaluable for diagnosing forms of harm, perhaps *slow metabolism* names an aspiration toward watery infrastructures involving many biotic co-conspirators.

See also: **Animals, Clean, Commons, Design, Indigenous Activism, Nonlinear, Sustainability**

Notes

1 The city of Santa Barbara is built on the unceded lands of the Chumash people.
2 I would like to acknowledge Rebecca Baker, Sage Gerson, and Surojit Kayal, who contributed helpful observations during this site visit.
3 U.S. Geological Service, "How Much Water Does the Typical Hydraulically Fractured Well Require?," accessed July 21, 2021, https://www.usgs.gov/faqs/how-much-water-does-typical-hydraulically-fractured-well-require?qt-news_science_products=0#qt-news_science_products; Jennifer Weston, "Water Is Life: The Rise of the Mní Wičóni Movement," *Cultural Survival*, March 3, 2017, https://www.culturalsurvival.org/publications/cultural-survival-quarterly/water-life-rise-mni-wiconi-movement.
4 Stacy Alaimo, "States of Suspension: Trans-corporeality at Sea," *Interdisciplinary Studies in Literature and the Environment* 19, no. 3 (2012): 476–93; Astrida Neimanis, *Bodies of Water: Posthuman Feminist Phenomenology* (London: Bloomsbury, 2017).
5 Thomas Welche, email to the author, July 26, 2021.
6 Timothy Choy, "A Commentary: Breathing Together Now," *Engaging Science, Technology, and Society* 6 (2020): 586–90; Natasha Myers, "Becoming Sensor in the Planthroposcene: An Interview with Natasha Myers," interview by Meredith Evans, *Fieldsights*, July 9, 2020, https://culanth.org/fieldsights/becoming-sensor-an-interview-with-natasha-myers.
7 Hannah Landecker, "The Metabolism of Philosophy, in Three Parts," in *Dialectic and Paradox: Configurations of the Third in Modernity*, ed. Berhard Malkmus and Ian Cooper (Bern: Peter Lang, 2013), 193–94.
8 Surojit Kayal, "Metabolic City: Ecology, Infrastructure, and Emergence in the East Kolkata Wetlands" (presentation, ASLE Conference, 2021); East Kolkata Wetlands Management Authority, "East Kolkata Wetlands" (last modified June 27, 2022), http://ekwma.in/ek/.

WIND

What color is the wind?

DAVID McDERMOTT HUGHES

What color is the wind? That's an absurd question to ask about an invisible force. But wind is coded and associated with a certain color of person: white people. In North America, wind has blown *for* white or European people. Not every breeze for every pale face. But it has engaged with key groups at crucial junctures: in capitalism, imperialism, and the current moment of transition from fossil fuels to **renewable** energy.

Consider colonialism. The trade winds brought Columbus to the Caribbean. He was sailing on uncharted waters, but he knew the midocean breezes, which depend upon latitude more than anything else. Atlantic easterlies imported conquerors, investors, refugees, indentured laborers, and—in the only truly involuntary case—enslaved Africans to the Americas. Wind did not automatically make Europeans successful in the Western Hemisphere. For that, they needed guns, germs, steel, and domesticated **animals**. But before smallpox felled the first **Indigenous** person, it needed to arrive in the body of a sailor beneath a mast. And that sailor—or his captain, at any rate—had to possess a certain will to traverse. What gave that European the confidence to reach for the ever-receding horizon? Wind. To those who could read and channel it, it was the highway, the signpost, and the map all in one. "I propose," wrote Columbus to the Spanish Crown in 1493, "to make a new navigational chart, in which I will locate the entire sea and lands of the sea-ocean in the proper place under their wind."[1] Conquest followed.

That alliance of wind, watercraft, and warriors prevailed until the mid-nineteenth century. Many a schooner carried enslaved Africans below decks. Then, as coal and steamships replaced commercial sailing, sails retired from the worlds of conquest and commerce. They relaxed into the elite leisure of yacht clubs and regattas. Sailing today is largely a millionaire's **sport**. It is also more private than other expensive sports. With no training or certification, one can rent a surfboard, kayak, canoe, or tennis court. To sail, one must either take expensive lessons to qualify for expensive rentals or plunk down four figures for an entry-level dingy. The sport, in short, sifts for **class**—which means, in the United States, that it sifts for race as well. From Narragansett to Nantucket, the same wind blows differently for different people.

Relationships to wind are all the more uneven when it blows very hard. The poor man's regatta is a hurricane. Residents of apartments and trailers, in other words, experience the wind through wreckage, rather than recreation. In Zora Neale Hurston's best-known novel, God sends a hurricane to a huddled knot of men and women in the Everglades: "The wind came back with triple fury. . . . He meant to measure their puny might against His. They seemed to be staring at the dark, but their eyes were watching God." "The sea walk[s] the earth with a heavy heel" and destroys a settlement of African Americans living independently of white people.[2] Along the Gulf Coast, hurricanes periodically level poor people's aspirations to join the middle class—even more markedly among descendants of enslaved Africans. Segregation and redlining put African Americans in harm's way. As in New Orleans's Ninth Ward, darker-complected people live at lower elevations and possess fewer means of escape. Evacuation is more private than other moments of mass transit. Spike Lee's **documentary** *When the Levees Broke: A Requiem in Four Acts* (2006) captures the decades of neglect, disregard, and systematic selfishness that left **Black** men floating face down. Malevolent wind joins forces with Jim Crow.

Modern wind turbines arrive with this cultural, racially coded baggage. Those machines do not appeal to one group more than another. Rather, the whole debate takes place in terms more familiar to white people than to Black people. Yachtsmen frequently oppose offshore turbines as dangerous obstructions. And as eyesores: to many leisure-seeking coastal residents and visitors, sixty-meter-long carbon fiber blades seem unnatural and out of place. The Kennedys lobbied successfully to kill the Cape Wind project off Nantucket.[3] Meanwhile, another group of mostly white people lobbies for wind power. Environmentalists, engineers, and steel workers want even longer blades along much of the Eastern Seaboard. So, debates on wind power in the United States mostly involve white, comparatively affluent people, who tend to live in locales and practice professions that have often excluded minorities.

Memory deepens the exclusion. To survivors of Katrina and those who identify with them, wind does not appear as something beneficent. New Orleans, of course, is a city, and the urban life tends to shape one's view of wind. In a city, you notice air when it moves too fast. You don't ride the wind; you shelter from it. This is the racial difference, then: to a certain class of white people, wind turbines update an old technology, perhaps too radically, but using means and spaces they know. To a certain class of Black people, wind turbines are utterly foreign and unprecedented.

So the environmental **justice** movement and the forceful urban constellation of Black- and Brown-led activism has said virtually nothing for or against wind power. These movements understand the climate crisis and the energy transition. Indeed, in cities like New York, grassroots groups are starting to demand solar panels on neighborhood roofs—for backup, resilient **electricity**, and autonomy from private utilities. To my knowledge, nobody has made a similar demand for wind turbines. Of course, one would also have to demand an innovative **design** with blades shorter than sixty meters. Try turning the axis ninety degrees, from horizontal to vertical. Vertical axis turbines spin compactly—like a top—if somewhat less efficiently. They require more maintenance—that is, jobs.[4] They can be installed side by side on a roofline, making them more efficient per acre of scarce urban space. Few advocates of wind power are thinking along these lines, yet. They should do so now. The movement for wind (and against fossil fuels) needs urban people and people of color. It needs people who have not and will not skipper a sailboat. Let's start a new eolian tradition, of the neighborhood drawing power—in multiple senses—from airflow just above it. Wind once conquered; now it should liberate.

See also: **Airplane**, **Local**, **Natural Gas**, **Organize**, **Settler Colonialism**, **Shipping**

Notes

1 Cristóbal Colón, *Diaro de a bordo*, ed. Luis Arranz Márquez (Madrid: EDAF, 2006), 67.
2 Zora Neale Hurston, *Their Eyes Were Watching God* (New York: Perennial, 1990), 160, 163.
3 David McDermott Hughes, *Who Owns the Wind?* (New York: Verso, 2021), 225–26.
4 Paul Gipe, *Wind Energy for the Rest of Us* (Bakersfield, CA: Wind-works.org, 2016), 137–173.

YOUTH

Why is youth culture indispensable to energy transition?

DEREK GLADWIN

One of the most effective campaigns for energy-climate **justice** in the mid-2020s is spearheaded by international youth culture. While Swedish activist **Greta** Thunberg has become a famous spokesperson for the fight against climate change, she also exemplifies how youth culture values energy transition as a fundamental part of climate justice and how youth can mobilize knowledge and **action** on a global scale. Greta is, however, not alone in this movement. Other youth in the Global North and the Global South, including Isra Hirsi, Autumn Peltier, Bruno Rodriguez, Eyal Weintraub, Leah Namugerwa, and Helena Gualinga, have promoted social action and educated others about what is at stake in the climate emergency.

Contemporary youth are a driving force for energy transition as they advocate for energy-climate justice by leading instead of following. Physicist Antoine Bret conceptualizes this relationship as the energy-climate continuum: the intricate and inseparable links between fossil fuel energy and climate change.[1] Youth are the next generation of energy users, and their knowledge and perspectives about energy issues have begun to eclipse previous generations' privileged authority over decisions about, or visions of, energy futures. Today's youth are catalyzing a new kind of energy politics that engages in building worlds beyond fossil fuels by imagining what low-carbon energy futures might look like and how to achieve them.

Listening to and following youth on issues of climate and energy can occur at a planetary **scale**, as with celebrity activists like Greta, but it can also proliferate change locally through mobilizing schools and communities. For example, the International Youth Deliberation on Energy Futures (IYDEF) offered an interactive and relational energy literacy curriculum that was created for students from high schools around the world. Involving twenty-two schools in eighteen countries, this program brought together hundreds of students in the final three years of their secondary **education**; the students met together in an online, collaborative classroom for the duration of a school year (October 2019–April 2020).[2]

Initiated and led by energy culture researchers and facilitated by teachers located in their respective countries,[3] this program aimed to address a fundamental question: what can happen when global youth gather in educational contexts to explore the obstacles and opportunities involved in energy transition and the creation of socially just climate futures? Education is inherently a practice of imagining futures and understanding how knowledges and epistemologies shift and adapt to support different forms of worldmaking. Futurist and computational poet Nick Montfort describes this process as "future-making," or "consciously trying" to imagine multiple progressive futures.[4] When engaged in future-making, students explored vital questions: Who gets to imagine our futures? What are the best ways to do so?

Educating youth about energy transition in the IYDEF also allowed them to confront the acceleration of hopelessness, particularly by considering concrete actions at structural and sociocultural levels, including both collective and individual actions. When learning about energy transition, students recognized not only that other futures might exist, but also that they themselves play an indispensable role in imagining and creating those futures in the present. The shift came as students were provided space to conceptualize alternative futures and concrete changes to fossil fuel use in diverse contexts, including their own roles in these futures both locally and globally. By listening to one another and working together across their diverse contexts, students learned to collaborate and build collective action on energy transition. While many students began the program with a broad motivation to "save the planet," this collaborative educational experience focused on energy transition provided them with practical ways of living on the **planet** and changing their relationships to energy in the present.[5]

Beyond the IYDEF, other examples demonstrate how youth are energizing the transition. The Fridays for Future international initiative that Greta catalyzed involves over fourteen million participants, and it is still organized and led by youth.[6] In collaboration with the United Nations Energy Compact, the government of Denmark recently announced the launch of the Student Energy Solutions Movement (as an extension of the successful international group Student Energy, launched in 2009), which promises "10,000 youth-led clean energy projects by 2030."[7] In Canada, Generation Power Youth invites Indigenous youth to become critical voices in the transition to **renewable** energy.[8]

Youth culture might be the most renewable fuel driving energy transition. On questions of energy-climate justice, initiatives like those described above aim to energize social and cultural action. Matthew Hoffman, codirector of the Environmental Governance Lab at the University of Toronto, warns that "climate despair is the new climate denial, dulling the sense of urgency and blunting the momentum for action," and triggering a paralysis that "we can least afford."[9] The politics of hope relies upon momentum, but it can be difficult to generate or sustain, particularly amid pandemics, drought, flooding, fires, and mass species **extinction**. The UN secretary-general António Guterres admits that "the planet is broken."[10] Fortunately, youth momentum materializes out of the desire to fix a broken planet through practical efforts to live and build just futures in the present.

See also: **Communication, Trans-, Transition**

Notes

1 Antoine Bret, *The Energy-Climate Continuum: Lessons from Basic Science and History* (New York: Springer, 2014), 3.
2 Just Powers, University of Alberta, "International Youth Deliberation on Energy Futures," accessed June 8, 2021, https://www.justpowers.ca/projects/international-youth-deliberation-on-energy-futures/.
3 Researchers and curriculum developers of IYDEF included Derek Gladwin (University of British Columbia), Eva-Lynn Jagoe (University of Toronto), Carrie Karsgaard (University of Alberta), Jordan Kinder (McGill University), Lynette Shultz (University of Alberta), Mark Simpson (University of Alberta), Imre Szeman (University of Waterloo), and Sheena Wilson (University of Alberta).
4 Nick Montfort, *The Future* (Cambridge, MA: MIT Press, 2017), 4.
5 Carrie Karsgaard, Derek Gladwin, and Lynette Shultz, "Designing the International Youth Deliberation on Energy Futures: The Global Classroom," *Journal of Environmental Education* (under review).
6 Fridays for Future, "Who We Are," accessed July 3, 2021, https://fridaysforfuture.org/what-we-do/who-we-are/.
7 Student Energy, "Launching the Movement," accessed July 3, 2021, https://studentenergy.org/about/solutionsmovement/.
8 Generation Power, Indigenous Clean Energy Social Enterprise, "Generation Power Youth," accessed July 3, 2021, https://www.generationpower.ca/get-involved/generation-power-youth.

9 Matthew Hoffman, "Reasons for Hope on Climate Change in 2021," *YES! Magazine*, January 12, 2021, https://www.yesmagazine.org/environment/2021/01/12/climate-change-hope-momentum.

10 António Guterres, "The UN Secretary-General Speaks on the State of the Planet" (United Nations, last modified August 28, 2022), https://www.un.org/en/climatechange/un-secretary-general-speaks-state-planet.

ACKNOWLEDGMENTS

This project was born in the bewildering pandemic year of 2020, when Imre emailed Jennifer to ask whether it was time for a follow-up to our 2017 edited collection, *Fueling Culture: 101 Words for Energy and Environment.* That book was a work of infrastructural capacity building for conversations in the energy humanities that were then beginning to emerge. The fact that we had, give or take, 101 chapters for *Fueling Culture* came as a late and happy surprise, as also turned out to be true for *Power Shift*, which appears more than a decade after the first Petrocultures Research Group conference at the University of Alberta in 2012. Our first thanks, then, go to the burgeoning communities of thinkers and artists for whom and with whom we have done this work of learning how to think and what to do about energy. The generous and brilliant colleagues we've met at subsequent Petrocultures conferences, After Oil Schools, and other energy humanities events are our inspiration, our audience, our collaborators, our comrades. We hope that this book finds new readers who will become cherished interlocutors we haven't yet met.

A book always involves the work of many more people than those listed on its cover. This is especially true for a book like this one, which collects the voices and perspectives of more than one hundred people. We thank our contributors for their willingness to lend their ideas to *Power Shift* and for their patience while this project moved from concept to reality over a four-year period. It's a truism that happens to be true: we could not have done this without you!

The tireless labor of several research assistants was invaluable to bringing this book to fruition. For their organizational and technological skills, critical acumen, and sheer smarts, a huge thanks to Zhizhui Chen, Alessandra Gianino, Liam Jagoe, Tatiana Rios, and Valerie Uher. Jennifer Wenzel gratefully acknowledges support from the Columbia Climate School's Climate and Society Departmental Research Assistant Program, which funded Alessandra Gianino's position.

Derek Krissoff was the director of West Virginia University Press when we started *Power Shift*; Than Saffel was there at the end as the Press's director. Both showed a genuine commitment to this project. We also want to thank members of the WVUP editorial team, Marguerite Avery, Kristen Bettcher, and Natalie Homer.

Our friends and colleagues read the introduction both carefully and with care and helped us draw out the ideas that were lurking in the back of the argument in early drafts. Darin Barney, Eva-Lynn Jagoe, Graeme Macdonald, Mark Simpson, and Caleb Wellum: thanks for this, and for so, so much more.

We're grateful for the continuing patience, support, and wisdom of our partners, Eva-Lynn Jagoe and Joey Slaughter. And, as ever, for the inspiring memory of Patsy Yaeger: the spark.

CONTRIBUTORS

Mohamed Adow is the founder and director of Power Shift Africa.

Fallon Samuels Aidoo teaches real estate and historic preservation at Tulane University.

Allisa Ali lives and works in Montreal, Quebec, Canada.

Leah Aronowsky is a historian of science at the Columbia Climate School.

Zainab Ashraf is a coordinator for UW RAISE at the University of Waterloo.

Sākihitowin Awāsis teaches in geography and environment at Western University.

Stacey Balkan teaches environmental humanities at Florida Atlantic University.

Ana Isabel Baptista teaches environmental policy and sustainability management at the New School.

Daniel A. Barber teaches architecture at the University of Technology Sydney.

Darin Barney is Grierson Chair in Communication Studies at McGill University.

Kylie Benton-Connell is a doctoral candidate in anthropology at the New School.

Kai Bosworth is a human geographer at Virginia Commonwealth University.

Dominic Boyer teaches energy humanities and anthropology at Rice University.

Guy Brodsky is passionate about decarbonization, distributed energy resources, and climate justice.

Rebecca Byrnes is a lawyer and deputy director of the Fossil Fuel Treaty Initiative.

Angela Carter teaches political science at Memorial University.

Mijin Cha teaches environmental studies at the University of California, Santa Cruz.

Deedee Chao is a graduate student at the University of California, Davis.

Joshua Clover is an abolition communist.

Stephen Collis is a poet who teaches at Simon Fraser University.

Kyle Conway teaches communication at the University of Ottawa.

Brian Cozen teaches environmental communication at California State University, Fresno.

Emerson Cram teaches in communication and in gender, women's and sexuality studies at the University of Iowa.

Cara Daggett teaches political science at Virginia Tech.

Brigt Dale teaches and directs research at Nordland Research Institute, Norway.

Robert Danisch teaches communication arts at the University of Waterloo.

Ashley Dawson is a writer and activist for fossil abolition.

Erin Dean teaches anthropology and environmental humanities at Carnegie Mellon University.

Christiaan De Beukelaer works on culture and climate at the University of Melbourne.

Shane Denson teaches media theory and aesthetics at Stanford University.

Pauline Destrée teaches anthropology at Durham University.

Katy Didden is a poet and teaches creative writing at Ball State University.

Marianna Dudley teaches environmental humanities at the University of Bristol.

Todd Dufresne teaches philosophy at Lakehead University.

Keller Easterling is a designer, writer, and professor at Yale University.

Danielle Endres is professor and director of the Environmental Humanities Program at the University of Utah.

Jaimey Hamilton Faris teaches critical theory and art history at the University of Hawai'i, Mānoa.

Marcela da Silveira Feital is an environmental sociologist at the State University of Campinas, Brazil.

Kesha Fevrier teaches in geography and planning and in Black studies at Queen's University.

Darren Fleet teaches environmental communication and media studies at Simon Fraser University.

Stephanie Foote is the Jackson and Nichols Professor of English at West Virginia University.

Carolyn Fornoff teaches Latin American studies at Cornell University.

Grace Franklin is a doctoral candidate at the University of Southern California.

Michael B. Gerrard teaches and directs the Sabin Center for Climate Change Law at Columbia Law School.

Alfonso Giuliani researches economics at CES and at PHARE, Université Paris 1 Panthéon-Sorbonne.

Derek Gladwin teaches in education at the University of British Columbia.

Walter Gordon is the Karr Family Provostial Fellow at Stanford University.

Gökçe Günel teaches anthropology at Rice University.

Shane Gunster teaches communication at Simon Fraser University.

Anushree Gupta is a doctoral researcher in liberal arts at IIT Hyderabad.

Max Haiven is Canada Research Chair in the Radical Imagination at Lakehead University.

Ted Hamilton teaches English at Bucknell University.

Liz Harmer is author of the novels *The Amateurs* and *Strange Loops*.

Matthew S. Henry is a social scientist at the National Renewable Energy Laboratory.

Amber Hickey teaches art history at the University of Tennessee at Chattanooga.

Cymene Howe teaches anthropology and codirects the Science and Technology Studies Program at Rice University.

Cameron Hu is a Mellon Postdoctoral Fellow at the Center for the Humanities at Wesleyan University.

David McDermott Hughes teaches anthropology at Rutgers University.

Joy Hutchinson is a doctoral candidate in public health sciences at the University of Waterloo.

Eva-Lynn Jagoe teaches comparative literature at University of Toronto.

Amy Janzwood teaches in political science at the Bieler School of the Environment at McGill University.

Lesley Johnston teaches at the University of Waterloo.

Melody Jue teaches English at the University of California, Santa Barbara.

Carrie Karsgaard teaches in education at Cape Breton University.

Aalok Khandekar studies science and technology studies at the Indian Institute of Technology Hyderabad.

Jordan B. Kinder teaches communication studies at Wilfrid Laurier University.

Donald V. Kingsbury teaches in political science and Latin American studies at the University of Toronto.

Rachel Krueger is a policy analyst at Public Safety Canada.

Stéphane La Branche is a climate sociologist and coordinator for the International Panel of Experts on Behavior Change.

Carlos Larrea is research professor at the Universidad Andina Simón Bolivar in Quito, Ecuador.

Aarti Latkar is a doctoral fellow at the Institute for Culture and Society at Aarhus University in Denmark.

Sherry Lee teaches musicology at the University of Toronto.

Stephanie LeMenager teaches English and environmental studies at the University of Oregon.

Emily MacCallum is a doctoral candidate in musicology and environmental studies at the University of Toronto.

Graeme Macdonald teaches literature at the University of Warwick.

Alicia Massie is a doctoral candidate in communication at Simon Fraser University.

Misty Matthews-Roper is a doctoral candidate at the School of Environment, Resources and Sustainability at the University of Waterloo.

Tanner Mirrlees teaches communication and digital media at Ontario Tech University.

Lisa Moore teaches creative writing at Memorial University in Newfoundland and Labrador.

Dustin Mulvaney teaches environmental studies at San Jose State University.

Swaralipi Nandi teaches English at Loyola College, Hyderabad.

Rodrigo Nunes teaches philosophy at the Pontifical Catholic University of Rio de Janeiro.

David E. Nye is Professor Emeritus of American Studies at the University of Southern Denmark.

Sarah O'Brien researches social anthropology at the University of Manchester, UK.

Zeynep Oguz teaches anthropology of development at the University of Edinburgh.

Jessica O'Reilly teaches international studies at Indiana University Bloomington.

Anne Pasek is Canada Research Chair in Media, Culture, and the Environment at Trent University.

Emily Pawley is Walter E. Beach '56 Chair in Sustainability Studies and teaches history at Dickinson College.

Andrew Pendakis teaches theory and rhetoric at Brock University.

Drew Pendergrass is a doctoral candidate in environmental engineering at Harvard University.

Helen Petrovsky teaches aesthetics at the Institute of Philosophy of the Russian Academy of Sciences.

Kristin D. Phillips teaches anthropology at Emory University.

Philomena Polefrone is the advocacy associate manager at the American Booksellers Association.

Thomas Pringle teaches cinema and media studies at the University of Southern California.

Andrew Revkin is a longtime environmental journalist, author, and podcaster.

Emily Rich is a doctoral candidate in English at the University of California, Davis.

Kimberly Skye Richards teaches writing at the University of British Columbia.

Daniela Russ is a historical sociologist at the University of Leipzig, Germany.

Deena Rymhs teaches American Indian studies at the University of Illinois Urbana-Champaign.

Meg Samuelson teaches literature at the University of Adelaide.

Siddharth Sareen works on the governance of energy transitions at the Fridtjof Nansen Institute.

Jeremy J. Schmidt teaches environmental geography at Queen Mary University of London.

Gianfranco Selgas is British Academy Postdoctoral Fellow and teaches Latin American studies at University College London.

Hiroki Shin is an energy historian and Vice-Chancellor's Illuminate Fellow at Queen's University Belfast.

Mark Simpson teaches in English and film studies at the University of Alberta.

Adam Sobel is a working climate scientist who also thinks and writes about broader issues.

Sanaz Sohrabi is a filmmaker and researcher of visual culture.

John Szabo is a DAAD scholar at DIW Berlin and a fellow at the Institute for World Economics, CERS, HUN-REN.

Imre Szeman is director of the Institute for Environment, Conservation and Sustainability at the University of Toronto Scarborough.

Isaac Thornley is a doctoral candidate in environmental studies at York University.

Petra Tschakert is a geographer who teaches at Curtin University in Australia.

Rebecca Tuhus-Dubrow is a writer based in Southern California.

Bhushan Tuladhar is an environmental engineer and Chief of Party for USAID Clean Air at FHI 360 Nepal.

Valerie Uher is a doctoral candidate at the University of Waterloo.

Josefin Wangel teaches at the Swedish University of Agricultural Sciences.

Caleb Wellum teaches history at the University of Toronto Mississauga.

Jennifer Wenzel teaches postcolonial studies and environmental/energy humanities at Columbia University.

Parke Wilde works on nutrition science and policy at Tufts University.

Rhys Williams works in energy and infrastructure humanities at the University of Glasgow.

Sheena Wilson teaches media, communications, and cultural studies at the University of Alberta.

James Wilt is a doctoral candidate in environment and geography at the University of Manitoba.

Anna Zalik teaches geography and environmental studies at York University.

ALTERNATIVE INDEX

Economics

Extractions

Infra/structures

Knowledge

Narratives

Nonhumans

Politics

Technologies

Time

Transitions

Values

INDEX